Human Herpesviruses: Diagnosis and Clinical Management

Human Herpesviruses: Diagnosis and Clinical Management

Edited by Aydan Jones

hayle medical

New York

Hayle Medical,
750 Third Avenue, 9th Floor,
New York, NY 10017, USA

Visit us on the World Wide Web at:
www.haylemedical.com

ISBN: 978-1-64647-539-1

Cataloging-in-Publication Data

Human herpesviruses : diagnosis and clinical management / edited by Aydan Jones.
p. cm.
Includes bibliographical references and index.
ISBN 978-1-64647-539-1
1. Human herpesvirus-6 infections. 2. Herpesvirus diseases--Diagnosis. 3. Herpesvirus diseases--Treatment.
4. Perinatology. I. Jones, Aydan.
QR201.H48 H86 2023
616.911 2--dc23

Table of Contents

Preface

Herpesvirus refers to a type of virus that belongs to the herpesviridae family. It has been classified into eight different types which include herpes simplex type I (HSV-1), herpes simplex type II (HSV-2), varicella-zoster virus (VZV/HHV-3), Epstein-Barr virus (EBV/HHV-4), cytomegalovirus (CMV/HHV-5), herpesvirus type 6 (HBLV/HHV-6), herpesvirus type 7 (HHV-7) and Kaposi's sarcoma herpesvirus (KSHV/HHV-8). Both direct and serologic techniques are used to diagnose the human herpesvirus 6 HHV-6 infection. One of the most prominent techniques used for quantifying viral DNA in blood, organs and other body fluids is real-time PCR. There is currently no officially approved treatment for herpes infections. However, supportive therapy including sufficient hydration and acetaminophen for fever is recommended in all patients with symptomatic HHV-6 infection. This book is compiled in such a manner, that it will provide in-depth knowledge about the diagnosis and management of human herpesviruses. It will help new researchers by foregrounding their knowledge about this virus. The book will provide comprehensive knowledge to medical students and professionals.

Significant researches are present in this book. Intensive efforts have been employed by authors to make this book an outstanding discourse. This book contains the enlightening chapters which have been written on the basis of significant researches done by the experts.

Finally, I would also like to thank all the members involved in this book for being a team and meeting all the deadlines for the submission of their respective works. I would also like to thank my friends and family for being supportive in my efforts.

Editor

1

KSHV LANA—The Master Regulator of KSHV Latency

Timsy Uppal [1], Sagarika Banerjee [2], Zhiguo Sun [2], Subhash C. Verma [1] and Erle S. Robertson [2,*]

[1] Department of Microbiology and Immunology, University of Nevada, Reno, School of Medicine, 1664 N Virginia Street, MS 320, Reno, NV 89557, USA;
E-Mails: tuppal@medicine.nevada.edu (T.U.); scverma@medicine.nevada.edu (S.C.V.)

[2] Department of Microbiology and the Tumor Virology Program of the Abramson Cancer Center, Perelman School of Medicine at the University of Pennsylvania, 201E Johnson Pavilion, 3610 Hamilton Walk, Philadelphia, PA 19104, USA;
E-Mails: sagarika@mail.med.upenn.edu (S.B.); sunzhig@upenn.edu (Z.S.)

* Author to whom correspondence should be addressed; E-Mail: erle@mail.med.upenn.edu;

External Editor: Zhi-Ming Zheng

Abstract: Kaposi's sarcoma associated herpesvirus (KSHV), like other human herpes viruses, establishes a biphasic life cycle referred to as dormant or latent, and productive or lytic phases. The latent phase is characterized by the persistence of viral episomes in a highly ordered chromatin structure and with the expression of a limited number of viral genes. Latency Associated Nuclear Antigen (LANA) is among the most abundantly expressed proteins during latency and is required for various nuclear functions including the recruitment of cellular machineries for viral DNA replication and segregation of the replicated genomes to daughter cells. LANA achieves these functions by recruiting cellular proteins including replication factors, chromatin modifying enzymes and cellular mitotic apparatus assembly. LANA directly binds to the terminal repeat region of the viral genome and associates with nucleosomal proteins to tether to the host chromosome. Binding of LANA to TR recruits the replication machinery, thereby initiating DNA replication within the TR. However, other regions of the viral genome can also initiate replication as determined by Single Molecule Analysis of the Replicated DNA (SMARD) approach. Recent, next generation sequence analysis of the viral transcriptome shows the expression of additional genes during latent phase. Here, we discuss the newly annotated latent genes and the role of major latent proteins in KSHV biology.

Keywords: Kaposi's sarcoma-associated herpesvirus; KSHV; KS; minichromosome; DNA replication; reactivation; signaling

---

1. Introduction

Kaposi sarcoma (KS) is one of the most common virally induced cancers among HIV-infected patients. KS is an endothelial cell lineage tumor that is caused by Kaposi's sarcoma-associated herpesvirus (KSHV) or eighth human herpesvirus (HHV-8), sub-classified as a gamma herpesvirus and first discovered in 1994 by Chang and Moore's group from patients with KS by a subtraction-PCR based method called Representational Difference Analysis (RDA) [1,2]. Clinically, KS exists in several forms; *Classic indolent KS*, which is the prevalent form in HIV-negative elderly male patients of Mediterranean and Middle Eastern origin, *African endemic KS* that is a relatively aggressive form and most common in HIV-negative children, *Iatrogenic KS or post-transplant KS*, associated with patients undergoing immunosuppressive therapy after renal transplantation, and *AIDS-related epidemic KS*, which develops in HIV-infected individuals and is among the leading causes of death in AIDS patients [3]. Though KSHV is found in all forms of KS, infection with KSHV is necessary but not sufficient for the development of the KS [4–6].

After the discovery of KSHV in KS tissues, KSHV sequences were rapidly identified in two other lymphoproliferative disorders: (a) Primary Effusion Lymphoma (PEL) and the (b) plasmablastic variant of Multicentric Castleman's Disease (MCD) [7,8]. Spindle cells expressing endothelial markers including CD31, CD34, CD36 and factor XIII are shown to be the cell type of origin in KS, whereas PEL and MCD are considered to be of B-cell lineage as they express B-cell surface markers. Additionally, KS and MCD are polyclonal in nature while there is a pattern of B-cell monoclonality in all PEL samples [9,10]. In immunocompetent individuals, CD19+ B cells appear to be the main target of KSHV infection [11]. Recently, a newly characterized KSHV-associated condition, KICS (KSHV Inflammatory Cytokine Syndrome) has been reported in patients with HIV and KSHV co-infection with elevated levels of interleukin-6 and KSHV, though the cell origin of this condition is still a matter of discussion [12]. Although the incidence of KS in HIV-infected individuals has declined with the introduction and widespread use of highly active antiretroviral therapy (HAART), it is anticipated that KSHV will remain one of the co-morbid agent in persons with or at high risk of acquiring HIV infection [13]. Therefore, understanding the molecular pathogenesis of KSHV infection and KSHV-HIV co-infection is crucial for development of drug therapies to treat and control the associated pathological processes.

2. Epidemiology

Serologic and epidemiologic studies have revealed that KSHV does not lead to a ubiquitous infection and KSHV-infected individuals are found throughout the world. There is an interesting but enigmatic association between KS prevalence and KSHV seroprevalence, though there are major variations in KS prevalence geographically [14,15].

The advent of serological assays to detect antibodies against KSHV has enabled the study of the distribution of KSHV seroprevalence among different risk groups. Currently, in the most frequently used serological assays such as immunofluorescence assay or ELISA, serum antibody to KSHV is detected using either KSHV-infected cells or recombinant capsid proteins/KSHV lysate as antigens, respectively [16–18]. Among the general population, the seroprevalence of KSHV infection in northern Europe, Asia and America is <10%, but the overall KSHV seropositivity is >50% in most of sub-Saharan Africa. The Mediterranean regions have intermediate seroprevalence that ranges from 30% in Sicily to 3% in northern Italy [19].

Genotypes of KSHV are categorized based on highly variable but tracer K1 and K15 genes located next to the terminal repeat in the viral genome. As a result, analysis of the divergence of the K1 gene has led to the identification of 5 genotypes (A through E) and several subtypes [20,21]. In general, the sequence variation between different subgroups is <3% at the nucleotide level, however two hyper-variable K1 regions-VR1 and VR2 display up to 60% variability [22]. Subgroups A1-4 and subtype C are prevalent in North Europe, America, some regions of Asia and Middle East and subtype A is predominant in AIDS-associated cases but not in non-AIDS patients. Subgroups B1-4 are primarily found in sub-Saharan Africa whereas subgroups D and E are abundant in Australia, the Pacific and Brazilian Amerindians [23,24].

Although the transmission modes and the risk factors of KSHV have not been clarified yet, both sexual and non-sexual modes can transmit KSHV [25,26]. In KSHV-endemic regions like Africa, saliva-mediated transmission is the most common mode of transmission among children, as high KSHV copy numbers are detected in the saliva of seropositive patients while sexual transmission may be predominant among homosexual men in the areas of low prevalence where the risk of KSHV transmission rises with the number of sexual partners [27,28]. Organ transplantation can also transmit KSHV and cause viral infection and is documented to occur both in infected organ donor and transplant recipients due to virus reactivation [29,30]. KSHV transmission through blood transfusion or among intravenous drug users is rare but evidence of both modes of transmission has been reported [31].

3. Life Cycles of KSHV and Control of Latency

Full genome sequence of KSHV virus from KS lesions and PEL cells reveal that the viral genome exists as a linear double-stranded DNA of *ca.* 170 kb in the virion particles. The genome consists of a central ~137 kb long unique coding region (LUR) with 53.5% GC content, which is flanked by multiple, non-coding terminal repeat units having 84.5% GC content [32]. The number of repeats in the TRs varies from 32–50 copies [33]. LUR is the viral protein-coding region, which encodes approximately 90 ORFs, (some with homology to the cellular genes), 12 microRNAs and several ncRNAs [34]. Usually the complete viral particle consisting of a viral capsid and an envelope, 150–200 nm in diameter, is not observed in the KS samples by electron microscope, as the genome is maintained in latent form without reactivation [35]. However, they can be observed in PEL cells induced by Na-butyrate (NaB) or 12 *O*-tetradecanoylphorbol-13-acetate (TPA) [36]. Following infection, viral DNA is delivered to the nucleus where it is circularized using host enzymatic machinery, generating a chromatinized nuclear episome [37]. Studies to identify the processes through which the incoming histone-free virion DNA gets chromatinized in the nucleus are still in its infancy.

Like the other members of the γ-herpesvirus family, KSHV exhibits two different phases of infection: *persistent latent infection* and a *transient lytic reactivation* that are distinguished by their viral gene expression patterns. During latent infection, which is the default pathway of KSHV infection *in vivo* and *in vitro*, the viral genome is maintained as a circular episome within the host cell nucleus with highly restricted protein expression in order to maintain the genome in the dividing cells and to limit host immune responses while enhancing cell survival and virus persistence [38]. The latent phase is reversible and certain environmental and physiological factors including oxidative stress, hypoxia or inflammatory cytokines may periodically reactivate the hidden latent virus to enter the second program of viral gene expression and lytic reactivation [39–43]. In this phase, expressions of the remaining viral open reading frames (ORFs) are activated in a temporally regulated cascade, leading to three classes of lytic genes: immediate early (IE), early (E) and late (L) genes [44]. The host cell machinery is redirected to manufacture and assemble the infectious progeny virions, ultimately causing cell apoptosis due to virus production and shut off of host cell protein synthesis. In KS tissue, KSHV is found predominantly in the latent phase as assessed by immunohistochemical and genetic analysis of KS tissue, while there is limited expression of KSHV viral genes especially vIL-6 in PEL cells. In contrast, MCD is associated with KSHV in its lytic state, a feature that is unusual among herpesvirus-associated tumors [45].

The dogma of the classic transcriptional program shifted to a more complex expression pattern after technological advances enabled the discovery that KSHV induces diverse gene expression profiles in different endothelial cell types, lymphatic endothelial cells (LECs) and blood endothelial cells (BECs) [46]. The gene expression pattern analyzed in these cells infected with recombinant virus rKSHV.219 clearly demonstrated that infected BECs exhibit conventional latent gene transcripts during latency, whereas LECs have a unique virus latency program with a widespread expression of numerous lytic genes, including RTA, K-bZIP and ORF45 that do not produce any infectious virions [46,47]. During this “dysregulated lytic program”, expression of KSHV ORF45 leads to selective activation of mTORC1 by ERK2-mediated activation of RSK1, which sensitizes these LECs to rapamycin-induced killing. However, the exact benefits of this different transcriptional program to KSHV during persistent infection are yet to be identified.

Recent analysis of the KSHV genome by Arias *et al.* using next-generation sequencing, mRNA-Seq and Ribo-Seq has provided a plethora of information about the genomic landscape and peptide coding potential of KSHV during the productive (lytic) cycle [48]. Using a tightly-controlled and highly inducible epithelial iSLK-219 cell line, several hidden genomic and functional features have been uncovered, leading to the generation of a novel revised annotation of the KSHV genome: KSHV 2.0 [48]. This includes 45% more coding capacity with expanded illustration of 49 viral transcripts, 70 ORFs, non-coding RNAs, polyadenylation sites, splice junctions and initiation/termination codons of main ORFs [48]. The coding capacity of KSHV has been attributed to multiple strategies, including splicing, mRNA editing and the usage of alternative start sites leading to multiple small and upstream ORFs (sORFs and uORFs) [48]. In summary, these comprehensive and high resolution approaches have identified the underappreciated complexity of the KSHV gene expression during the different stages of the viral life cycle [48].

Central to KSHV infection is the ability of the virus to establish life-long, non-productive latent infection, which can later be reactivated in the host cell. In general, this oncogenic γ-herpesvirus faces a number of problems to establish and control the latency in the infected host cell without losing the genome [49]. To begin with, the incoming epigenetic naïve and linear KSHV genome must be circularized and chromatinized following infection, to generate the KSHV epigenome and trigger the transcriptionally silent latent cycle. Further, one or more of the expressed latent genes must allow stable latent dsDNA replication and KSHV genome segregation to maintain the genome in new daughter cells. Additionally, successful latency establishment also requires hijacking of several viral and cellular pathways in order to repress KSHV reactivation and to escape host immune surveillance. Thus, KSHV's existence and persistence in the host cell requires a dynamic balance between the two gene expression programs, although how this balance is maintained still remains unclear. Here we discuss in detail the latent maintenance of the dsDNA KSHV genome in the infected host cell with particular attention to the role of one of the major latent antigens, latency-associated nuclear antigen (LANA) which is constitutively expressed in all latently infected KSHV cells and is an important contributor to efficient KSHV transcriptional regulation.

4. The Latency Program of KSHV and the Key Players

Examination of latent infection in PEL cell lines that maintain 98%–99% of latently infected cells has led to the characterization of a major latency locus that is abundantly and consistently transcribed in cell lines derived from the KSHV-infected patients and restrict transcription to only a few of the 90 KSHV genes (Figure 1) [50,51]. This latency locus includes four genes, encoding ORF73/LANA, ORF72/v-Cyclin, ORF71/v-FLIP (*Fas-associated death domain-like interleukin-1β-converting enzyme-inhibitory protein*) and K12/Kaposin family of protein (Kaposin A, B and C) along with 12 microRNAs that can be processed to yield 18 miRNAs (at last count) [52–54]. Viral interferon regulatory factor-3 (v-IRF-3) and mRNA of several other viral genes including viral G protein-coupled receptor, vGPCR encoded by ORF74, K14, vIL-6 and Processivity factor encoded by ORF59 have also been detected in most KS and PEL-tumor models [55]. Together, these latency-associated viral proteins are required for constant latent infection and survival of the infected cell. Latency transcript cluster, including, LANA, v-Cyclin and v-FLIP are located adjacent to each other and are transcribed from a constitutively active promoter, LANA promoter or LTc [49]. These three genes are separated from the K12 gene which is transcribed from a second promoter, the kaposin promoter, or LTd, located just downstream of LANA. The kaposin promoter encodes a spliced transcript encoding the kaposin proteins and a bicistronic RNA for v-Cyclin and v-FLIP [56]. This promoter also governs the expression of 12 KSHV pre-miRNAs. The ORF74-K14 transcript initiates from 5' UTR of the LANA-vCyclin-vFLIP and is expressed in latently infected cells along with LANA-vCyclin-vFLIP genes expression.

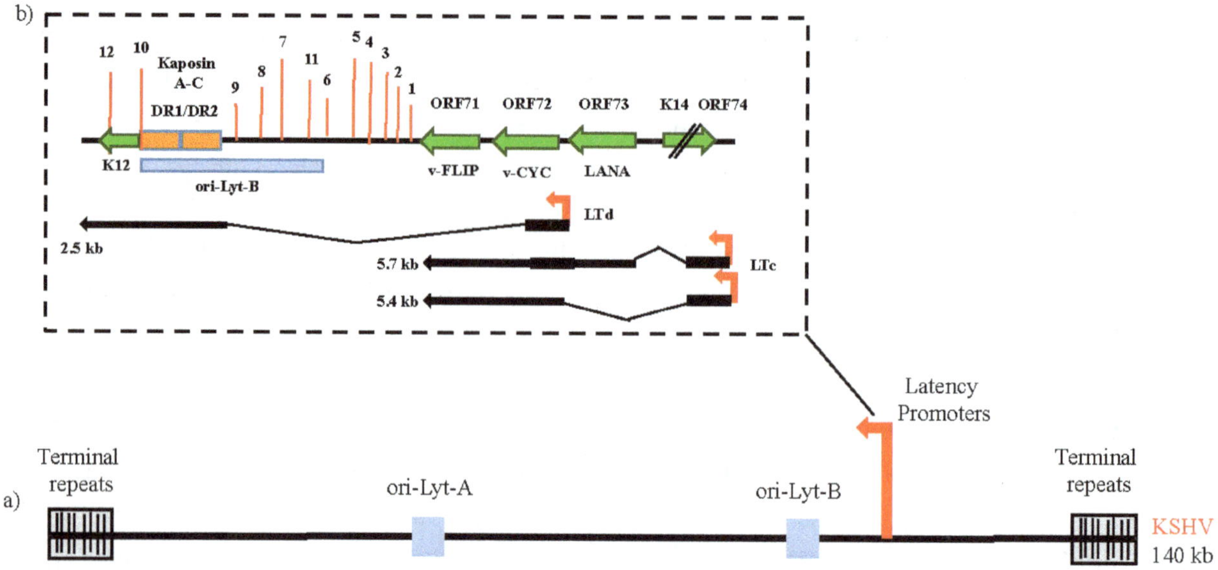

Figure 1. (**a**) Schematic representation of the linear KSHV genome with terminal repeat sequences, origin of lytic replication (ori-Lyt-A and ori-Lyt-B) and latency promoter; (**b**) The major latency locus of KSHV is shown in an expanded view. **Top** panel: The four major open reading frames (ORFs)-ORF73/LANA, ORF72/v-CYC, ORF 71/v-FLIP and ORF K12/Kaposins A-C along with the adjacent K14 and ORF74 genes are shown by green arrows. Position of 12 pre-miRNA sequences is shown as vertical red lines. **Bottom** panel: The schematic diagram of the transcripts directed by the Kaposin promoter (LTd) and the LANA promoter (LTc) is shown.

Another unlinked latency promoter is found to encode the vIRF3 gene, a member of IRF superfamily [49]. This protein has been identified as the LANA-2 protein encoded by the ORF K10.5 of KSHV and is expressed in KSHV-infected hematopoietic tissues including PEL and MCD, but not in KS spindle cells. This protein is known to inhibit the function of certain cellular IRFs and block the interferon induction. Wies *et al.* have shown that vIRF3 is also required for proliferation and survival of PEL cells infected with KSHV as *in vitro* knockdown of vIRF3 expression in PEL cells reduced the cell proliferation and increased the activity of caspase-3 and/or caspase-7, triggering programmed cell death [57].

Additionally, a third latency locus has been identified which drives the expression of the ORF-K1 protein [58]. K1 is a 46-kDa transmembrane signaling protein that imitates signaling through the B cell receptor [59]. In many KSHV-infected cell lines, this gene is transcribed at very low levels during latency and is upregulated during lytic reactivation [60]. More sensitive methods using microarrays and proteomics have identified several viral transcripts and peptide motifs that further provide valuable knowledge regarding viral latent gene expression. In the following sections, we describe some of these key KSHV latent genes and their role in KSHV-associated latency.

4.1. ORF73/LANA (Enables Replicative Immortality)

As mentioned above, KSHV establishes stable latent infections that play an essential role in KSHV-induced malignancies and pathogenesis. One of the key aspects of KSHV-associated oncogenesis is the ability of the latent viral genome to persist as extra-chromosomal episomes in the dividing cells that express a dynamic pattern of viral genes to drive host cell proliferation and survival. As the KSHV viral genome does not encode its own centromeric proteins, it is probable that it follows an alternative way to maintain and replicate its episome from parental cells to progeny cells. In KSHV, LANA, a major KSHV-encoded latent protein is considered to be critical for the maintenance, replication and efficient segregation of the viral genome from generation to generation. In order to carry out this function, LANA, a multifunctional nuclear protein (1162 amino acid in length and 220–230 kDa in size), binds directly to the conserved TR sequences of the KSHV genome through its *C*-terminal domain and docks onto the host chromosome through its *N*-terminal chromatin-binding domain (CBD), thus enabling the KSHV genome to hitch a ride on the host chromosome during mitosis and maintain a stable copy number in the latently infected cells. Also, among all the latent proteins, LANA is the most consistently expressed antigen in the KSHV-infected cells and is always detected as a dot-like staining pattern by immunohistochemistry. LANA is found to bind and interact with multiple cellular proteins, including *tumor suppressors*-p53, pRb and von Hippel Lindau (VHL), *transcription factors*- ATF4/CREB2 and STAT3, *chromatin-binding proteins*-HP1, H2A/H2B, MeCP2, Brd4 as well as *signal transducers*-GSK-3β, in order to inhibit apoptosis and stimulate spindle cell proliferation. LANA has also been proposed to bind to several viral promoters and suppress viral lytic gene transcription and thereby influence the maintenance of latency. Thus LANA is a highly versatile oncogenic protein that plays a central role in the pathogenesis of KSHV infection.

4.2. ORF 72/v-Cyclin (Sustains Cell Proliferation)

ORF72 encodes the functional viral homologue of cellular cyclin D. Like its cellular homologue, v-Cyclin acts like a constitutive activator of cellular cyclin-dependent kinase 6 (CDK6) and regulates cell cycle and cell proliferation [61]. The v-Cyclin mediates phosphorylation and inhibition of its cellular counterpart pRb protein, Histones H1, CDK inhibitor (cdki) and p27 (Kip1) through the formation of v-Cyclin-CDK6 complex [62]. The exact role of this viral protein in regulating KSHV life cycle is not fully understood but studies indicate that v-Cyclin mediates phosphorylation of nucleophosmin (NPM), through its association with CDK6 and facilitates NPM-LANA interaction and recruitment of HDAC1 to promote KSHV latency [63]. Additionally, due to its close functional relationship with murine gammaherpesvirus 68 (MHV68) v-Cyclin, KSHV v-Cyclin is believed to function as a modulator of the latent-lytic switch [64].

4.3. ORF71/v-FLIP (Resists Cell Death)

ORF71 encodes the KSHV homologue of cellular FLICE (Fas-associated death domain (FADD)-like interleukin-1 beta-converting enzyme) inhibitory protein, v-FLIP or K13. The v-FLIP activates one of the key cellular survival pathways, the NF-κB pathway, in latently infected PEL cells to promote cell proliferation and survival during latency [65,66]. It does so by binding to the inhibitor of kB-kinase γ (IKK γ) thereby leading to the activation of the NF-κB pathway [67,68]. NF-κB pathway activation by v-FLIP has been linked to KSHV lytic replication as a KSHV mutant lacking the v-FLIP gene is shown to inhibit ORF 50/RTA lytic gene expression [69].

4.4. K12/Kaposins

The Kaposin locus located a few kilobases away from the v-FLIP gene, upstream of the ORF73 is a complex and poorly understood locus that encodes for at least three proteins, namely, Kaposin A, B and C [54]. Kaposin A is a small hydrophobic latent oncogenic protein with inefficient transforming potential in rodent fibroblasts whereas Kaposin B is a small soluble nuclear protein, which acts like a scaffold or adaptor protein and activates p38/MAPK signaling pathway after binding to a key p38 substrate called mitogen-activated protein kinase 2 (MK2) [70]. All these proteins are shown to contribute to the pro-inflammatory KS tumor microenvironment.

4.5. Viral microRNAs

As shown in Figure 1, the kaposin locus also encodes 12 pre-miRNAs, out of which 10 miRNA (miR-K1-9 and -K11) are encoded between kaposin and the ORF71/K13 region while the other two miRNAs, miR-K10 and -K12 are located in the coding and 3' untranslated region of K12, respectively [71]. These KSHV-encoded pre-miRNAs further produce 18 mature miRNAs that are highly conserved in all the KSHV isolates. All miRNAs are expressed in latently infected cells and deletion of most viral miRNAs leads to a modest enhancement of KSHV reactivation, suggesting an important role of these miRNAs in regulating latent gene expression [72,73]. Among these miRNA, miR-K1 represses the expression of IκBα-an inhibitor of the pro-survival NF-κB pathway and inhibits the activation of lytic viral promoters. KSHV miR-K10 affects the TNF-like weak inducer of apoptosis (TWEAKR) and inhibits cell apoptosis by suppressing pro-inflammatory responses, which might contribute to KSHV latent infection [74]. Several other miRNAs such as miR-K3, -K4, -K5 and -K9 target nuclear factor I/B, Rbl2 protein, Bcl-2 associated factor, BCLAF1 and activates 3'UTR region of RTA protein respectively, in order to induce viral reactivation. As these miRNAs are expressed in latency, they could potentially target both cellular and viral miRNAs and contribute to the neovascular phenotype of KS [75].

5. Role of LANA in KSHV Episome Maintenance and Partitioning during Cell Division

In latently infected cells, the KSHV genome persists as an episome, a closed circular extra-chromosomal genome. In each KSHV-infected PEL cell, the copy number of the KSHV episome seems to be stable (~50–100 copies/cell) [76,77]. To accomplish this, the KSHV episome must replicate and then efficiently segregate to the daughter cells after mitosis to avoid the loss in copy number. During mitosis, the replicated viral episomes are segregated to the daughter cells mediated by LANA. LANA is a multifunctional protein required for episomal maintenance and segregation as deletion of KSHV LANA resulted in the complete loss of episomal genomes and failure to establish latent infection [78,79]. To successfully partition the episomes into dividing daughter cells, LANA simultaneously binds to the viral episome at the TR region and the host chromatin [80,81]. The *C*-terminal domain of LANA binds cooperatively to two sites (LANA binding sites) within each TR element, which is necessary for episome persistence and efficient segregation of KSHV episomes to the daughter cell [82,83]. In KSHV infected cells, LANA tethers to the condensed mitotic chromatin through interactions with several host proteins. These include H2A/B, histone H1, MeCP2, Brd4, NuMA, Bub1 and CENP-F [84–90].

A condensed chromatin consists of a DNA double helix wrapped around core histones (nucleosomes) compacted by linker histone (H1) along with other nucleosomal proteins. The histone proteins are the chief protein components of chromatin, acting as spools around which DNA winds. Histones are a family of small, positively charged proteins termed H2A, H2B, H3, and H4 and H1. The histones H2A, H2B, H3 and H4 are the core histones, while histone H1 is the linker histone. Two molecules each of H2A, H2B, H3, and H4 form the histone octameric nucleosome, which is bound and wrapped with DNA [91]. The linker histone H1 binds the nucleosome at the entry and exit sites of the DNA, thus locking the DNA into place and allowing the formation of the higher order structure [92]. It has been reported that LANA interacts directly with chromosomes via histones H2A and H2B (H2A/B) in the cells and uses them as a docking station to tether viral episomes to the cellular chromatin. By using affinity purification and mass spectrometry, Barbera *et al.* found that the *N*-terminus of LANA (LANA 1-32) can bind to the core histone proteins, H2A, H2B, H3 and H4 as well as Ku70, Ku80 and PARP1. The metaphase spread assays performed in knockout mouse embryonic fibroblasts (MEFs) showed that Ku70, Ku80, PRAP1 do not mediate the association between LANA and the chromosome. Since nucleosomes are comprised of two H2A/H2B dimers and one H3-H4 tetramer wrapped by DNA, LANA binding to any of the histone components would result in precipitation of all four core histones of the nucleosome, therefore further studies using immunoprecipitation and GST pull down assays showed that the *N* terminus LANA 1-32 and the full-length LANA can both specifically bind to H2A and H2B rather than H3, H4, and this binding can be found throughout the cell cycle. The roles of H2A/H2B in LANA-host chromosome association were further confirmed in *Xenopus laevis* sperm chromatin, which is naturally devoid of H2A/H2B. These experiments showed that LANA cannot bind to *Xenopus laevis* sperm chromatin, but this chromatin binding can be rescued after assembly of nucleosomes containing H2A/H2B [84].

The structure of LANA bound to the nucleosome resolved by X-ray diffraction showed that the *N*-terminus of LANA directly binds nucleosome core particles; the nucleosomal surface functions as a docking station for LANA [84]. Also a hairpin formed by the KSHV *N*-terminus of LANA is seen to interact with host chromatin through binding to an acidic patch formed by H2A/H2B dimer within the nucleosome. Analysis of the molecular surfaces of both, the LANA peptide and the H2A-H2B dimer demonstrate an appropriate charge complementarity, indicating that the LANA *N*-terminal region has evolved to recognize this region of the nucleosome core particles with high specificity [84]. It is very interesting that LANA cannot bind to either H2A or H2B alone, but is found to be associated with the H2A/H2B dimer [93]. Besides the H2A/B dimer of the nucleosome core particles, the association between LANA and the linker histone H1 was also found in body cavity-based lymphoma (BCBL) cells [90]. A recent report showed that LANA can associate with H2AX, an isoform of H2A, to contribute to the persistence of the KSHV genome in KSHV-positive cells [94]. These results strongly suggested that H2A/H2B is not only important but also essential for LANA chromosome association.

The centromere is a chromosomal apparatus that is required for chromosome segregation by ensuring the delivery of one copy of each chromosome to each daughter at cell division [95]. In mitosis, some kinetochore proteins such as Bub1 and CENP-F assemble at the surface of centromere and act as the docking site of the spindle microtubule binding [96]. As viral genomes lack centromeres, the KSHV episomes must tether to the host chromosomes to ensure that they are partitioned to daughter cells. A recent report showed that KSHV achieves this through LANA, which simultaneously binds to KSHV viral genomes and the centromeres of mitotic chromosomes via the formation of complexes with CENP-F and Bub1 [89]. Both the *N*-terminal and *C*-terminal domains of LANA strongly bind to Bub1 and CENP-F. The *N*-terminus of LANA showed greater efficiency of binding to CENP-F, whereas the basic *C*-terminal domain of LANA has a minor binding affinity to CENP-F as compared to the *N*-terminal domain. The dynamic association of LANA and Bub1/CENP-F demonstrated by immunofluorescence assays showed strong co-localization of LANA and Bub1 at each phase during mitosis and interphase. Also, the co-localization of LANA, Bub1 and the KSHV episomes was shown using fluorescence *in-situ* hybridization (FISH) assay. A dramatic decrease in the copy number of KSHV episomes was found in Bub1 knockdown KSHV-infected cell lines [89]. However, no significant changes in the KSHV episome copy number was observed in CENP-F1 knock-down KSHV positive cell lines [89]. The Bub1 knockdown KSHV-positive cells showed significant reduction in the number of KSHV episomes, which may be due to an inefficient passage of KSHV episomes to the progeny nuclei during mitosis and failure to maintain the viral genome in the absence of Bub1. This suggests that the interaction of CENP-F with LANA is probably redundant, compared to Bub1, although important in the context of LANA's function to tether KSHV genome to the host chromosome to ensure persistence of the viral genome. As a component of the spindle assembly checkpoint (ASC), Bub1 is important for the formation of mitotic checkpoint complex (MCC) [97,98]. The interaction of LANA with Bub1 might interfere with the function of Bub1 and the correct spindle formation, although this interaction is critical for segregation into the daughter cells. A recent study showed that LANA can promote the degradation of Bub1 in a ubiquitin dependent pathway which finally resulted in chromosomal instability (CIN) in KSHV-infected tumor cells [99].

Nuclear mitotic apparatus (NuMA) protein (238-kDa in human) is a component of the mitotic apparatus used for the segregation of dividing nuclei [100,101]. During the metaphase and anaphase stage of mitotic cell division, NuMA is distributed to the spindle poles to organize microtubule movement and to stabilize the mitotic spindle [102–104]. NuMA has also been shown to associate with small nuclear ribonucleoproteins and splicing factors involved in recycling and phosphorylation of RNA-processing factors and thus has been implicated in the regulation of DNA replication and transcription [105–107]. As a nuclear matrix protein, NuMA is thought to support the nuclear shape in differentiating cells [108]. In addition, a role for NuMA in DNA anchoring has also been proposed based on its interaction with matrix attachment regions, which anchor DNA on the nuclear matrix [109]. NuMA is also known to interact with a number of essential mitotic components, including microtubules [110,111], dynein/dynactin [112] and has crucial functions related to cell cycle progression [101,109,113,114]. NuMa has been shown to play a vital role in the maintenance of KSHV genome in the host cell in a cell cycle dependent manner [85]. During the interphase, NuMa may serve as the nuclear matrix and support the KSHV genome maintenance and segregation into the new daughter cells. We have shown that NuMA and LANA can interact with each other during interphase and their interaction is temporally lost as the cells enter the mitotic phase [85]. During mitosis, NuMA forms complexes with dynein/dynactin and microtubules, which is important for the segregation of replicated viral DNA. It has been shown that the amino and carboxy-terminal domains of LANA have different functions and can co-operate to tether KSHV genome to the human chromosomes [115]. It has been reported that NuMA binds to LANA in the carboxy-terminus between amino acids 840 and 963, which is adjacent to the TR-binding region and also involves interactions with dynein/dynactin and microtubules [85]. Additionally, the involvement of dynein/dynactin and microtubules in this interaction was proved by blocking the association of NuMA with these proteins, which resulted in the loss of KSHV episomes in the daughter cells.

Methyl CpG binding protein 2 (MeCP2) is a nuclear protein of about 75–80 kDa, that preferentially binds to methylated CpG dinucleotides [116–118]. It has been suggested that MeCP2 plays a key role in the transcriptional silencing of genes in CpG-methylated regions, activation of euchromatic genes, and mRNA splicing [119–121]. MeCP2 may also alter higher-order chromatin architecture [122,123]. It has been shown that MeCP2 interacts with the *N*-terminus of LANA and this interaction is vital for tethering of LANA to the host chromosomes [124]. In a separate study, it has been reported that the association of LANA and MeCP2 is modulated by the chromatin-binding motif of LANA located at the *N*-terminus [86]. In the same study, the authors also reported that co-expression of fluorescently tagged LANA and MeCP2 in murine cells resulted in remarkable relocalization of LANA from its diffuse nucleoplasmic distribution to being concentrated at the chromocenters, which corresponds to major accumulations of pericentric heterochromatin. Although the methyl-CpG-binding domain in LANA is sufficient to localize MeCP2 to chromocenters, relocalization of LANA requires both the methyl-CpG-binding domain and transcription repression domain. MeCP2 has been reported to enhance the transactivation of human E2F1 promoter through LANA and this effect is dependent on the chromatin-binding motif and methyl-CpG-binding domain. These findings indicate that multiple interactions are required for LANA to stably associate with chromatin and may occur as a two-step process in which nucleosome binding by the chromatin-binding motif facilitates the interaction of LANA with sequence- or context-specific cofactors such as MeCP2. These multivalent interactions possibly allow LANA to stabilize MeCP2 on low-affinity sites and facilitate KSHV to reprogram selected aspects of host gene expression [86].

DEK is a ubiquitous nuclear protein of ~43 kDa and has been shown to predominately associate with chromatin [125]. The protein was first identified in a chromosomal translocation with the NUP214 nucleoporin protein in a subset of acute myeloid leukemias [126]. It is also a ubiquitously expressed DNA-binding phosphoprotein that recognized Ets-binding sites in the human immunodeficiency virus type-2 enhancer [127–129], and is a constituent of splicing complexes [130]. It has been shown that DEK interacts with the carboxy-terminal domain of LANA between the amino acids 986 to 1043. Unlike the punctate localization of LANA on chromosomes of infected cells, DEK has broad distributions on human chromosomes [124]. It has been suggested that DEK associates with histones and plays a supporting role in LANA-mediated tethering to the host chromosome [131].

Bromodomain and extra-terminal (BET) proteins are a class of highly conserved bromodomain (BRD) containing proteins involved in fundamental cellular processes, such as meiosis, embryonic development, cell cycle regulation, and transcription, and have elevated activity in human leukemia [132–137]. Through their bromodomains, BRD2/RING3 interacts with acetylated histone H4 and BRD4 interacts with acetylated histones H3 and H4, thereby providing a docking station for other proteins to attach to the chromatin [138–141]. Interaction of BRD2/RING3 has been shown to promote G1/S transition [142]. It has been shown that LANA recruits BRD2/RING3 to chromatin [143,144]. LANA binds to RING3 through the extra-terminal (ET) domain, characteristic of fsh-related proteins, suggesting a highly conserved function in terms of protein-protein interactions [145–147]. In *in vitro* assays, it has been shown that BRD2/RING3 mediates phosphorylation of Ser/Thr residues within the *C*-terminal domain of LANA between amino acids 951 and 1107 [143]. RING3 localizes to the euchromatin regions in the interphase nucleus even in the absence of the KSHV viral genome and is released to the cytoplasm during mitosis. In KSHV-infected cells, most of the RING3 proteins co-localize with LANA suggesting that RING3 may contribute to KSHV genome persistence by local euchromatic microenvironment around the viral episomes tethered to the heterochromatic region of the chromosome [144]. The BRD2/RING3 binding domain of LANA was mapped between amino acids 1007 and 1055 [88]. Mutants of LANA capable of supporting replication and dimerization were able to interact with BRD2/RING3, suggesting that RING3 may be important for interaction of the *C*-terminal domain of LANA with heterochromatin [88]. It has also been seen that the DNA binding domain of LANA interacts with the extra-terminal domain of Brd4. Since Brd4 is associated with mitotic chromosomes throughout mitosis and is co-localized with LANA and KSHV episomes on host mitotic chromosomes, Brd4 might also be an interacting protein that mediates LANA tethering onto the mitotic chromosomes [87,88]. A model summarizing the role of cellular proteins in viral genome segregation is presented in Figure 2.

6. Role of LANA in KSHV DNA Replication

During latency the KSHV genome exists as a closed circular episome tethered to the host chromosomal DNA and replicates in harmony with the host cellular DNA (once per cell cycle). In general, KSHV-infected PEL cells maintain 50–100 copies of KSHV episome per cell and the copy number remains the same after multiple cell divisions. Viral DNA replication is a fundamental process to maintain a constant number of latent viral genome in the proliferating cells. In the absence of DNA replication, the KSHV genome is rapidly lost from the dividing cells leading to failure in the maintenance of persistent viral infection. LANA, a sequence-specific DNA binding protein, is shown to play an indispensable role in KSHV episomal DNA replication and segregation of the newly constructed genome copies to daughter nuclei. Replication of the KSHV genome is thought to be executed using cellular replication factors and LANA is documented to assist through its interaction with a variety of host cellular proteins.

6.1. LANA-TR Mediated DNA Replication of KSHV Episome

The terminal repeats (TRs) in the KSHV genome contain a DNA replication origin *cis*-element called ori-P that consists of two LANA-binding sites (LBS): a higher affinity site/LBS1 and a lower affinity site/LBS2, followed by an adjacent 32-bp GC-rich segment (reviewed in [33]). LANA, being a *trans*-acting protein, binds to the TRs forming a LANA DNA replication element. A 31 bp sequence upstream of LBS1/LBS2 is mapped as a replicator element (RE) and is critical for the initiation of replication. Plasmids containing a single copy of the TR with both LBS1/LBS2 and RE are considered to be replication sufficient in a LANA dependent manner; however, at least two copies of the TR are required for stable episomal persistence. Based on colocalization studies, LANA has been shown to co-localize with the artificial episomes (containing two TR units) along with mitotic chromosomes, thereby providing a model in which LANA docks the KSHV genome to the host chromatin during host cell division. LANA promotes viral DNA replication by directly binding at LANA-binding sites and recruiting the components of cellular pre-replication complexes (pre-RC), which include origin recognition complexes ORC1-6 (ORCs), Cdc6, Cdt1 and mini-chromosomal maintenance proteins (MCMs) to the origin of replication [148,149].

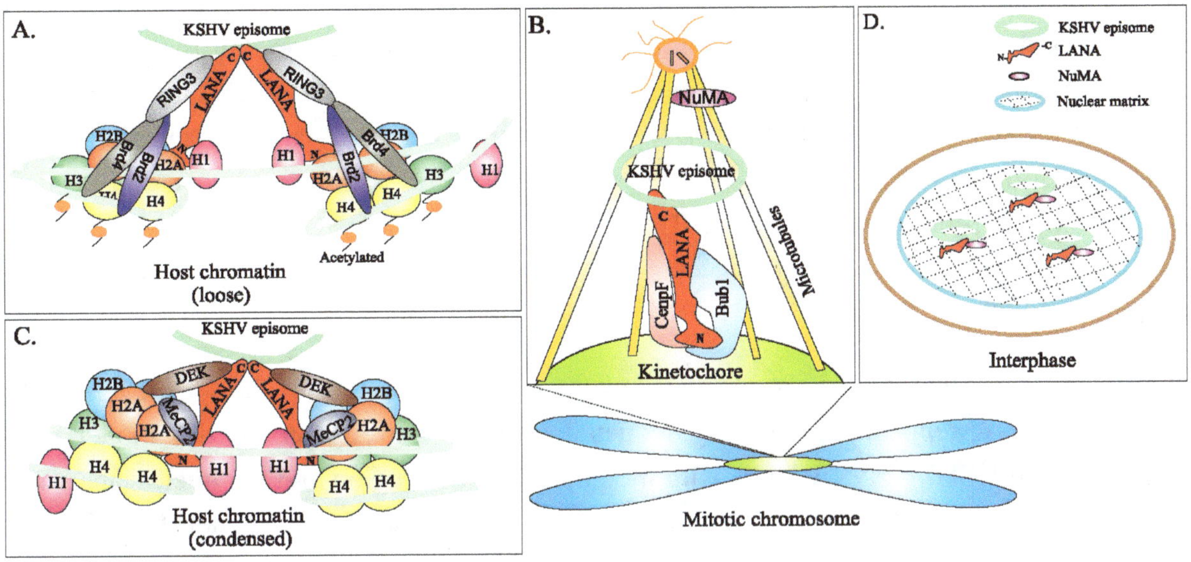

Figure 2. Schematic model showing the association of LANA with host proteins that aid in tethering to the host chromatin and segregation. The interaction of host cellular proteins with LANA during interphase (panel **D**) and the Mitotic phase (M-phase) is shown. During interphase, NuMA binds to the nuclear matrix and tethers KSHV genome by interacting with the carboxy-terminus of LANA, whereas during the M-phase, this interaction is lost and NuMA interacts with the microtubules and localizes to the spindle poles. Panel **A** shows the condensed chromatin structure during M-phase. LANA interacts with the core histone proteins (H2A, H2B, H3 and H4) through the *N*-terminal domain and binds to the KSHV episome through the *C*-terminal domain. Nucleosome binding by the chromatin-binding motif facilitates interaction of LANA with MeCP2 at the *N*-terminal domain of LANA. DEK protein interacts with the *C*-terminal domain of LANA and also to histones H2A, H2B, H3 and H4 facilitating tethering of LANA to the host chromatin. Panel **B** shows that both the *N*- and *C*-terminus of LANA strongly bind to the kinetochore proteins Bub1 and CENP-F that ensures delivery of KSHV episomes to the daughter cells during chromatid segregation. Panel **C** shows a loose chromatin structure, where BRD2/RING3 interacts with acetylated histone H4 and BRD4 interacts with acetylated histones H3 and H4, whereas LANA binds to the RING3 through the *C*-terminal domain, thus showing the contribution of host BET proteins in KSHV genome persistence.

According to the amino acid sequence, LANA is divided into three different protein domains: a proline-rich *N*-terminal domain containing the sequence motif that tethers LANA to host chromosomes, a long glutamic acid-rich internal repeat central domain and a carboxy-terminal domain (amino acids 770–1162 fragment). The association of LANA (LBS1/2 within the TRs) with the pre-RC complex units occurs in a cell cycle dependent manner, facilitated by both the *N*- and *C*-terminal domains of LANA. As only a few viral proteins are expressed during latent infection and these do not posses enzymatic activity required for DNA replication, it is suggested that the LANA-TR complex recruits the cellular DNA replication machinery to replicate viral DNA. LANA's role in latent DNA replication is regulated through the formation of the LANA-TR complex and further recruitment of the host cell replication machinery in a coordinated manner. Consistent with these studies, mutations within amino acids 4–32 of the *N*-terminal region of LANA and the region that regulates binding of LANA to LANA-binding sites of the TRs, led to a rapid loss of the KSHV genome in human cells, indicating that these regions are critical for DNA replication. Also the *C*-terminal domain of LANA is essential for episomal replication, as it interacts with several chromosome binding and origin recognition complex proteins such as Brd2/RING3/CBP/ORC2 and HBO1 and there is a decrease in the DNA replication efficiency when the expression levels of these proteins are knocked down by siRNAs [150].

The nuclear matrix region has been identified as the replication initiation site for the host cell-cycle dependent viral DNA replication. Studies conducted by Ueda's group proposed a model for latent DNA replication of KSHV, LANA binds with the nuclear matrix region and the nuclear matrix region serves as the site of the replication factory [33]. According to the proposed model, LANA associates with ori-P through LANA binding sites on TR and recruits it to the nuclear matrix region during the late G1 phase. This LANA-bound ori-P then serves as the launching pad for the recruitment of other cellular proteins, Cdc6, Cdt1 and MCMs (to establish a complete pre-RC complex) to the nuclear matrix region for the initiation of viral DNA replication (Figure 3).

Due to the absence of DNA polymerase/helicase activity by LANA, it is believed that LANA-mediated KSHV DNA replication depends on the enzymes that display these activities. One such enzyme, Topoisomerase II (TopoIIβ), is known to control the topology of DNA and to initiate DNA replication by activating double-stranded breaks on DNA. DNA affinity chromatography and proteomics analysis using KSHV TR DNA and the LANA binding site (as affinity column) recognized topoisomerase IIβ (TopoIIβ) as an important LANA-interacting protein. Our group showed that LANA interacts with TopoIIβ and forms complexes in KSHV-infected cells [151]. These studies confirmed that LANA recruits TopoIIβ to ori-P through its *N*-terminal domain for replication initiation. Further, a selective inhibitor of TopoIIβ, ellipticine, negatively regulated replication of TR confirming its role in DNA replication [151].

A recent report showed that the interaction of LANA with the replication factor C (RFC) complex is critical for KSHV episomal replication and genome persistence [152]. The RFC complex consists of a five-subunit (Rfc1, Rfc2, Rfc3, Rfc4 and Rfc5) protein complex that is needed for DNA replication [153]. The RFC complex is an AAA+ clamp loading ATPase and catalyzes the loading of DNA polymerase clamp loader, PCNA (proliferating cell nuclear antigen) on to the DNA. LANA recruits PCNA and LANA-enhanced PCNA loading is considered to be necessary for viral replication and persistent infection, as LANA mutants lacking RFC interaction negatively regulated LANA-mediated latent DNA replication in infected cells. These findings suggested that PCNA loading is a rate-limiting step in DNA replication and that LANA enhancement of PCNA loading permits efficient virus replication and persistence.

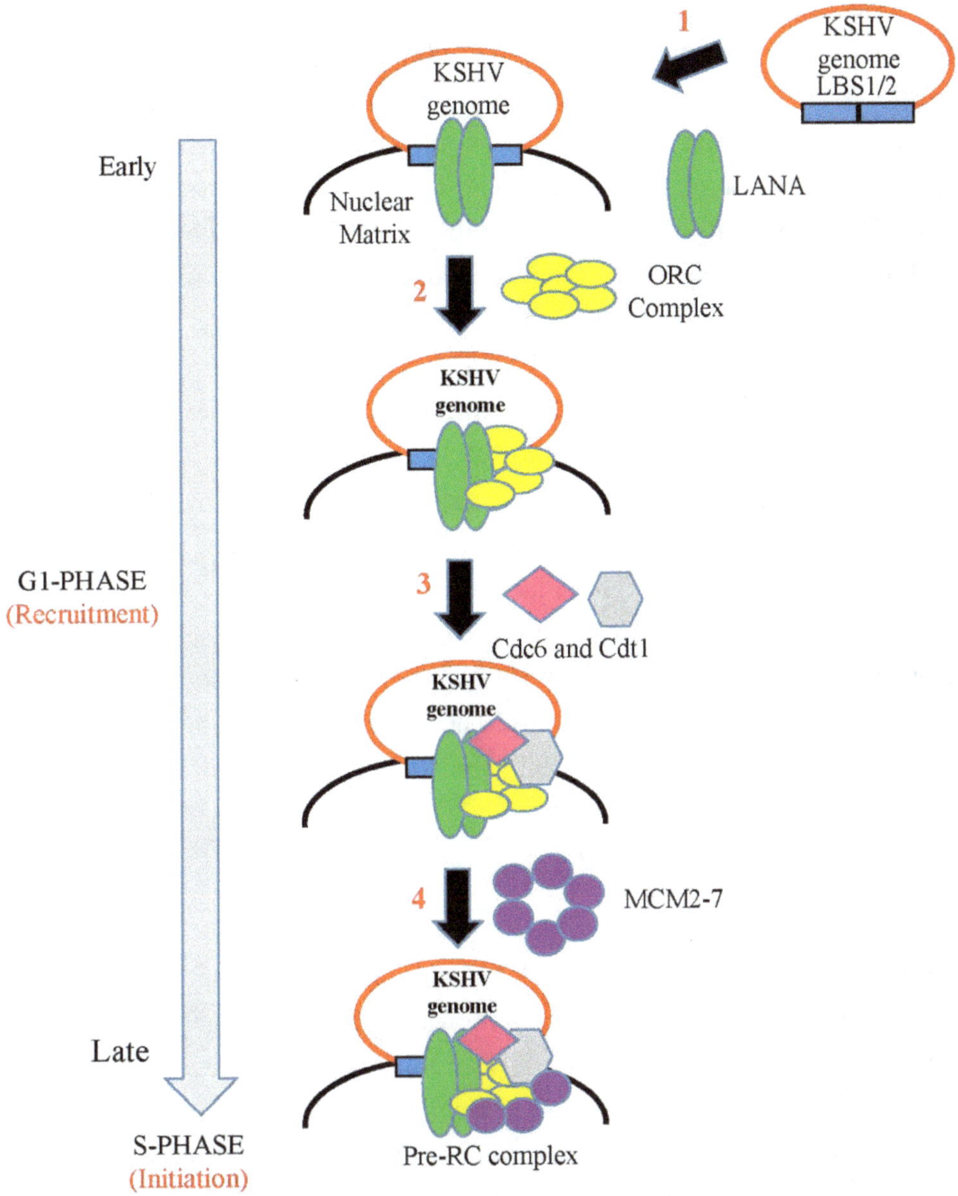

Figure 3. A model of the KSHV latent DNA replication. (**1**) LANA binds to the LANA-binding sites (LBS1 and LBS2) or replication origin of the terminal repeats (TR) region of the KSHV genome and recruits it to the nuclear matrix region; (**2**) LANA then recruits the host cellular machinery factors such as Origin Replication Complexes (ORCs) to the replication origin which is followed by (**3**) sequential loading of Cdc6, Cdt1; and (**4**) heterohexameric complex Mcm2-7 to the origins to form pre-replicative complex (pre-RC) during late G1 phase followed by replication of DNA during early S phase.

6.2. Epigenetic Regulation of LANA-TR Mediated KSHV DNA Replication

It is important to note that epigenetic modifications including DNA methylation, chromatin modifications and nucleosome positioning, also control KSHV DNA replication during latency [43,154]. Chromatin modifications are shown to regulate DNA replication by controlling the recruitment of replication proteins as well as the access of the replication machinery to the KSHV genome. Chromatin structure at the KSHV TR consists of four nucleosomes and two LANA binding sites [86]. The GC rich regions of TR have been shown to have a high tendency to form repressive heterochromatin that alters the binding of the cellular proteins required for pre-RC complex formation [43]. During the G1/S phase of the cell cycle, chromatin structure at the TR is altered and hidden DNA becomes more accessible to replication machinery components. Also, there is significant alteration in histone modification patterns at TR. The TR typically possesses high levels of acetylation at histones H3 and H4, in contrast to the internal region of the genome, which is abundant in activating histone marks (H3K4me3). During G1/S phase, there is a significant reduction in these activating histone marks while histone hyper acetylation remains constant. Levels of H3K4me3 appear to be correlated with the recruitment of MCM3 by the LANA-ORC complex at the latent origin of replication. LANA is also critical to the recruitment of HBO1, a member of the Myst family histone acetyl-transferases that interacts with ORC1 and MCM3 [148].

We recently showed that LANA upregulates the expression of Survivin, an inhibitor of apoptosis to increase the proliferation of KSHV-infected B cells [155]. Our previous studies showed that LANA recruits Aurora kinase B (AK-B), which induces the phosphorylation of Survivin at residue T34 leading to an enhanced activity of p300, which inhibits histone deacetylase 1 (HDAC-1) activity [156]. This in turns leads to an increase in acetylation of histone H3 on the viral genome that further increases the viral copy number in KSHV-infected B cells. The result is a boost of KSHV DNA replication in latently infected B-cells. These studies highlight the innovative role of LANA in the regulation of latent viral replication prior to mitosis.

6.3. Non-TR Mediated KSHV DNA Replication

Studies focused on the identification of LANA binding sites on KSHV genome identified TR-associated LANA binding and replication initiation sites. Our continued interrogation of the KSHV genome, led to the discovery of an additional DNA replication origin, called ori-A, using the single molecule analysis of replicated DNA (SMARD) technique [157,158]. In this technique, replicating DNA is labeled with IdU and CldU in order to determine the position, direction and the regions of replication forks on the replicated molecules. Our results indicated that replication of DNA can initiate throughout the KSHV genome. SMARD also showed that the utilization of multiple replication initiation sites occurs across large regions of the genome rather than at a specified sequence. The replication origin of the terminal repeats showed a slight preference for their usage, indicating that LANA dependent origin at the terminal repeats (TR) plays a role in genome duplication. Furthermore, ChIP assays for ORC2 and MCM3, which are part of the pre-replication complex (pre-RC), determined multiple genomic sites with these proteins. These suggested that initiation of replication is likely to be affected by the genomic context rather than by the DNA sequence.

7. LANAs Interaction with Various Cellular and Viral Pathways/Promoters

7.1. Analysis and Implication of Multiple LANA-Binding Sites within the KSHV Genome

Apart from regulating viral genome copy number and gene expression, LANA is also known to modulate host cell gene expression by interacting with different transcription factors and chromatin regulatory proteins, and through direct binding to the regulatory regions of the cellular genes [88,148,159–168]. Deep Sequencing and genome-wide analysis has been done to identify the LANA binding sites in KSHV latently infected cells. The results have revealed several LANA binding sites throughout the viral and host-cell genome with a small subset of LANA binding sites near transcription start sites (TSS) and gene promoters [81,169,170] of some cellular genes, suggesting a role for LANA in the regulation of host genes during KSHV infection. Interestingly, LANA binding on host chromatin is seen to be cell type specific and most LANA binding sites are found in endothelial cells [81,169,170].

Two-independent LANA-specific-ChIP-Seq experiments performed using the KSHV-positive BCBL-1 PEL cells have identified a total of 256 LANA binding sites with 17.5% of the peaks (45/256) being situated within ±2 kb of the TSS. In addition, two GC-rich DNA motifs were recognized in these LANA binding sites, with motif 1 being identical to the sequence of KSHV LBS1 and motif 2 displaying less similarity to KSHV LBS1/LBS2. Among the 256 total LANA binding peaks detected, LANA was shown to be associated with the genes within the p53 and tumor necrosis factor (TNF) regulatory networks, both of which help in the survival of latently infected cells. LANA was seen to bind near the TSS [81,170] of numerous genes that interact with p53, suggesting that LANA may be physically and functionally involved with p53 for the co-regulation of a common network of cellular genes important for cell cycle control in response to stress or DNA damage. Based on the comparison of LANA-specific-ChIP-Seq data with IFN-γ–inducible STAT1 DNA binding ChIP-Seq data, Lu *et al.* found that most LANA binding sites either co-localize or lie adjacent to STAT1 binding sites at the promoter regions of STAT1-dependent IFN-γ inducible genes, namely TAP1, PSMB9 and PARL [81]. The most enriched LANA binding sites were seen in the promoter regions of TAP1 and PSMB9, both of which are associated with antigen processing and presentation to the major histocompatibility complex class I (MHC-I) [171]. Indeed, LANA was found to compete with STAT1 for binding to the TAP1/PSMB9 promoter region, thus having the potency to attenuate activation of IFN-γ–inducible genes modulating the host cell antiviral immune response [81].

To better understand the interaction of LANA with cellular chromatin, Don Ganem's group carried out LANA ChIP-Seq experiments in the KSHV-infected epithelial cell line iSLK-219 (with tight latency and highly inducible lytic reactivation) and KSHV-infected primary lymphatic endothelial cells LEC-219s (phenotypic similarities to KS tumors). These studies suggested that the transcriptionally active promoter sites of host genes targeted by LANA are identical to the LANA binding site 1 (LBS1) motif in KSHV DNA [170]. A total of 267 reproducible LANA-binding sites were observed in iSLK-219 cells and 2481 in LEC-219s. Whereas most of the LANA binding sites in LEC-219 were unique to the LEC-219 cell line, 67% (179 peaks) of the LANA binding sites mapped in iSLK-219 cells were also detected in LEC-219s [170]. While 41.8% of LANA-enriched peaks were located within 1 kb of a TSS, nearly 11.6% of the peaks were also mapped between −1 kb and −10 kb of a TSS (distal promoter). The same study also identified 29 LANA-binding sites in the host genome of BCBL-1 cells, though out of these novel LANA-binding sites only 8 were previously detected and reported in the study by Lu *et al.*, using the same cell line [170]. Additionally, 4 gene promoters (SGMS1, SBF2, IQGAP3 and NIPAL2) were found to bind to LANA in 5 different cell lines studied, namely KSHV-infected endothelial cells (LEC-219), primary blood endothelial cells (BECs), Burkitt's lymphoma B cells (BJAB), PEL cell line (BCBL-1) and epithelial cells (iSLK-219) [170]. Due to the preferential enrichment of LANA at a TSS, the chromatin that is associated with the proximal promoters occupied by LANA was also examined. ChIP-Seq analysis of histone activating marks (H3K4me3) in iSLK-219 cells displayed the co-localization of LANA and H3K4me3 at proximal promoters, suggesting the possibility of LANA association with active chromatin [170]. Most of the LANA-associated TSS were highly populated with activating H3K4me3 histone marks whereas only a small subset of LANA-enriched regions showed increased levels of the repressive (H3K27me3) histone marks [170]. However, the presence of LANA on the proximal promoter did not impact their transcriptional activity or regulated the deposition of H3K4me3 marks. These observed results were further supported by those reported earlier by Hu *et al.*, where LANA binding to TSS correlated with H3K4me3, but not with the H3K27me3 marks and 86% of all LANA bound promoters were found to be transcriptionally active [169]. Taken together, these results suggested that LANA binding to the host gene promoters in the KSHV-infected cells is not sufficient to modulate the host gene expression, indicating LANA's indirect role in the regulation of gene expression during infection [169]. The ChIP-Seq results showed that the association of LANA with the host and viral genome was disrupted during late lytic cycle of KSHV infection [170].

LANA Chip-Seq studies performed by Hu *et al.* using BCBL-1 and TIVE-LTC cells (Telomerase-Immortalized human umbilical Vein Endothelial cells) showed that LANA binds to many more promoters in TIVE-LTC cells compared to BCBL-1 (also supported by the study of Mercier *et al*). Additionally, LANA was found to bind to host cellular DNA directly using two distinct motifs [169]. Fewer LANA-binding sites (58 of 2180 LANA binding sites in BCBL-1 cells, and 205 of 2951 LANA-binding sites in TIVE-LTC cells) showed sequence homology to the known LANA binding sites (LBS1/LBS2) [169]. A large number of LANA binding sites instead had a novel motif $(TCCAT)_3$ whose affinity for LANA was lower than LBS1, but was comparable to LBS2 [169]. In contrast to the report by Hu *et al.*, Mercier *et al.* identified the majority (58.8%) of the LANA ChIP-peaks to have sequences resembling LBS1 [169].

In a recent study by Hu *et al.*, LANA occupancy on the human genome in BCBL-1 and TIVE-LTC cells was studied and 26 genes were found enriched between the two cell types [169]. Three genes, PARL, NIPAL2, IQGAP3 were BCBL-1 specific while four genes, MRPL53, NFYC, CCDC90B and HIST2BE were TIVE-LTC specific. Also WDR74 displayed similar binding profiles in both cell lines [169]. Survivin and Id-1, regulated by LANA, [169,172] contained LANA peaks within the promoters in both cell types. The gene ontology analysis showed that in BCBL-1 cells the putative LANA targets were related to phosphorous metabolic processes, regulation of cellular enzymatic activity and regulation of cellular response to stress, whereas in TIVE-LTC cells the LANA targets detected are involved in regulation of macromolecular metabolic process, nutrient levels, and angiogenesis which is a hallmark of KS [169]. However, it was observed that 14% of the cellular promoters that are bound by LANA were not expressed in BCBL-1 cells during latency [169]. In KSHV-infected LEC cells, only 3.7% of the host gene promoters bound by LANA were differentially regulated [170]. This suggested that LANA binding to a host gene promoter alone has no direct impact on gene expression. However, certain genes like IQGAP3 gene can be directly regulated by LANA [81,170]. Thus, it appears that although LANA binding to the host gene promoter is not responsible for the host gene expression, it still remains an important permissive factor that allows co-ordination with other factors to regulate host gene expression.

7.2. LANA Associates with Several Viral Promoters and Causes Epigenetic Modifications

Upon *de novo* KSHV infection, there is a constant expression of the key latent genes along with a transient yet robust expression of a handful of lytic genes. A series of studies have shown that the immediate-early lytic protein, Replication and Transcription Activator (RTA), encoded by KSHV ORF50 (691 amino acid residues/120 kDa) is the only viral lytic protein that is necessary and sufficient to disrupt latency and promote the complete lytic cascade. RTA protein acts as the master latent-to-lytic switch that triggers KSHV to enter into the productive transcriptional program required for viral spread and KS pathogenesis (reviewed in [173]). RTA auto-activates its own promoter and trans-activates the expression of multiple downstream lytic genes including ORF K8, K5, K2, K12, ORF 6, ORF 57, ORF 74, K9, ORF 59, K3, ORF 37, K1, K8.1A, ORF 21, vIL-6, PAN RNA, vIRF1, ORF-K1 and small viral capsid proteins ORF 65, either alone through an RTA-responsive element or in combination with other viral regulatory genes [69,174–176]. KSHV RTA was also shown to interact with host cellular proteins and modulate cellular as well as viral gene expression. However, establishment of latency disrupts a full-blown expression of RTA and other lytic genes, though the molecular events behind this rapid inhibition of the lytic promoters are still poorly understood.

Earlier study showed a significant increase in the immediate-early gene expression following transfection of a LANA-deletion mutant of KSHV, suggesting that LANA plays a critical role in transcription repression of the lytic genes [177]. LANA, the major latent protein, is involved in the repression of the basal level of RTA promoter activity and other RTA-responsive lytic promoters as: (1) LANA interacts with several transcription repressors and histone-modifying enzymes associated with lytic genes silencing (reviewed in [173]); (2) LANA promotes establishment of latent histone modification patterns on the RTA promoter and represses the RTA-mediated gene activation (reviewed in [69]); and (3) The mutual LANA and RTA feedback regulatory mechanism promotes the establishment of KSHV latency [162].

The transcription repression activity of LANA occurs through its interaction with transcriptional repressors (heterochromatin protein HP1α, methyl-CpG-binding protein MeCP2, histone deacetylase co-repressor mSin3 complex and DNA methyltransferases, DNMTs) and chromatin-remodeling proteins (H3K9me3 histone methyltransferase SUV39H1, H3K9 demethylase KDM3A, histone acetyltransferase CBP and chromatin transcription complex FACT) [86,161,178–182]. Previous reports have indicated that interaction between LANA and the recombination signal sequence binding protein Jκ (RBP-Jκ), a major transcriptional repressor of the Notch signaling pathway, is essential to suppress the transcription of RTA [183,184]. By directly interacting with RBP-Jκ protein, LANA not only recruits co-repressors to down-regulate the expression of RTA gene but also represses RTA auto-activation activity. The fact that both positive and negative regulators of RTA gene expression use the same RBP-Jκ dependent mechanism, suggests that the switch between latency and lytic reactivation is fine-tuned by the levels of LANA and RTA proteins in KSHV-infected cells. Furthermore, RTA also induces LANA expression providing a negative feedback in keeping viral lytic reactivation under check.

DNA methylation and post-translational histone modifications play a central role in the regulation of gene expression [185]. DNA methylation of functionally conserved immediate early gene of herpesvirus genome by *de novo* methyltransferases, DNMT3a/DNMT3b results in establishment of methylation marks on the immediate early gene promoter (but not at the latency locus) [186], followed by gene silencing (reviewed in [34]). Treatment of BCBL-1 cells with histone deacetylase inhibitors, including sodium butyrate (NaB) and Trichostatin A, causes a rapid dissociation of LANA from the RTA promoter, initiating transcription activation of the RTA gene [187]. Interestingly, expression of LANA is also regulated by post-translational modifications such as arginine methylation, phosphorylation and SUMOylation [178]. Furthermore, phosphorylation of LANA by several kinases including glycogen synthase kinase (GSK-3), DNA-PK/Ku and Pim family kinase members, Pim-1 and Pim-3, has been reported to promote viral reactivation by negative modulation of LANA function [188–190]. LANA is also identified as a substrate for protein arginine methyltransferase 1 (PRMT1) and methylation at the R20 site is found to influence strong binding of LANA to the KSHV genome and repression of lytic genes [191]. Hence, these studies clearly show that LANA plays a role in lytic gene silencing during the establishment of latency.

Several independent groups have characterized the epigenetic modification of KSHV episomes during latency using genome-wide ChIP-seq assays, which revealed that latency-associated genes such as LANA are distinctively associated with activating histone marks (acetylated H3 and H3K4me3) while most lytic genes are enriched with either bivalent chromatin (acH3, H3K4me3 and H3K27me3) or active chromatin (acH3/H3K4me3-rich) [192–195]. In addition, two histone-modifying enzymes, EZH2, a H3K27me3 histone methyltransferase of the Polycomb Repressive Complex 2 (PRC2) and JMJD2A, a H3K9me3 histone demethylase, are associated with the latent KSHV genome [196,197]. The regulated removal of the gene-silencing epigenetic mark, trimethylation of lysine 27 of histone H3 (H3K27me3), by either transient expression of UTX or JMJD3 (the H3K27 demethylase) or by blocking with EZH2 (H3K7 methyltransferase) of PRC2 complex, disrupts latency and induces lytic reactivation [194]. Intriguingly, LANA can associate with JMJD3 and EZH2, and this interaction maintains H3K27me3-enriched heterochromatin on lytic genes to repress their expression during latency [194].

7.3. LANA Associates with Host Proteins and Signaling Pathways

LANA is shown to interact with various cellular proteins, indicating that LANA may target these proteins to promote KSHV-mediated tumorigenesis. Although KSHV is equipped to manipulate and deregulate several cellular proteins and associated signaling pathways (reviewed in [38,198]), it is not yet understood how this interaction inhibits apoptosis and leads to cell proliferation and cell transformation. Deeper understanding of the interplay of LANA and cellular factors in KSHV-infected cells may provide valuable information on the precise mechanisms of KS infection that could enable the development of drug therapies for KSHV-induced oncogenesis.

At the gene transcriptional level, LANA has been shown to inhibit KSHV lytic reactivation as it blocks the expression of KSHV RTA, which is critical for the KSHV latency to lytic switch [176]. LANA has been reported to display both transcription repression and activation activities by interacting with multiple cellular transcriptional factors, including the components of mSin3 complex, CBP, RING3, GSK-3β, p53 and pRb (for gene silencing) and E2F, Sp1, RBP-Jκ, ATF4/CREB2, CBP, Id-1 and Ets-1 (for gene activation) [160–162,172,199–205]. Here we discuss some of these key cellular proteins along with newly identified LANA-cellular protein interactions.

7.3.1. LANA Interaction with p53 and pRb

LANA has long been known to bind to and block p53-transcriptional activity and inhibit p53-induced cell death [156,166]. In human cells, constitutive expression of LANA was shown to induce chromosomal instability by suppressing the transcription of p53 from its endogenous promoter. Our previous studies have indicated that in KSHV-infected cells, p53 can be degraded by recruitment of the cellular EC_5S ubiquitin complex, targeted by the SOCS-box (suppressor of cytokine signaling) motif of LANA [206]. Recently, our group reported that LANA can also upregulate the levels of Aurora A, a centrosome-associated Ser/Thr oncogenic kinase, which promotes phosphorylation of p53 and LANA-mediated p53 ubiquitylation and degradation [207]. LANA was also shown to interact with the retinoblastoma (Rb) protein and enhance the transcriptional activation of the E2F-responsive gene/cyclic E promoter [208]. Co-expression of LANA and H-Ras has been shown to transform primary rat embryonic fibroblast (REF) cells *in vitro* [166]. LANA expression in lymphoid cells is found to overcome the cyclin-dependent kinase inhibitor, p16INK4a, and BRD4- and BRD2/RING3-induced G1 cell cycle arrest and stimulate E2F-mediated S-phase entry [62,88].

7.3.2. LANA Interaction with GSK-3β (Wnt Signaling Pathway)

LANA, also interacts with glycogen synthase kinase (GSK-3β), a kinase involved in phosphorylation and subsequent degradation of many proteins involved in cell cycle regulation, such as β-catenin, proto-oncoprotein c-Myc, c-Jun and cyclin D genes [209]. Interaction of LANA with GSK-3β, an important modulator of the Wnt signaling pathway, leads to an overall inactivation and nuclear sequestering of GSK-3β due to extracellular signal-regulated kinase 1/2 (ERK1/2)-mediated phosphorylation of GSK-3 at Ser9, which stabilizes cytosolic β-catenin and makes it available for transcriptional activation of target genes [210]. LANA also interacts and stabilizes c-Myc, another known GSK-3β substrate whose activity is regulated by ERK1/2 [211]. LANA decreases c-Myc phosphorylation on Thr58, an event that promotes Myc-induced apoptosis; and increases phosphorylation of c-Myc at Ser62, an event that transcriptionally activates c-Myc [211].

7.3.3. LANA Interacts with ANG and Promotes Angiogenesis

LANA has been reported to interact with the 14-kDa multifunctional protein angiogenin (ANG), which is considered to play a critical role in establishment of KSHV latency, anti-apoptosis and tumor angiogenesis [212]. Also, in KSHV-latently infected telomerase-immortalized microvascular endothelial (TIME) and BCBL-1 PEL cell lines, LANA and ANG were found to co-interact with Annexin A2, a protein that plays significant roles in cell proliferation, apoptosis and exo-/endocytosis [213]. Suppression of Annexin A2 or ANG expression in PEL cells was found to increase cell death, whereas depletion of Annexin A2 led to a concomitant decrease in both ANG and LANA protein expression, indicating that these three proteins are integrated and functionally important to promote viability of latently infected cells.

7.3.4. LANA Interacts with BMP-Activated Smad1

Recently, in the KSHV-transformed mesenchymal precursor (KMM) cells, KSHV LANA was shown to affect the bone morphogenetic protein (BMP) signaling pathway and convert it to an oncogenic BMP-Smad1-Id pathway, which might contribute to the pathogenesis of KSHV-induced malignancies [200]. BMP signaling pathways are involved in both cell promotion and inhibition of tumor progression. Also, inhibitor of DNA (Id) proteins are major downstream targets of BMP signaling and negative-regulators of basic helix-loop-helix (HLH) proteins [214]. In these studies, the *N*-terminus of LANA (1–432 amino acid residues) is found to interact with MH2-M, the LANA-binding domain of BMP-activated pSmad1 [199]. LANA-Smad1 interaction led to an increase in the Smad1 loading on the Id promoter which then upregulates the Id protein expression in the KS lesions. Chemical inhibition of the BMP-Smad1-Id pathway by Dorsomorphin and WSS25 inhibits the growth of the KSHV-induced tumors, indicating that small inhibitors targeting this pathway can serve as potential therapeutic agents for the treatment of KS.

7.3.5. LANA Interacts with Host KAP1 and Facilitates the Establishment of KSHV Latency

Two studies have recently reported that LANA interacts with a new host nuclear protein, called KAP1 (KRAB-associated protein 1) both *in vivo* and *in vitro* and down-regulates the lytic gene expression to facilitate KSHV latency [215–217]. KAP1 is a cellular transcriptional repressor that controls chromatin remodeling and KSHV biphasic life cycles. LANA is reported to bind to both the *N*- and *C*-terminal domains of KAP1, which interacts with SETDB1, H3K9me3 methyltransferase and recruits them to the lytic promoter region of the KSHV genome [218]. In cells harboring latent KSHV, shRNA knockdown of KAP-1 resulted in the induction of lytic genes and 5-fold increase of RTA-mediated lytic reactivation. Thus, KAP1 is believed to play an important role in the complete shutdown of the transient lytic gene expression during the early stages of KS infection.

7.3.6. LANA Interacts with Daxx and Contributes to VEGFR–Mediated Angiogenesis

The death-associated protein Daxx is a multifunctional protein that binds to several cellular host proteins and regulates a variety of cellular processes, including transcriptional repression and apoptosis [219]. Co-immunoprecipitation analysis of human HeLa cells harboring KSHV LANA identified Daxx as a potential LANA-binding protein [202]. These results indicate that LANA associates with Daxx and colocalizes in the nucleus of KSHV-infected BCBL-1 cells. The Glu/Asp-rich domain within LANA (321–344 amino acid region of LANA) was shown to bind to the central 63–440 amino acid region within Daxx, containing two paired amphipathic helices and its following 20 amino acids. Also, Daxx was found to repress Ets-1 dependent VEGF receptor-1 gene expression. However, there is an indirect regulation of the Ets-1 responsive promoters through a LANA-Daxx interaction [202]. Co-IP and immunoblotting experiments indicate that LANA-binding to Daxx disrupts the association between Daxx and Ets-1, which in turn, contributes to an increased level of VEGR receptor in KS tissues [202]. Vascular endothelial growth factor (VEGF) and its receptors are highly expressed in KS lesions and are known to play a key role in angiogenesis.

8. Concluding Remarks

The viral protein LANA encoded by KSHV during latent infection is critical for the maintenance of latency, episome replication, segregation, and also for regulating certain viral and cellular genes. LANA has been reported to interact with many host cellular proteins important for DNA replication, transcriptional regulation and cell cycle control. The interaction between LANA and host proteins leads to maintenance of the KSHV episome and transformation of the host cell. Strategies that can interrupt the interaction between LANA and host cell proteins may provide effective treatment and prevention of KSHV-associated diseases. LANA is also known to regulate several host cellular pathways. ChIP-Seq data revealed that LANA binds directly to the transcription start site of many host cellular genes. Interestingly LANA was found to bind to the genes within the p53 and TNF networks, both of which promote survival of latently infected cells. However, only a small percentage of LANA-bound cellular promoters showed direct LANA-driven gene regulation. Therefore, LANA binding to the promoter region of host genes may facilitate participation of other factors to regulate the LANA-driven host gene expression. Further investigations of LANA's role in KSHV latent DNA replication, manipulation of host and viral gene expression, and association with several cellular and viral proteins are required to facilitate our understanding of LANA-mediated KSHV latency and related oncogenesis.

Acknowledgments

We sincerely apologize to authors whose important work could not be cited here owing to space limitations. This work was supported by the public health grants from National Institute of Health (grant numbers CA174459 and AI105000) to Subhash C. Verma and (grant numbers CA137894, CA171979, CA174439, CA177423, P30-DK-050306 and P01-CA-174439) to Erle S. Robertson.

References

1. Sturzl, M.; Zietz, C.; Monini, P.; Ensoli, B. Human herpesvirus-8 and Kaposi's sarcoma: Relationship with the multistep concept of tumorigenesis. *Adv. Cancer Res.* **2001**, *81*, 125–159.
2. Chang, Y.; Cesarman, E.; Pessin, M.S.; Lee, F.; Culpepper, J.; Knowles, D.M.; Moore, P.S. Identification of herpesvirus-like DNA sequences in AIDS-associated Kaposi's sarcoma. *Science* **1994**, *266*, 1865–1869.
3. Boshoff, C.; Chang, Y. Kaposi's sarcoma-associated herpesvirus: A new DNA tumor virus. *Annu. Rev. Med.* **2001**, *52*, 453–470.
4. Whitby, D.; Howard, M.R.; Tenant-Flowers, M.; Brink, N.S.; Copas, A.; Boshoff, C.; Hatzioannou, T.; Suggett, F.E.; Aldam, D.M.; Denton, A.S.; *et al.* Detection of Kaposi sarcoma associated herpesvirus in peripheral blood of HIV-infected individuals and progression to Kaposi's sarcoma. *Lancet* **1995**, *346*, 799–802.

5. Gao, S.J.; Kingsley, L.; Hoover, D.R.; Spira, T.J.; Rinaldo, C.R.; Saah, A.; Phair, J.; Detels, R.; Parry, P.; Chang, Y.; *et al*. Seroconversion to antibodies against Kaposi's sarcoma-associated herpesvirus-related latent nuclear antigens before the development of Kaposi's sarcoma. *N. Engl. J. Med.* **1996**, *335*, 233–241.
6. Martin, J.N.; Ganem, D.E.; Osmond, D.H.; Page-Shafer, K.A.; Macrae, D.; Kedes, D.H. Sexual transmission and the natural history of human herpesvirus 8 infection. *N. Engl. J. Med.* **1998**, *338*, 948–954.
7. Cesarman, E.; Chang, Y.; Moore, P.S.; Said, J.W.; Knowles, D.M. Kaposi's sarcoma-associated herpesvirus-like DNA sequences in AIDS-related body-cavity-based lymphomas. *N. Engl. J. Med.* **1995**, *332*, 1186–1191.
8. Soulier, J.; Grollet, L.; Oksenhendler, E.; Cacoub, P.; Cazals-Hatem, D.; Babinet, P.; d'Agay, M.F.; Clauvel, J.P.; Raphael, M.; Degos, L.; *et al*. Kaposi's sarcoma-associated herpesvirus-like DNA sequences in multicentric Castleman's disease. *Blood* **1995**, *86*, 1276–1280.
9. Gill, P.S.; Tsai, Y.C.; Rao, A.P.; Spruck, C.H., 3rd; Zheng, T.; Harrington, W.A., Jr.; Cheung, T.; Nathwani, B.; Jones, P.A. Evidence for multiclonality in multicentric Kaposi's sarcoma. *Proc. Natl. Acad. Sci. USA* **1998**, *95*, 8257–8261.
10. Gessain, A.; Duprez, R. Spindle cells and their role in Kaposi's sarcoma. *Int. J. Biochem. Cell Biol.* **2005**, *37*, 2457–2465.
11. Ambroziak, J.A.; Blackbourn, D.J.; Herndier, B.G.; Glogau, R.G.; Gullett, J.H.; McDonald, A.R.; Lennette, E.T.; Levy, J.A. Herpes-like sequences in HIV-infected and uninfected Kaposi's sarcoma patients. *Science* **1995**, *268*, 582–583.
12. Uldrick, T.S.; Wang, V.; O'Mahony, D.; Aleman, K.; Wyvill, K.M.; Marshall, V.; Steinberg, S.M.; Pittaluga, S.; Maric, I.; Whitby, D.; *et al*. An Interleukin-6-Related Systemic Inflammatory Syndrome in Patients Co-Infected with Kaposi Sarcoma-Associated Herpesvirus and HIV but without Multicentric Castleman Disease. *Clin. Infect. Dis.* **2010**, *51*, 350–358.
13. Holkova, B.; Takeshita, K.; Cheng, D.M.; Volm, M.; Wasserheit, C.; Demopoulos, R.; Chanan-Khan, A. Effect of highly active antiretroviral therapy on survival in patients with AIDS-associated pulmonary Kaposi's sarcoma treated with chemotherapy. *J. Clin. Oncol.* **2001**, *19*, 3848–3851.
14. Martin, J.N. The epidemiology of KSHV and its association with malignant disease. In *Human Herpesviruses: Biology, Therapy, and Immunoprophylaxis*; Arvin, A., Campadelli-Fiume, G., Mocarski, E., Moore, P.S., Roizman, B., Whitley, R., Yamanishi, K., Eds.; Cambridge University Press: Cambridge, UK, 2007.
15. Dourmishev, L.A.; Dourmishev, A.L.; Palmeri, D.; Schwartz, R.A.; Lukac, D.M. Molecular genetics of Kaposi's sarcoma-associated herpesvirus (human herpesvirus-8) epidemiology and pathogenesis. *Microbiol. Mol. Biol. Rev.* **2003**, *67*, 175–212.
16. Gao, S.J.; Kingsley, L.; Li, M.; Zheng, W.; Parravicini, C.; Ziegler, J.; Newton, R.; Rinaldo, C.R.; Saah, A.; Phair, J.; *et al*. KSHV antibodies among Americans, Italians and Ugandans with and without Kaposi's sarcoma. *Nat. Med.* **1996**, *2*, 925–928.
17. Kedes, D.H.; Operskalski, E.; Busch, M.; Kohn, R.; Flood, J.; Ganem, D. The seroepidemiology of human herpesvirus 8 (Kaposi Kaposi's sarcoma-associated herpesvirus): Distribution of infection in KS risk groups and evidence for sexual transmission. *Nat. Med.* **1996**, *2*, 918–924.

18. Simpson, G.R.; Schulz, T.F.; Whitby, D.; Cook, P.M.; Boshoff, C.; Rainbow, L.; Howard, M.R.; Gao, S.J.; Bohenzky, R.A.; Simmonds, P.; *et al.* Prevalence of Kaposi's sarcoma associated herpesvirus infection measured by antibodies to recombinant capsid protein and latent immunofluorescence antigen. *Lancet* **1996**, *348*, 1133–1138.
19. Dukers, N.H.; Rezza, G. Human herpesvirus 8 epidemiology: What we do and do not know. *AIDS* **2003**, *17*, 1717–1730.
20. Hayward, G.S. KSHV strains: The origins and global spread of the virus. *Semin. Cancer Biol.* **1999**, *9*, 187–199.
21. Hayward, G.S.; Zong, J.C. Modern evolutionary history of the human KSHV genome. *Curr. Top. Microbiol. Immunol.* **2007**, *312*, 1–42.
22. Stebbing, J.; Bourboulia, D.; Johnson, M.; Henderson, S.; Williams, I.; Wilder, N.; Tyrer, M.; Youle, M.; Imami, N.; Kobu, T.; *et al.* Kaposi's sarcoma-associated herpesvirus cytotoxic T lymphocytes recognize and target Darwinian positively selected autologous K1 epitopes. *J. Virol.* **2003**, *77*, 4306–4314.
23. Poole, L.J.; Zong, J.C.; Ciufo, D.M.; Alcendor, D.J.; Cannon, J.S.; Ambinder, R.; Orenstein, J.M.; Reitz, M.S.; Hayward, G.S. Comparison of genetic variability at multiple loci across the genomes of the major subtypes of Kaposi's sarcoma-associated herpesvirus reveals evidence for recombination and for two distinct types of open reading frame K15 alleles at the right-hand end. *J. Virol.* **1999**, *73*, 6646–6660.
24. Zong, J.; Ciufo, D.M.; Viscidi, R.; Alagiozoglou, L.; Tyring, S.; Rady, P.; Orenstein, J.; Boto, W.; Kalumbuja, H.; Romano, N.; *et al.* Genotypic analysis at multiple loci across Kaposi's sarcoma herpesvirus (KSHV) DNA molecules: Clustering patterns, novel variants and chimerism. *J. Clin. Virol.* **2002**, *23*, 119–148.
25. Vitale, F.; Viviano, E.; Perna, A.M.; Bonura, F.; Mazzola, G.; Ajello, F.; Romano, N. Serological and virological evidence of non-sexual transmission of human herpesvirus type 8 (HHV8). *Epidemiol. Infect.* **2000**, *125*, 671–675.
26. Sitas, F.; Newton, R. Kaposi's sarcoma in South Africa. *J. Natl. Cancer Inst. Monogr.* **2001**, *28*, 1–4.
27. Bagni, R.; Whitby, D. Kaposi's sarcoma-associated herpesvirus transmission and primary infection. *Curr. Opin. HIV AIDS* **2009**, *4*, 22–26.
28. Vieira, J.; Huang, M.L.; Koelle, D.M.; Corey, L. Transmissible Kaposi's sarcoma-associated herpesvirus (human herpesvirus 8) in saliva of men with a history of Kaposi's sarcoma. *J. Virol.* **1997**, *71*, 7083–7087.
29. Farge, D.; Lebbe, C.; Marjanovic, Z.; Tuppin, P.; Mouquet, C.; Peraldi, M.N.; Lang, P.; Hiesse, C.; Antoine, C.; Legendre, C.; *et al.* Human herpes virus-8 and other risk factors for Kaposi's sarcoma in kidney transplant recipients. Groupe Cooperatif de Transplantation d'Ile de France (GCIF). *Transplantation* **1999**, *67*, 1236–1242.
30. Garcia-Astudillo, L.A.; Leyva-Cobian, F. Human herpesvirus-8 infection and Kaposi's sarcoma after liver and kidney transplantation in different geographical areas of Spain. *Transpl. Immunol.* **2006**, *17*, 65–69.

31. Hladik, W.; Dollard, S.C.; Mermin, J.; Fowlkes, A.L.; Downing, R.; Amin, M.M.; Banage, F.; Nzaro, E.; Kataaha, P.; Dondero, T.J.; *et al.* Transmission of human herpesvirus 8 by blood transfusion. *N. Engl. J. Med.* **2006**, *355*, 1331–1338.
32. Russo, J.J.; Bohenzky, R.A.; Chien, M.C.; Chen, J.; Yan, M.; Maddalena, D.; Parry, J.P.; Peruzzi, D.; Edelman, I.S.; Chang, Y.; *et al.* Nucleotide sequence of the Kaposi sarcoma-associated herpesvirus (HHV8). *Proc. Natl. Acad. Sci. USA* **1996**, *93*, 14862–14867.
33. Ohsaki, E.; Ueda, K. Kaposi's Sarcoma-Associated Herpesvirus Genome Replication, Partitioning, and Maintenance in Latency. *Front. Microbiol.* **2012**, *3*, 7.
34. Toth, Z.; Brulois, K.; Jung, J.U. The chromatin landscape of Kaposi's sarcoma-associated herpesvirus. *Viruses* **2013**, *5*, 1346–1373.
35. Fukumoto, H.; Kanno, T.; Hasegawa, H.; Katano, H. Pathology of Kaposi's Sarcoma-Associated Herpesvirus Infection. *Front. Microbiol.* **2011**, *2*, 175.
36. Yu, Y.; Black, J.B.; Goldsmith, C.S.; Browning, P.J.; Bhalla, K.; Offermann, M.K. Induction of human herpesvirus-8 DNA replication and transcription by butyrate and TPA in BCBL-1 cells. *J. Gen. Virol.* **1999**, *80*, 83–90.
37. Lieberman, P.M. Keeping it quiet: Chromatin control of gammaherpesvirus latency. *Nat. Rev. Microbiol.* **2013**, *11*, 863–875.
38. Cai, Q.; Verma, S.C.; Lu, J.; Robertson, E.S. Molecular biology of Kaposi's sarcoma-associated herpesvirus and related oncogenesis. *Adv. Virus Res.* **2010**, *78*, 87–142.
39. Davis, D.A.; Rinderknecht, A.S.; Zoeteweij, J.P.; Aoki, Y.; Read-Connole, E.L.; Tosato, G.; Blauvelt, A.; Yarchoan, R. Hypoxia induces lytic replication of Kaposi sarcoma-associated herpesvirus. *Blood* **2001**, *97*, 3244–3250.
40. Ye, F.; Zhou, F.; Bedolla, R.G.; Jones, T.; Lei, X.; Kang, T.; Guadalupe, M.; Gao, S.J. Reactive oxygen species hydrogen peroxide mediates Kaposi's sarcoma-associated herpesvirus reactivation from latency. *PLoS Pathog.* **2011**, *7*, e1002054.
41. Xie, J.; Ajibade, A.O.; Ye, F.; Kuhne, K.; Gao, S.J. Reactivation of Kaposi's sarcoma-associated herpesvirus from latency requires MEK/ERK, JNK and p38 multiple mitogen-activated protein kinase pathways. *Virology* **2008**, *371*, 139–154.
42. Li, Q.; He, M.; Zhou, F.; Ye, F.; Gao, S.J. Activation of Kaposi's sarcoma-associated herpesvirus (KSHV) by inhibitors of class III histone deacetylases: Identification of sirtuin 1 as a regulator of the KSHV life cycle. *J. Virol.* **2014**, *88*, 6355–6367.
43. Pantry, S.N.; Medveczky, P.G. Epigenetic regulation of Kaposi's sarcoma-associated herpesvirus replication. *Semin. Cancer Biol.* **2009**, *19*, 153–157.
44. Jenner, R.G.; Alba, M.M.; Boshoff, C.; Kellam, P. Kaposi's sarcoma-associated herpesvirus latent and lytic gene expression as revealed by DNA arrays. *J. Virol.* **2001**, *75*, 891–902.
45. Polizzotto, M.N.; Uldrick, T.S.; Hu, D.; Yarchoan, R. Clinical Manifestations of Kaposi Sarcoma Herpesvirus Lytic Activation: Multicentric Castleman Disease (KSHV-MCD) and the KSHV Inflammatory Cytokine Syndrome. *Front. Microbiol.* **2012**, *3*, 73.
46. Chang, H.H.; Ganem, D. A unique herpesviral transcriptional program in KSHV-infected lymphatic endothelial cells leads to mTORC1 activation and rapamycin sensitivity. *Cell Host Microbe* **2013**, *13*, 429–440.
47. Yogev, O.; Boshoff, C. Redefining KSHV latency. *Cell Host Microbe* **2013**, *13*, 373–374.

48. Arias, C.; Weisburd, B.; Stern-Ginossar, N.; Mercier, A.; Madrid, A.S.; Bellare, P.; Holdorf, M.; Weissman, J.S.; Ganem, D. KSHV 2.0: A comprehensive annotation of the Kaposi's sarcoma-associated herpesvirus genome using next-generation sequencing reveals novel genomic and functional features. *PLoS Pathog.* **2014**, *10*, e1003847.
49. Speck, S.H.; Ganem, D. Viral latency and its regulation: Lessons from the gamma-herpesviruses. *Cell Host Microbe* **2010**, *8*, 100–115.
50. Burysek, L.; Pitha, P.M. Latently expressed human herpesvirus 8-encoded interferon regulatory factor 2 inhibits double-stranded RNA-activated protein kinase. *J. Virol.* **2001**, *75*, 2345–2352.
51. Dittmer, D.; Lagunoff, M.; Renne, R.; Staskus, K.; Haase, A.; Ganem, D. A cluster of latently expressed genes in Kaposi's sarcoma-associated herpesvirus. *J. Virol.* **1998**, *72*, 8309–8315.
52. Kedes, D.H.; Lagunoff, M.; Renne, R.; Ganem, D. Identification of the gene encoding the major latency-associated nuclear antigen of the Kaposi's sarcoma-associated herpesvirus. *J. Clin. Investig.* **1997**, *100*, 2606–2610.
53. Rainbow, L.; Platt, G.M.; Simpson, G.R.; Sarid, R.; Gao, S.J.; Stoiber, H.; Herrington, C.S.; Moore, P.S.; Schulz, T.F. The 222- to 234-kilodalton latent nuclear protein (LNA) of Kaposi's sarcoma-associated herpesvirus (human herpesvirus 8) is encoded by orf73 and is a component of the latency-associated nuclear antigen. *J. Virol.* **1997**, *71*, 5915–5921.
54. Sadler, R.; Wu, L.; Forghani, B.; Renne, R.; Zhong, W.; Herndier, B.; Ganem, D. A complex translational program generates multiple novel proteins from the latently expressed kaposin (K12) locus of Kaposi's sarcoma-associated herpesvirus. *J. Virol.* **1999**, *73*, 5722–5730.
55. Wen, K.W.; Damania, B. Kaposi sarcoma-associated herpesvirus (KSHV): Molecular biology and oncogenesis. *Cancer Lett.* **2010**, *289*, 140–150.
56. Pearce, M.; Matsumura, S.; Wilson, A.C. Transcripts encoding K12, v-FLIP, v-cyclin, and the microRNA cluster of Kaposi's sarcoma-associated herpesvirus originate from a common promoter. *J. Virol.* **2005**, *79*, 14457–14464.
57. Wies, E.; Mori, Y.; Hahn, A.; Kremmer, E.; Sturzl, M.; Fleckenstein, B.; Neipel, F. The viral interferon-regulatory factor-3 is required for the survival of KSHV-infected primary effusion lymphoma cells. *Blood* **2008**, *111*, 320–327.
58. Ganem, D. KSHV and the pathogenesis of Kaposi sarcoma: Listening to human biology and medicine. *J. Clin. Investig.* **2010**, *120*, 939–949.
59. Lee, H.; Guo, J.; Li, M.; Choi, J.K.; deMaria, M.; Rosenzweig, M.; Jung, J.U. Identification of an immunoreceptor tyrosine-based activation motif of K1 transforming protein of Kaposi's sarcoma-associated herpesvirus. *Mol. Cell. Biol.* **1998**, *18*, 5219–5228.
60. Lagunoff, M.; Majeti, R.; Weiss, A.; Ganem, D. Deregulated signal transduction by the K1 gene product of Kaposi's sarcoma-associated herpesvirus. *Proc. Natl. Acad. Sci. USA* **1999**, *96*, 5704–5709.
61. Van Dross, R.; Yao, S.; Asad, S.; Westlake, G.; Mays, D.J.; Barquero, L.; Duell, S.; Pietenpol, J.A.; Browning, P.J. Constitutively active K-cyclin/cdk6 kinase in Kaposi sarcoma-associated herpesvirus-infected cells. *J. Natl. Cancer Inst.* **2005**, *97*, 656–666.
62. Direkze, S.; Laman, H. Regulation of growth signalling and cell cycle by Kaposi's sarcoma-associated herpesvirus genes. *Int. J. Exp. Pathol.* **2004**, *85*, 305–319.

63. Sarek, G.; Jarviluoma, A.; Moore, H.M.; Tojkander, S.; Vartia, S.; Biberfeld, P.; Laiho, M.; Ojala, P.M. Nucleophosmin phosphorylation by v-cyclin-CDK6 controls KSHV latency. *PLoS. Pathog.* **2010**, *6*, e1000818.
64. Liang, X.; Paden, C.R.; Morales, F.M.; Powers, R.P.; Jacob, J.; Speck, S.H. Murine gamma-herpesvirus immortalization of fetal liver-derived B cells requires both the viral cyclin D homolog and latency-associated nuclear antigen. *PLoS Pathog.* **2011**, *7*, e1002220.
65. Ye, F.C.; Zhou, F.C.; Xie, J.P.; Kang, T.; Greene, W.; Kuhne, K.; Lei, X.F.; Li, Q.H.; Gao, S.J. Kaposi's sarcoma-associated herpesvirus latent gene vFLIP inhibits viral lytic replication through NF-kappaB-mediated suppression of the AP-1 pathway: A novel mechanism of virus control of latency. *J. Virol.* **2008**, *82*, 4235–4249.
66. Guasparri, I.; Keller, S.A.; Cesarman, E. KSHV vFLIP is essential for the survival of infected lymphoma cells. *J. Exp. Med.* **2004**, *199*, 993–1003.
67. Israel, A. The IKK complex, a central regulator of NF-kappaB activation. *Cold Spring Harb. Perspect. Biol.* **2010**, *2*, a000158.
68. Field, N.; Low, W.; Daniels, M.; Howell, S.; Daviet, L.; Boshoff, C.; Collins, M. KSHV vFLIP binds to IKK-gamma to activate IKK. *J. Cell Sci.* **2003**, *116*, 3721–3728.
69. Ye, F.; Lei, X.; Gao, S.J. Mechanisms of Kaposi's Sarcoma-Associated Herpesvirus Latency and Reactivation. *Adv. Virol.* **2011**, doi:10.1155/2011/193860.
70. McCormick, C.; Ganem, D. The kaposin B protein of KSHV activates the p38/MK2 pathway and stabilizes cytokine mRNAs. *Science* **2005**, *307*, 739–741.
71. Lin, Y.T.; Kincaid, R.P.; Arasappan, D.; Dowd, S.E.; Hunicke-Smith, S.P.; Sullivan, C.S. Small RNA profiling reveals antisense transcription throughout the KSHV genome and novel small RNAs. *RNA* **2010**, *16*, 1540–1558.
72. Feldman, E.R.; Kara, M.; Coleman, C.B.; Grau, K.R.; Oko, L.M.; Krueger, B.J.; Renne, R.; van Dyk, L.F.; Tibbetts, S.A. Virus-encoded microRNAs facilitate gammaherpesvirus latency and pathogenesis *in vivo*. *MBIO* **2014**, *5*, e00981–e00914.
73. Moody, R.; Zhu, Y.; Huang, Y.; Cui, X.; Jones, T.; Bedolla, R.; Lei, X.; Bai, Z.; Gao, S.J. KSHV microRNAs mediate cellular transformation and tumorigenesis by redundantly targeting cell growth and survival pathways. *PLoS Pathog.* **2013**, *9*, e1003857.
74. Suffert, G.; Malterer, G.; Hausser, J.; Viiliainen, J.; Fender, A.; Contrant, M.; Ivacevic, T.; Benes, V.; Gros, F.; Voinnet, O.; *et al.* Kaposi's sarcoma herpesvirus microRNAs target caspase 3 and regulate apoptosis. *PLoS Pathog.* **2011**, *7*, e1002405.
75. Qin, Z.; Jakymiw, A.; Findlay, V.; Parsons, C. KSHV-Encoded MicroRNAs: Lessons for Viral Cancer Pathogenesis and Emerging Concepts. *Int. J. Cell Biol.* **2012**, *2012*, 603961.
76. Lacoste, V.; Judde, J.G.; Bestett, G.; Cadranel, J.; Antoine, M.; Valensi, F.; Delabesse, E.; Macintyre, E.; Gessain, A. Virological and molecular characterisation of a new B lymphoid cell line, established from an AIDS patient with primary effusion lymphoma, harbouring both KSHV/HHV8 and EBV viruses. *Leuk. Lymphoma* **2000**, *38*, 401–409.
77. Ueda, K.; Sakakibara, S.; Ohsaki, E.; Yada, K. Lack of a mechanism for faithful partition and maintenance of the KSHV genome. *Virus Res.* **2006**, *122*, 85–94.
78. Collins, C.M.; Medveczky, P.G. Genetic requirements for the episomal maintenance of oncogenic herpesvirus genomes. *Adv. Cancer Res.* **2002**, *84*, 155–174.

79. Ye, F.C.; Zhou, F.C.; Yoo, S.M.; Xie, J.P.; Browning, P.J.; Gao, S.J. Disruption of Kaposi's sarcoma-associated herpesvirus latent nuclear antigen leads to abortive episome persistence. *J. Virol.* **2004**, *78*, 11121–11129.
80. Wong, L.Y.; Wilson, A.C. Kaposi's sarcoma-associated herpesvirus latency-associated nuclear antigen induces a strong bend on binding to terminal repeat DNA. *J. Virol.* **2005**, *79*, 13829–13836.
81. Lu, F.; Tsai, K.; Chen, H.S.; Wikramasinghe, P.; Davuluri, R.V.; Showe, L.; Domsic, J.; Marmorstein, R.; Lieberman, P.M. Identification of host-chromosome binding sites and candidate gene targets for Kaposi's sarcoma-associated herpesvirus LANA. *J. Virol.* **2012**, *86*, 5752–5762.
82. Ballestas, M.E.; Kaye, K.M. Kaposi's sarcoma-associated herpesvirus latency-associated nuclear antigen 1 mediates episome persistence through cis-acting terminal repeat (TR) sequence and specifically binds TR DNA. *J. Virol.* **2001**, *75*, 3250–3258.
83. Cotter, M.A., 2nd; Subramanian, C.; Robertson, E.S. The Kaposi's sarcoma-associated herpesvirus latency-associated nuclear antigen binds to specific sequences at the left end of the viral genome through its carboxy-terminus. *Virology* **2001**, *291*, 241–259.
84. Barbera, A.J.; Chodaparambil, J.V.; Kelley-Clarke, B.; Joukov, V.; Walter, J.C.; Luger, K.; Kaye, K.M. The nucleosomal surface as a docking station for Kaposi's sarcoma herpesvirus LANA. *Science* **2006**, *311*, 856–861.
85. Si, H.; Verma, S.C.; Lampson, M.A.; Cai, Q.; Robertson, E.S. Kaposi's sarcoma-associated herpesvirus-encoded LANA can interact with the nuclear mitotic apparatus protein to regulate genome maintenance and segregation. *J. Virol.* **2008**, *82*, 6734–6746.
86. Matsumura, S.; Persson, L.M.; Wong, L.; Wilson, A.C. The latency-associated nuclear antigen interacts with MeCP2 and nucleosomes through separate domains. *J. Virol.* **2010**, *84*, 2318–2330.
87. You, J.; Srinivasan, V.; Denis, G.V.; Harrington, W.J., Jr.; Ballestas, M.E.; Kaye, K.M.; Howley, P.M. Kaposi's sarcoma-associated herpesvirus latency-associated nuclear antigen interacts with bromodomain protein Brd4 on host mitotic chromosomes. *J. Virol.* **2006**, *80*, 8909–8919.
88. Ottinger, M.; Christalla, T.; Nathan, K.; Brinkmann, M.M.; Viejo-Borbolla, A.; Schulz, T.F. Kaposi's sarcoma-associated herpesvirus LANA-1 interacts with the short variant of BRD4 and releases cells from a BRD4- and BRD2/RING3-induced G1 cell cycle arrest. *J. Virol.* **2006**, *80*, 10772–10786.
89. Xiao, B.; Verma, S.C.; Cai, Q.; Kaul, R.; Lu, J.; Saha, A.; Robertson, E.S. Bub1 and CENP-F can contribute to Kaposi's sarcoma-associated herpesvirus genome persistence by targeting LANA to kinetochores. *J. Virol.* **2010**, *84*, 9718–9732.
90. Cotter, M.A., 2nd; Robertson, E.S. The latency-associated nuclear antigen tethers the Kaposi's sarcoma-associated herpesvirus genome to host chromosomes in body cavity-based lymphoma cells. *Virology* **1999**, *264*, 254–264.
91. Luger, K.; Mader, A.W.; Richmond, R.K.; Sargent, D.F.; Richmond, T.J. Crystal structure of the nucleosome core particle at 2.8 A resolution. *Nature* **1997**, *389*, 251–260.
92. Zhou, Y.B.; Gerchman, S.E.; Ramakrishnan, V.; Travers, A.; Muyldermans, S. Position and orientation of the globular domain of linker histone H5 on the nucleosome. *Nature* **1998**, *395*, 402–405.

93. Barbera, A.J.; Chodaparambil, J.V.; Kelley-Clarke, B.; Luger, K.; Kaye, K.M. Kaposi's sarcoma-associated herpesvirus LANA hitches a ride on the chromosome. *Cell Cycle* **2006**, *5*, 1048–1052.
94. Jha, H.C.; Upadhyay, S.K.; MahadeshPrasad, A.J.; Lu, J.; Cai, Q.; Saha, A.; Robertson, E.S. H2AX phosphorylation is important for LANA-mediated Kaposi's sarcoma-associated herpesvirus episome persistence. *J. Virol.* **2013**, *87*, 5255–5269.
95. Doxsey, S.; Zimmerman, W.; Mikule, K. Centrosome control of the cell cycle. *Trends Cell Biol.* **2005**, *15*, 303–311.
96. Guse, A.; Carroll, C.W.; Moree, B.; Fuller, C.J.; Straight, A.F. *In vitro* centromere and kinetochore assembly on defined chromatin templates. *Nature* **2011**, *477*, 354–358.
97. Johnson, V.L.; Scott, M.I.; Holt, S.V.; Hussein, D.; Taylor, S.S. Bub1 is required for kinetochore localization of BubR1, Cenp-E, Cenp-F and Mad2, and chromosome congression. *J. Cell Sci.* **2004**, *117*, 1577–1589.
98. Saitoh, S.; Kobayashi, Y.; Ogiyama, Y.; Takahashi, K. Dual regulation of Mad2 localization on kinetochores by Bub1 and Dam1/DASH that ensure proper spindle interaction. *Mol. Biol. Cell* **2008**, *19*, 3885–3897.
99. Sun, Z.; Xiao, B.; Jha, H.C.; Lu, J.; Banerjee, S.; Robertson, E.S. Kaposi's Sarcoma-Associated Herpesvirus-Encoded LANA Can Induce Chromosomal Instability through Targeted Degradation of the Mitotic Checkpoint Kinase Bub1. *J. Virol.* **2014**, *88*, 7367–7378.
100. Compton, D.A.; Szilak, I.; Cleveland, D.W. Primary structure of NuMA, an intranuclear protein that defines a novel pathway for segregation of proteins at mitosis. *J. Cell Biol.* **1992**, *116*, 1395–1408.
101. Harborth, J.; Wang, J.; Gueth-Hallonet, C.; Weber, K.; Osborn, M. Self assembly of NuMA: Multiarm oligomers as structural units of a nuclear lattice. *EMBO J.* **1999**, *18*, 1689–1700.
102. Du, Q.; Stukenberg, P.T.; Macara, I.G. A mammalian Partner of inscuteable binds NuMA and regulates mitotic spindle organization. *Nat. Cell Biol.* **2001**, *3*, 1069–1075.
103. Du, Q.; Macara, I.G. Mammalian Pins is a conformational switch that links NuMA to heterotrimeric G proteins. *Cell* **2004**, *119*, 503–516.
104. Kisurina-Evgenieva, O.; Mack, G.; Du, Q.; Macara, I.; Khodjakov, A.; Compton, D.A. Multiple mechanisms regulate NuMA dynamics at spindle poles. *J. Cell Sci.* **2004**, *117*, 6391–6400.
105. Gobert, G.N.; Hueser, C.N.; Curran, E.M.; Sun, Q.Y.; Glinsky, V.V.; Welshons, W.V.; Eisenstark, A.; Schatten, H. Immunolocalization of NuMA and phosphorylated proteins during the cell cycle in human breast and prostate cancer cells as analyzed by immunofluorescence and postembedding immunoelectron microscopy. *Histochem. Cell Biol.* **2001**, *115*, 381–395.
106. Sparks, C.A.; Fey, E.G.; Vidair, C.A.; Doxsey, S.J. Phosphorylation of NUMA occurs during nuclear breakdown and not mitotic spindle assembly. *J. Cell Sci.* **1995**, *108*, 3389–3396.
107. Zeng, C.; He, D.; Berget, S.M.; Brinkley, B.R. Nuclear-mitotic apparatus protein: A structural protein interface between the nucleoskeleton and RNA splicing. *Proc. Natl. Acad. Sci. USA* **1994**, *91*, 1505–1509.
108. Merdes, A.; Cleveland, D.W. The role of NuMA in the interphase nucleus. *J. Cell Sci.* **1998**, *111*, 71–79.

109. Luderus, M.E.; den Blaauwen, J.L.; de Smit, O.J.; Compton, D.A.; van Driel, R. Binding of matrix attachment regions to lamin polymers involves single-stranded regions and the minor groove. *Mol. Cell. Biol.* **1994**, *14*, 6297–6305.
110. Du, Q.; Taylor, L.; Compton, D.A.; Macara, I.G. LGN blocks the ability of NuMA to bind and stabilize microtubules. A mechanism for mitotic spindle assembly regulation. *Curr. Biol.* **2002**, *12*, 1928–1933.
111. Haren, L.; Merdes, A. Direct binding of NuMA to tubulin is mediated by a novel sequence motif in the tail domain that bundles and stabilizes microtubules. *J. Cell Sci* **2002**, *115*, 1815–1824.
112. Merdes, A.; Ramyar, K.; Vechio, J.D.; Cleveland, D.W. A complex of NuMA and cytoplasmic dynein is essential for mitotic spindle assembly. *Cell* **1996**, *87*, 447–458.
113. Yang, C.H.; Snyder, M. The nuclear-mitotic apparatus protein is important in the establishment and maintenance of the bipolar mitotic spindle apparatus. *Mol. Biol. Cell* **1992**, *3*, 1259–1267.
114. Gueth-Hallonet, C.; Weber, K.; Osborn, M. NuMA: A bipartite nuclear location signal and other functional properties of the tail domain. *Exp. Cell Res.* **1996**, *225*, 207–218.
115. Verma, S.C.; Robertson, E.S. Molecular biology and pathogenesis of Kaposi sarcoma-associated herpesvirus. *FEMS Microbiol. Lett.* **2003**, *222*, 155–163.
116. Aravind, L.; Landsman, D. AT-hook motifs identified in a wide variety of DNA-binding proteins. *Nucleic Acids Res.* **1998**, *26*, 4413–4421.
117. Meehan, R.R.; Lewis, J.D.; Bird, A.P. Characterization of MeCP2, a vertebrate DNA binding protein with affinity for methylated DNA. *Nucleic Acids Res.* **1992**, *20*, 5085–5092.
118. Nan, X.; Campoy, F.J.; Bird, A. MeCP2 is a transcriptional repressor with abundant binding sites in genomic chromatin. *Cell* **1997**, *88*, 471–481.
119. Young, J.I.; Hong, E.P.; Castle, J.C.; Crespo-Barreto, J.; Bowman, A.B.; Rose, M.F.; Kang, D.; Richman, R.; Johnson, J.M.; Berget, S.; *et al.* Regulation of RNA splicing by the methylation-dependent transcriptional repressor methyl-CpG binding protein 2. *Proc. Natl. Acad. Sci. USA* **2005**, *102*, 17551–17558.
120. Chahrour, M.; Jung, S.Y.; Shaw, C.; Zhou, X.; Wong, S.T.; Qin, J.; Zoghbi, H.Y. MeCP2, a key contributor to neurological disease, activates and represses transcription. *Science* **2008**, *320*, 1224–1229.
121. Yasui, D.H.; Peddada, S.; Bieda, M.C.; Vallero, R.O.; Hogart, A.; Nagarajan, R.P.; Thatcher, K.N.; Farnham, P.J.; Lasalle, J.M. Integrated epigenomic analyses of neuronal MeCP2 reveal a role for long-range interaction with active genes. *Proc. Natl. Acad. Sci. USA* **2007**, *104*, 19416–19421.
122. Georgel, P.T.; Horowitz-Scherer, R.A.; Adkins, N.; Woodcock, C.L.; Wade, P.A.; Hansen, J.C. Chromatin compaction by human MeCP2. Assembly of novel secondary chromatin structures in the absence of DNA methylation. *J. Biol. Chem.* **2003**, *278*, 32181–32188.
123. Nikitina, T.; Ghosh, R.P.; Horowitz-Scherer, R.A.; Hansen, J.C.; Grigoryev, S.A.; Woodcock, C.L. MeCP2-chromatin interactions include the formation of chromatosome-like structures and are altered in mutations causing Rett syndrome. *J. Biol. Chem.* **2007**, *282*, 28237–28245.
124. Krithivas, A.; Fujimuro, M.; Weidner, M.; Young, D.B.; Hayward, S.D. Protein interactions targeting the latency-associated nuclear antigen of Kaposi's sarcoma-associated herpesvirus to cell chromosomes. *J. Virol.* **2002**, *76*, 11596–11604.

125. Waldmann, T.; Scholten, I.; Kappes, F.; Hu, H.G.; Knippers, R. The DEK protein—An abundant and ubiquitous constituent of mammalian chromatin. *Gene* **2004**, *343*, 1–9.
126. Von Lindern, M.; Fornerod, M.; van Baal, S.; Jaegle, M.; de Wit, T.; Buijs, A.; Grosveld, G. The translocation (6;9), associated with a specific subtype of acute myeloid leukemia, results in the fusion of two genes, dek and can, and the expression of a chimeric, leukemia-specific dek-can mRNA. *Mol. Cell. Biol.* **1992**, *12*, 1687–1697.
127. Faulkner, N.E.; Hilfinger, J.M.; Markovitz, D.M. Protein phosphatase 2A activates the HIV-2 promoter through enhancer elements that include the pets site. *J. Biol. Chem.* **2001**, *276*, 25804–25812.
128. Fu, G.K.; Markovitz, D.M. Purification of the pets factor. A nuclear protein that binds to the inducible TG-rich element of the human immunodeficiency virus type 2 enhancer. *J. Biol. Chem.* **1996**, *271*, 19599–19605.
129. Fu, G.K.; Grosveld, G.; Markovitz, D.M. DEK, an autoantigen involved in a chromosomal translocation in acute myelogenous leukemia, binds to the HIV-2 enhancer. *Proc. Natl. Acad. Sci. USA* **1997**, *94*, 1811–1815.
130. McGarvey, T.; Rosonina, E.; McCracken, S.; Li, Q.; Arnaout, R.; Mientjes, E.; Nickerson, J.A.; Awrey, D.; Greenblatt, J.; Grosveld, G.; *et al.* The acute myeloid leukemia-associated protein, DEK, forms a splicing-dependent interaction with exon-product complexes. *J. Cell Biol.* **2000**, *150*, 309–320.
131. Alexiadis, V.; Waldmann, T.; Andersen, J.; Mann, M.; Knippers, R.; Gruss, C. The protein encoded by the proto-oncogene DEK changes the topology of chromatin and reduces the efficiency of DNA replication in a chromatin-specific manner. *Genes Dev.* **2000**, *14*, 1308–1312.
132. Chua, P.; Roeder, G.S. Bdf1, a yeast chromosomal protein required for sporulation. *Mol. Cell. Biol.* **1995**, *15*, 3685–3696.
133. Florence, B.; Faller, D.V. You bet-cha: A novel family of transcriptional regulators. *Front. Biosci.* **2001**, *6*, D1008–D1018.
134. Jang, M.K.; Mochizuki, K.; Zhou, M.; Jeong, H.S.; Brady, J.N.; Ozato, K. The bromodomain protein Brd4 is a positive regulatory component of P-TEFb and stimulates RNA polymerase II-dependent transcription. *Mol. Cell* **2005**, *19*, 523–534.
135. Yang, Z.; Yik, J.H.; Chen, R.; He, N.; Jang, M.K.; Ozato, K.; Zhou, Q. Recruitment of P-TEFb for stimulation of transcriptional elongation by the bromodomain protein Brd4. *Mol. Cell* **2005**, *19*, 535–545.
136. Sinha, A.; Faller, D.V.; Denis, G.V. Bromodomain analysis of Brd2-dependent transcriptional activation of cyclin A. *Biochem. J.* **2005**, *387*, 257–269.
137. Houzelstein, D.; Bullock, S.L.; Lynch, D.E.; Grigorieva, E.F.; Wilson, V.A.; Beddington, R.S. Growth and early postimplantation defects in mice deficient for the bromodomain-containing protein Brd4. *Mol. Cell. Biol.* **2002**, *22*, 3794–3802.
138. You, J.; Croyle, J.L.; Nishimura, A.; Ozato, K.; Howley, P.M. Interaction of the bovine papillomavirus E2 protein with Brd4 tethers the viral DNA to host mitotic chromosomes. *Cell* **2004**, *117*, 349–360.
139. Pamblanco, M.; Poveda, A.; Sendra, R.; Rodriguez-Navarro, S.; Perez-Ortin, J.E.; Tordera, V. Bromodomain factor 1 (Bdf1) protein interacts with histones. *FEBS Lett.* **2001**, *496*, 31–35.

140. Kanno, T.; Kanno, Y.; Siegel, R.M.; Jang, M.K.; Lenardo, M.J.; Ozato, K. Selective recognition of acetylated histones by bromodomain proteins visualized in living cells. *Mol. Cell* **2004**, *13*, 33–43.
141. Dey, A.; Chitsaz, F.; Abbasi, A.; Misteli, T.; Ozato, K. The double bromodomain protein Brd4 binds to acetylated chromatin during interphase and mitosis. *Proc. Natl. Acad. Sci. USA* **2003**, *100*, 8758–8763.
142. Denis, G.V.; Vaziri, C.; Guo, N.; Faller, D.V. RING3 kinase transactivates promoters of cell cycle regulatory genes through E2F. *Cell Growth Differ.* **2000**, *11*, 417–424.
143. Platt, G.M.; Simpson, G.R.; Mittnacht, S.; Schulz, T.F. Latent nuclear antigen of Kaposi's sarcoma-associated herpesvirus interacts with RING3, a homolog of the Drosophila female sterile homeotic (fsh) gene. *J. Virol.* **1999**, *73*, 9789–9795.
144. Mattsson, K.; Kiss, C.; Platt, G.M.; Simpson, G.R.; Kashuba, E.; Klein, G.; Schulz, T.F.; Szekely, L. Latent nuclear antigen of Kaposi's sarcoma herpesvirus/human herpesvirus-8 induces and relocates RING3 to nuclear heterochromatin regions. *J. Gen. Virol.* **2002**, *83*, 179–188.
145. Haynes, S.R.; Dollard, C.; Winston, F.; Beck, S.; Trowsdale, J.; Dawid, I.B. The bromodomain: A conserved sequence found in human, Drosophila and yeast proteins. *Nucleic Acids Res.* **1992**, *20*, 2603.
146. Nomura, N.; Nagase, T.; Miyajima, N.; Sazuka, T.; Tanaka, A.; Sato, S.; Seki, N.; Kawarabayasi, Y.; Ishikawa, K.; Tabata, S. Prediction of the coding sequences of unidentified human genes. II. The coding sequences of 40 new genes (KIAA0041-KIAA0080) deduced by analysis of cDNA clones from human cell line KG-1. *DNA Res.* **1994**, *1*, 223–229.
147. Jones, M.H.; Numata, M.; Shimane, M. Identification and characterization of BRDT: A testis-specific gene related to the bromodomain genes RING3 and Drosophila fsh. *Genomics* **1997**, *45*, 529–534.
148. Stedman, W.; Deng, Z.; Lu, F.; Lieberman, P.M. ORC, MCM, and histone hyperacetylation at the Kaposi's sarcoma-associated herpesvirus latent replication origin. *J. Virol.* **2004**, *78*, 12566–12575.
149. Verma, S.C.; Choudhuri, T.; Kaul, R.; Robertson, E.S. Latency-associated nuclear antigen (LANA) of Kaposi's sarcoma-associated herpesvirus interacts with origin recognition complexes at the LANA binding sequence within the terminal repeats. *J. Virol.* **2006**, *80*, 2243–2256.
150. Dheekollu, J.; Chen, H.S.; Kaye, K.M.; Lieberman, P.M. Timeless-dependent DNA replication-coupled recombination promotes Kaposi's Sarcoma-associated herpesvirus episome maintenance and terminal repeat stability. *J. Virol.* **2013**, *87*, 3699–3709.
151. Purushothaman, P.; McDowell, M.E.; McGuinness, J.; Salas, R.; Rumjahn, S.M.; Verma, S.C. Kaposi's Sarcoma-Associated Herpesvirus-Encoded LANA Recruits Topoisomerase IIbeta for Latent DNA Replication of the Terminal Repeats. *J. Virol.* **2012**, *86*, 9983–9994.
152. Sun, Q.; Tsurimoto, T.; Juillard, F.; Li, L.; Li, S.; de Leon Vazquez, E.; Chen, S.; Kaye, K. Kaposi's sarcoma-associated herpesvirus LANA recruits the DNA polymerase clamp loader to mediate efficient replication and virus persistence. *Proc. Natl. Acad. Sci. USA* **2014**, *111*, 11816–11821.
153. Green, C.M.; Erdjument-Bromage, H.; Tempst, P.; Lowndes, N.F. A novel Rad24 checkpoint protein complex closely related to replication factor C. *Curr. Biol.* **2000**, *10*, 39–42.

154. Lu, F.; Zhou, J.; Wiedmer, A.; Madden, K.; Yuan, Y.; Lieberman, P.M. Chromatin remodeling of the Kaposi's sarcoma-associated herpesvirus ORF50 promoter correlates with reactivation from latency. *J. Virol.* **2003**, *77*, 11425–11435.
155. Lu, J.; Jha, H.C.; Verma, S.C.; Sun, Z.; Banerjee, S.; Dzeng, R.; Robertson, E.S. Kaposi's sarcoma-associated herpesvirus-encoded LANA contributes to viral latent replication by activating phosphorylation of survivin. *J. Virol.* **2014**, *88*, 4204–4217.
156. Lu, J.; Verma, S.C.; Murakami, M.; Cai, Q.; Kumar, P.; Xiao, B.; Robertson, E.S. Latency-associated nuclear antigen of Kaposi's sarcoma-associated herpesvirus (KSHV) upregulates survivin expression in KSHV-associated B-lymphoma cells and contributes to their proliferation. *J. Virol.* **2009**, *83*, 7129–7141.
157. Verma, S.C.; Lu, J.; Cai, Q.; Kosiyatrakul, S.; McDowell, M.E.; Schildkraut, C.L.; Robertson, E.S. Single molecule analysis of replicated DNA reveals the usage of multiple KSHV genome regions for latent replication. *PLoS Pathog.* **2011**, *7*, e1002365.
158. Verma, S.C.; Lan, K.; Choudhuri, T.; Cotter, M.A.; Robertson, E.S. An autonomous replicating element within the KSHV genome. *Cell Host Microbe* **2007**, *2*, 106–118.
159. Chen, W.; Hilton, I.B.; Staudt, M.R.; Burd, C.E.; Dittmer, D.P. Distinct p53, p53: LANA, and LANA complexes in Kaposi's Sarcoma-associated Herpesvirus Lymphomas. *J. Virol.* **2010**, *84*, 3898–3908.
160. Friborg, J., Jr.; Kong, W.; Hottiger, M.O.; Nabel, G.J. p53 inhibition by the LANA protein of KSHV protects against cell death. *Nature* **1999**, *402*, 889–894.
161. Krithivas, A.; Young, D.B.; Liao, G.; Greene, D.; Hayward, S.D. Human herpesvirus 8 LANA interacts with proteins of the mSin3 corepressor complex and negatively regulates Epstein-Barr virus gene expression in dually infected PEL cells. *J. Virol.* **2000**, *74*, 9637–9645.
162. Lan, K.; Kuppers, D.A.; Robertson, E.S. Kaposi's sarcoma-associated herpesvirus reactivation is regulated by interaction of latency-associated nuclear antigen with recombination signal sequence-binding protein Jkappa, the major downstream effector of the Notch signaling pathway. *J. Virol.* **2005**, *79*, 3468–3478.
163. Lim, C.; Lee, D.; Seo, T.; Choi, C.; Choe, J. Latency-associated nuclear antigen of Kaposi's sarcoma-associated herpesvirus functionally interacts with heterochromatin protein 1. *J. Biol. Chem.* **2003**, *278*, 7397–7405.
164. Ohsaki, E.; Ueda, K.; Sakakibara, S.; Do, E.; Yada, K.; Yamanishi, K. Poly(ADP-ribose) polymerase 1 binds to Kaposi's sarcoma-associated herpesvirus (KSHV) terminal repeat sequence and modulates KSHV replication in latency. *J. Virol.* **2004**, *78*, 9936–9946.
165. Shamay, M.; Krithivas, A.; Zhang, J.; Hayward, S.D. Recruitment of the de novo DNA methyltransferase Dnmt3a by Kaposi's sarcoma-associated herpesvirus LANA. *Proc. Natl. Acad. Sci. USA* **2006**, *103*, 14554–14559.
166. Si, H.; Robertson, E.S. Kaposi's sarcoma-associated herpesvirus-encoded latency-associated nuclear antigen induces chromosomal instability through inhibition of p53 function. *J. Virol.* **2006**, *80*, 697–709.
167. Viejo-Borbolla, A.; Kati, E.; Sheldon, J.A.; Nathan, K.; Mattsson, K.; Szekely, L.; Schulz, T.F. A Domain in the *C*-terminal region of latency-associated nuclear antigen 1 of Kaposi's

sarcoma-associated Herpesvirus affects transcriptional activation and binding to nuclear heterochromatin. *J. Virol.* **2003**, *77*, 7093–7100.

168. Viejo-Borbolla, A.; Ottinger, M.; Bruning, E.; Burger, A.; Konig, R.; Kati, E.; Sheldon, J.A.; Schulz, T.F. Brd2/RING3 interacts with a chromatin-binding domain in the Kaposi's Sarcoma-associated herpesvirus latency-associated nuclear antigen 1 (LANA-1) that is required for multiple functions of LANA-1. *J. Virol.* **2005**, *79*, 13618–13629.
169. Hu, J.; Yang, Y.; Turner, P.C.; Jain, V.; McIntyre, L.M.; Renne, R. LANA binds to multiple active viral and cellular promoters and associates with the H3K4methyltransferase hSET1 complex. *PLoS Pathog.* **2014**, *10*, e1004240.
170. Mercier, A.; Arias, C.; Madrid, A.S.; Holdorf, M.M.; Ganem, D. Site-Specific Association with Host and Viral Chromatin by Kaposi's Sarcoma-Associated Herpesvirus LANA and Its Reversal during Lytic Reactivation. *J. Virol.* **2014**, *88*, 6762–6777.
171. Raghavan, M.; del Cid, N.; Rizvi, S.M.; Peters, L.R. MHC class I assembly: Out and about. *Trends Immunol.* **2008**, *29*, 436–443.
172. Tang, J.; Gordon, G.M.; Muller, M.G.; Dahiya, M.; Foreman, K.E. Kaposi's sarcoma-associated herpesvirus latency-associated nuclear antigen induces expression of the helix-loop-helix protein Id-1 in human endothelial cells. *J. Virol.* **2003**, *77*, 5975–5984.
173. Guito, J.; Lukac, D.M. KSHV Rta Promoter Specification and Viral Reactivation. *Front. Microbiol.* **2012**, *3*, 30.
174. Sun, R.; Lin, S.F.; Gradoville, L.; Yuan, Y.; Zhu, F.; Miller, G. A viral gene that activates lytic cycle expression of Kaposi's sarcoma-associated herpesvirus. *Proc. Natl. Acad. Sci. USA* **1998**, *95*, 10866–10871.
175. Bu, W.; Palmeri, D.; Krishnan, R.; Marin, R.; Aris, V.M.; Soteropoulos, P.; Lukac, D.M. Identification of direct transcriptional targets of the Kaposi's sarcoma-associated herpesvirus Rta lytic switch protein by conditional nuclear localization. *J. Virol.* **2008**, *82*, 10709–10723.
176. Deng, H.; Young, A.; Sun, R. Auto-activation of the rta gene of human herpesvirus-8/Kaposi's sarcoma-associated herpesvirus. *J. Gen. Virol.* **2000**, *81*, 3043–3048.
177. Li, Q.; Zhou, F.; Ye, F.; Gao, S.J. Genetic disruption of KSHV major latent nuclear antigen LANA enhances viral lytic transcriptional program. *Virology* **2008**, *379*, 234–244.
178. Campbell, M.; Izumiya, Y. Post-Translational Modifications of Kaposi's Sarcoma-Associated Herpesvirus Regulatory Proteins—SUMO and KSHV. *Front. Microbiol.* **2012**, *3*, 31.
179. Bachman, K.E.; Rountree, M.R.; Baylin, S.B. Dnmt3a and Dnmt3b are transcriptional repressors that exhibit unique localization properties to heterochromatin. *J. Biol. Chem.* **2001**, *276*, 32282–32287.
180. Verma, S.C.; Cai, Q.; Kreider, E.; Lu, J.; Robertson, E.S. Comprehensive analysis of LANA interacting proteins essential for viral genome tethering and persistence. *PLoS One* **2013**, *8*, e74662.
181. Sakakibara, S.; Ueda, K.; Nishimura, K.; Do, E.; Ohsaki, E.; Okuno, T.; Yamanishi, K. Accumulation of heterochromatin components on the terminal repeat sequence of Kaposi's sarcoma-associated herpesvirus mediated by the latency-associated nuclear antigen. *J. Virol.* **2004**, *78*, 7299–7310.

182. Kim, K.Y.; Huerta, S.B.; Izumiya, C.; Wang, D.H.; Martinez, A.; Shevchenko, B.; Kung, H.J.; Campbell, M.; Izumiya, Y. Kaposi's sarcoma-associated herpesvirus (KSHV) latency-associated nuclear antigen regulates the KSHV epigenome by association with the histone demethylase KDM3A. *J. Virol.* **2013**, *87*, 6782–6793.
183. Lu, J.; Verma, S.C.; Cai, Q.; Robertson, E.S. The single RBP-Jkappa site within the LANA promoter is crucial for establishing Kaposi's sarcoma-associated herpesvirus latency during primary infection. *J. Virol.* **2011**, *85*, 6148–6161.
184. Lu, J.; Verma, S.C.; Cai, Q.; Saha, A.; Dzeng, R.K.; Robertson, E.S. The RBP-Jkappa binding sites within the RTA promoter regulate KSHV latent infection and cell proliferation. *PLoS Pathog.* **2012**, *8*, e1002479.
185. Klose, R.J.; Bird, A.P. Genomic DNA methylation: The mark and its mediators. *Trends Biochem. Sci.* **2006**, *31*, 89–97.
186. Gray, K.S.; Forrest, J.C.; Speck, S.H. The de novo methyltransferases DNMT3a and DNMT3b target the murine gammaherpesvirus immediate-early gene 50 promoter during establishment of latency. *J. Virol.* **2010**, *84*, 4946–4959.
187. Lu, F.; Day, L.; Gao, S.J.; Lieberman, P.M. Acetylation of the latency-associated nuclear antigen regulates repression of Kaposi's sarcoma-associated herpesvirus lytic transcription. *J. Virol.* **2006**, *80*, 5273–5282.
188. Fujimuro, M.; Liu, J.; Zhu, J.; Yokosawa, H.; Hayward, S.D. Regulation of the interaction between glycogen synthase kinase 3 and the Kaposi's sarcoma-associated herpesvirus latency-associated nuclear antigen. *J. Virol.* **2005**, *79*, 10429–10441.
189. Cha, S.; Lim, C.; Lee, J.Y.; Song, Y.J.; Park, J.; Choe, J.; Seo, T. DNA-PK/Ku complex binds to latency-associated nuclear antigen and negatively regulates Kaposi's sarcoma-associated herpesvirus latent replication. *Biochem. Biophys. Res. Commun.* **2010**, *394*, 934–939.
190. Cheng, F.; Weidner-Glunde, M.; Varjosalo, M.; Rainio, E.M.; Lehtonen, A.; Schulz, T.F.; Koskinen, P.J.; Taipale, J.; Ojala, P.M. KSHV reactivation from latency requires Pim-1 and Pim-3 kinases to inactivate the latency-associated nuclear antigen LANA. *PLoS Pathog.* **2009**, *5*, e1000324.
191. Campbell, M.; Chang, P.C.; Huerta, S.; Izumiya, C.; Davis, R.; Tepper, C.G.; Kim, K.Y.; Shevchenko, B.; Wang, D.H.; Jung, J.U.; *et al.* Protein arginine methyltransferase 1-directed methylation of Kaposi sarcoma-associated herpesvirus latency-associated nuclear antigen. *J. Biol. Chem.* **2012**, *287*, 5806–5818.
192. Knipe, D.M.; Lieberman, P.M.; Jung, J.U.; McBride, A.A.; Morris, K.V.; Ott, M.; Margolis, D.; Nieto, A.; Nevels, M.; Parks, R.J.; *et al.* Snapshots: Chromatin control of viral infection. *Virology* **2013**, *435*, 141–156.
193. Gunther, T.; Grundhoff, A. The epigenetic landscape of latent Kaposi sarcoma-associated herpesvirus genomes. *PLoS Pathog.* **2010**, *6*, e1000935.
194. Toth, Z.; Maglinte, D.T.; Lee, S.H.; Lee, H.R.; Wong, L.Y.; Brulois, K.F.; Lee, S.; Buckley, J.D.; Laird, P.W.; Marquez, V.E.; *et al.* Epigenetic analysis of KSHV latent and lytic genomes. *PLoS Pathog.* **2010**, *6*, e1001013.
195. Toth, Z.; Brulois, K.; Lee, H.R.; Izumiya, Y.; Tepper, C.; Kung, H.J.; Jung, J.U. Biphasic euchromatin-to-heterochromatin transition on the KSHV genome following de novo infection. *PLoS Pathog.* **2013**, *9*, e1003813.

196. Rossetto, C.C.; Pari, G. KSHV PAN RNA associates with demethylases UTX and JMJD3 to activate lytic replication through a physical interaction with the virus genome. *PLoS Pathog.* **2012**, *8*, e1002680.
197. He, M.; Zhang, W.; Bakken, T.; Schutten, M.; Toth, Z.; Jung, J.U.; Gill, P.; Cannon, M.; Gao, S.J. Cancer angiogenesis induced by Kaposi sarcoma-associated herpesvirus is mediated by EZH2. *Cancer Res.* **2012**, *72*, 3582–3592.
198. Verma, S.C.; Lan, K.; Robertson, E. Structure and function of latency-associated nuclear antigen. *Curr. Top. Microbiol. Immunol.* **2007**, *312*, 101–136.
199. Liang, D.; Hu, H.; Li, S.; Dong, J.; Wang, X.; Wang, Y.; He, L.; He, Z.; Gao, Y.; Gao, S.J.; *et al.* Oncogenic herpesvirus KSHV Hijacks BMP-Smad1-Id signaling to promote tumorigenesis. *PLoS Pathog.* **2014**, *10*, e1004253.
200. Verma, S.C.; Borah, S.; Robertson, E.S. Latency-associated nuclear antigen of Kaposi's sarcoma-associated herpesvirus up-regulates transcription of human telomerase reverse transcriptase promoter through interaction with transcription factor Sp1. *J. Virol.* **2004**, *78*, 10348–10359.
201. Lim, C.; Sohn, H.; Gwack, Y.; Choe, J. Latency-associated nuclear antigen of Kaposi's sarcoma-associated herpesvirus (human herpesvirus-8) binds ATF4/CREB2 and inhibits its transcriptional activation activity. *J. Gen. Virol.* **2000**, *81*, 2645–2652.
202. Murakami, Y.; Yamagoe, S.; Noguchi, K.; Takebe, Y.; Takahashi, N.; Uehara, Y.; Fukazawa, H. Ets-1-dependent expression of vascular endothelial growth factor receptors is activated by latency-associated nuclear antigen of Kaposi's sarcoma-associated herpesvirus through interaction with Daxx. *J. Biol. Chem.* **2006**, *281*, 28113–28121.
203. Radkov, S.A.; Kellam, P.; Boshoff, C. The latent nuclear antigen of Kaposi sarcoma-associated herpesvirus targets the retinoblastoma-E2F pathway and with the oncogene Hras transforms primary rat cells. *Nat. Med.* **2000**, *6*, 1121–1127.
204. Fujimuro, M.; Hayward, S.D. The latency-associated nuclear antigen of Kaposi's sarcoma-associated herpesvirus manipulates the activity of glycogen synthase kinase-3beta. *J. Virol.* **2003**, *77*, 8019–8030.
205. Lim, C.; Gwack, Y.; Hwang, S.; Kim, S.; Choe, J. The transcriptional activity of cAMP response element-binding protein-binding protein is modulated by the latency associated nuclear antigen of Kaposi's sarcoma-associated herpesvirus. *J. Biol. Chem.* **2001**, *276*, 31016–31022.
206. Cai, Q.L.; Knight, J.S.; Verma, S.C.; Zald, P.; Robertson, E.S. EC5S ubiquitin complex is recruited by KSHV latent antigen LANA for degradation of the VHL and p53 tumor suppressors. *PLoS Pathog.* **2006**, *2*, e116.
207. Cai, Q.; Xiao, B.; Si, H.; Cervini, A.; Gao, J.; Lu, J.; Upadhyay, S.K.; Verma, S.C.; Robertson, E.S. Kaposi's sarcoma herpesvirus upregulates Aurora A expression to promote p53 phosphorylation and ubiquitylation. *PLoS Pathog.* **2012**, *8*, e1002566.
208. Hume, A.J.; Kalejta, R.F. Regulation of the retinoblastoma proteins by the human herpesviruses. *Cell Div.* **2009**, *4*, e1.
209. Cousins, E.; Nicholas, J. Molecular biology of human herpesvirus 8: Novel functions and virus-host interactions implicated in viral pathogenesis and replication. *Recent Results Cancer Res.* **2014**, *193*, 227–268.

210. Liu, J.; Martin, H.; Shamay, M.; Woodard, C.; Tang, Q.Q.; Hayward, S.D. Kaposi's sarcoma-associated herpesvirus LANA protein downregulates nuclear glycogen synthase kinase 3 activity and consequently blocks differentiation. *J. Virol.* **2007**, *81*, 4722–4731.
211. Liu, J.; Martin, H.J.; Liao, G.; Hayward, S.D. The Kaposi's sarcoma-associated herpesvirus LANA protein stabilizes and activates c-Myc. *J. Virol.* **2007**, *81*, 10451–10459.
212. Paudel, N.; Sadagopan, S.; Chakraborty, S.; Sarek, G.; Ojala, P.M.; Chandran, B. Kaposi's sarcoma-associated herpesvirus latency-associated nuclear antigen interacts with multifunctional angiogenin to utilize its antiapoptotic functions. *J. Virol.* **2012**, *86*, 5974–5991.
213. Paudel, N.; Sadagopan, S.; Balasubramanian, S.; Chandran, B. Kaposi's sarcoma-associated herpesvirus latency-associated nuclear antigen and angiogenin interact with common host proteins, including annexin A2, which is essential for survival of latently infected cells. *J. Virol.* **2012**, *86*, 1589–1607.
214. Yang, J.; Li, X.; Li, Y.; Southwood, M.; Ye, L.; Long, L.; Al-Lamki, R.S.; Morrell, N.W. Id proteins are critical downstream effectors of BMP signaling in human pulmonary arterial smooth muscle cells. *American J. Physiol. Lung Cell. Mol. Physiol.* **2013**, *305*, L312–L321.
215. Cai, Q.; Cai, S.; Zhu, C.; Verma, S.C.; Choi, J.Y.; Robertson, E.S. A Unique SUMO-2-Interacting Motif within LANA Is Essential for KSHV Latency. *PLoS Pathog.* **2013**, *9*, e1003750.
216. Sun, R.; Liang, D.; Gao, Y.; Lan, K. Kaposi's sarcoma-associated herpesvirus-encoded LANA interacts with host KAP1 to facilitate establishment of viral latency. *J. Virol.* **2014**, *88*, 7331–7344.
217. Zhang, L.; Zhu, C.; Guo, Y.; Wei, F.; Lu, J.; Qin, J.; Banerjee, S.; Wang, J.; Shang, H.; Verma, S.C.; *et al*. Inhibition of KAP1 enhances hypoxia-induced Kaposi's sarcoma-associated herpesvirus reactivation through RBP-Jkappa. *J. Virol.* **2014**, *88*, 6873–6884.
218. Chang, P.C.; Fitzgerald, L.D.; van Geelen, A.; Izumiya, Y.; Ellison, T.J.; Wang, D.H.; Ann, D.K.; Luciw, P.A.; Kung, H.J. Kruppel-associated box domain-associated protein-1 as a latency regulator for Kaposi's sarcoma-associated herpesvirus and its modulation by the viral protein kinase. *Cancer Res.* **2009**, *69*, 5681–5689.
219. Drane, P.; Ouararhni, K.; Depaux, A.; Shuaib, M.; Hamiche, A. The death-associated protein DAXX is a novel histone chaperone involved in the replication-independent deposition of H3.3. *Genes Dev.* **2010**, *24*, 1253–1265.

2

Contribution of Viral Mimics of Cellular Genes to KSHV Infection and Disease

Shuhei Sakakibara [1,2] **and Giovanna Tosato** [3,*]

[1] Department of Molecular Immunology, Research Institute for Microbial Diseases, Osaka University, Suita, Osaka 565-0871, Japan; E-Mail: sakakibara@ragtime.biken.osaka-u.ac.jp

[2] World Premier International Immunology Frontier Research Center, Osaka University, Suita, Osaka 565-0871, Japan

[3] Laboratory of Cellular Oncology, National Cancer Institute, National Institutes of Health, Bethesda, MD 20982, USA

* Author to whom correspondence should be addressed; E-Mail: tosatog@mail.nih.gov;

Abstract: Kaposi's sarcoma-associated herpesvirus (KSHV, also named Human herpesvirus 8 HHV-8) is the cause of Kaposi sarcoma (KS), the most common malignancy in HIV-infected individuals worldwide, primary effusion lymphoma (PEL) and multicentric Castleman disease (MCD). KSHV is a double-stranded DNA virus that encodes several homologues of cellular proteins. The structural similarity between viral and host proteins explains why some viral homologues function as their host counterparts, but sometimes at unusual anatomical sites and inappropriate times. In other cases, structural modification in the viral proteins can suppress or override the function of the host homologue, contributing to KSHV-related diseases. For example, viral IL-6 (vIL-6) is sufficiently different from human IL-6 to activate gp130 signaling independent of the α subunit. As a consequence, vIL-6 can activate many cell types that are unresponsive to cellular IL-6, contributing to MCD disease manifestations. Here, we discuss the molecular biology of KSHV homologues of cellular products as conduits of virus/host interaction with a focus on identifying new strategies for therapy of KS and other KSHV-related diseases.

Keywords: KSHV; vIL-6; vFLIP; NF-κB; Kaposi sarcoma; primary effusion lymphoma; multicentric Castleman disease; inflammation; tumor angiogenesis

---

1. Introduction

Kaposi's sarcoma-associated herpesvirus (KSHV) was identified in 1994 as a novel human herpesvirus that was shown to be the cause of Kaposi sarcoma (KS) [1]. Soon thereafter, HIV/AIDS-associated primary effusion lymphoma (PEL) and multicentric Castleman's disease (MCD) were linked to KSHV infection [2–5]. More recently, KICS (KSHV inflammatory cytokine syndrome) was described in HIV-infected patients [6]. All or some of these diseases can co-exist in individual AIDS patients. It is now recognized that while KSHV infection is necessary, it is not sufficient for development of these malignancies. For example, the epidemiology of KSHV shows that the frequency of KSHV infection worldwide is much higher that the frequency of KS, PEL, MCD and KICS [7,8]. KSHV cannot immortalize or transform primary cells of any lineage. In KS, the KSHV-infected "spindle" cells represent a minority of cells within KS lesions within a heterogeneous mixture of other cells, including endothelial cells and inflammatory cells. These other cell components appear critical for the development and progression of KS [9,10].

The disclosure of the entire DNA sequence of KSHV prompted investigation on several viral gene homologues of cellular genes, presumably pirated by the virus during evolution [11]. Since their discovery, many studies have focused on the identification of the roles of the KSHV-pirated genes in the development of KSHV-associated diseases. In this review, we discuss biochemical and functional features of these KSHV genes and gene products, and how we can use this knowledge to target viral gene products that play multiple roles in the viral life cycle, host cell transformation and tumorigenesis.

2. KSHV-related diseases: KS, MCD, KICS and PEL

2.1. KS

The Hungarian dermatologist Moritz Kaposi first described KS in the 1970s, before HIV and KSHV were discovered. Since the worldwide spread of HIV/AIDS, KS has been recognized as one of the AIDS-related diseases. Several epidemiologic types of KS are recognized; these include classic KS (usually arising in elderly men in regions surrounding the Mediterranean Sea), endemic KS (arising in HIV-negative individuals from Africa prior to the AIDS epidemic), epidemic KS (in HIV-infected individuals) and post-transplant KS (in transplant recipients). All these KS types are the same disease.

KS is a multi-focal endothelial tumor with a considerable inflammatory component and vascular proliferation. The multi-focal nature of KS is not due to metastatic spread from a primary tumor but rather has been attributed to blood colonization of independently infected circulating endothelial cells/endothelial precursors [12–14] or opportunistic spread of KSHV [15]. The KSHV-infected tumor cells, the "KS cells" are generally not clonal and represent a minority of cells within KS lesions. Endothelial cells are almost certainly the cells of origin of KS cells [16], but their phenotype and spindle cell morphology indicates that they are not comparable to the normal endothelium. KSHV infection of vascular endothelial cells causes the spindle cell morphology, which is attributable to expression of the KSHV-vFLIP (*ORF K13*) protein [17–19]. The KSHV-infected KS cells express the lymphatic endothelial cell markers VEGFR3, LYVE-1, VEGF-C, and Prox1, attributable to expression of KSHV-vIL-6 protein [20–22]. KSHV also induces endothelial-mesenchymal transition (EnMT) characterized by reduced expression of the endothelial cell markers CD31, VE-cadherin, CD34 and

Tie2, and expression of the mesenchymal markers αSMA (Acta2), NG-2 and PDGFRβ associated with increased cell motility [23,24]. This KSHV-induced transdifferentiation of endothelial cells is associated with activation of canonical Notch signaling (Figure 1), which provides a growth advantage to the KSHV-infected endothelial cells and is initiated by vFLIP (*ORF K13*) and vGPCR (*ORF74*) via incompletely defined pathways [23–25].

2.2. MCD and KICS

Multicentric Castleman disease (MCD) is a systemic lymphoproliferative disorder characterized by intermittent flares of severe inflammatory symptoms that include fever, night sweats, splenomegaly and lymphadenopathy associated with laboratory symptoms of hypoalbuminimia and anemia [26,27]. Characteristically, circulating levels of certain inflammatory cytokines, including IL-6 and IL-10, are elevated. The diagnosis of MCD is based on specific histologic features of the lesions [28]. This includes plasma cell infiltration of the mantle and inter-follicular zones of affected lymph nodes, which generates characteristic concentric layers that resemble the skin layers of onions, and increased vascularization of the interfollicular space.

With the spread of the AIDS epidemic, it was realized that MCD occurs at a higher rate in patients with HIV/AIDS and that in these patients MCD is almost universally associated with KSHV infection [4,5,29]. KSHV-LANA (latency-associated nuclear antigen)-expressing B cells, which are scattered towards the periphery of the affected follicle, are generally monotypic IgM/Igλ-expressing B cells [6,30]. vIL-6 is often detected in the circulation [6,31], particularly during disease flares, and circulating KSHV is usually present at high levels [26,32]. Recently, an MCD-related syndrome was identified and named KICS (KSHV Inflammatory Cytokine Syndrome): the clinical symptoms of KICS are indistinguishable from those in MCD, but enlarged lymph nodes are not observed and the histologic diagnosis of MCD is missing. Levels of IL-6, vIL-6 and IL-10, and KSHV viral load are comparably high in KICS and HIV-associated KSHV-MCD, and much higher than observed in KS [6,27].

KSHV gene products detected in MCD lesions include vIL-6 (*ORF K2*), PF-8 (*ORF59*), LANA (*ORF73*) and the vIRFs (*ORFs K9*, *K10/10.1*, *K10.5*, *K11* and *K11.1*), indicating that KSHV may be in its lytic phase, in at least a proportion of the infected cells [6,27,33–35].

There is no standard therapy for MCD. Siltuximab, a chimeric neutralizing monoclonal antibody against IL-6 has recently received FDA approval for use in HIV-negative and KSHV-negative MCD. Tocilizumab, a humanized neutralizing antibody against the IL-6R is approved in Japan as a therapy for KSHV-positive and KSHV-negative MCD. Several studies have shown that IL-6/IL-6R targeting (Figure 1) reduces MCD-associated lymph node swelling and fatigue [36,37]. Clinical benefit from the successful targeting of IL-6 or its receptor IL-6R in patients with MCD supports a contributing role of this cytokine in disease pathogenesis and symptomatogy. vIL-6, which is often measurable during MCD flares, is not usually neutralized by IL-6-neutralizing antibodies due to epitope differences [33]. Nonetheless, recent pre-clinical studies have shown that vIL-6 requires some level of IL-6/IL-6R signaling for activity, suggesting that IL-6/IL-6R targeting may also serve to reduce vIL-6 activity [38]. Yet targeting vIL-6 could be a treatment worth investigation in MCD when other treatments are ineffective. Rituximab, a humanized monoclonal antibody against the B-cell marker CD20 has shown efficacy in some cases of MCD (Figure 1) [39].

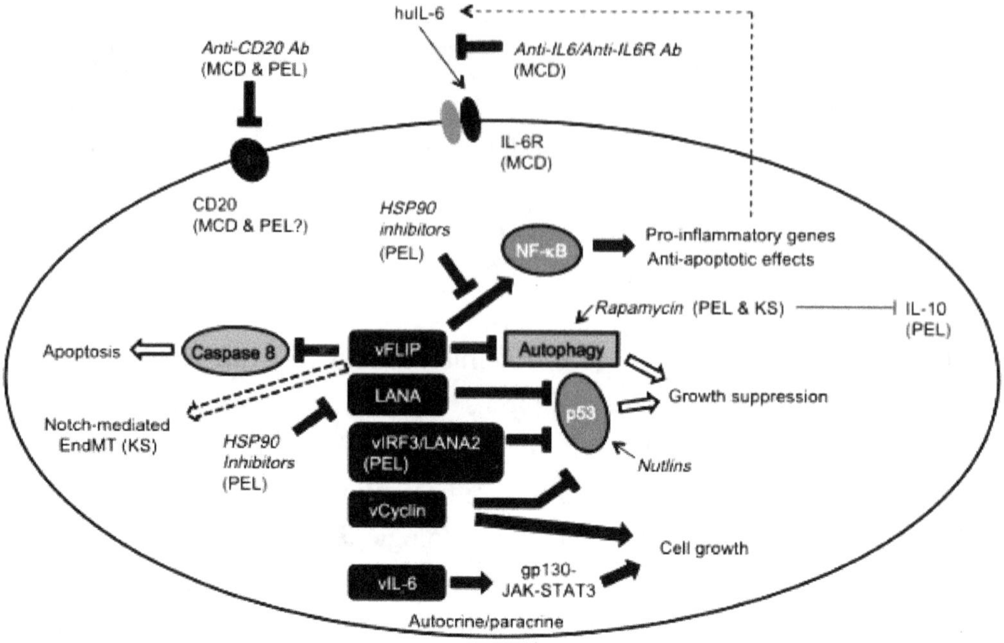

Figure 1. KSHV gene products and potential viral targets for the treatment of KSHV diseases. vFLIP induces pro-inflammatory genes and inhibits cell death by apoptosis and autophagy. vFLIP also enhances Notch-mediated EndMT. Several viral factors, including LANA, vIRF3/LANA2 and vCyclin inhibit p53 function and promote cell cycle progression. Autocrine/paracrine vIL-6 activates STAT3 via gp130. HSP90 inhibitors can target vFLIP and LANA, and could be effective in PEL. Rapamycin induces autophagy and inhibits vIL-10 secretion in PEL cells. Nutlins are p53 activators that could override p53 inhibition induced by several KSHV gene products. Anti-CD20 antibody has shown efficacy in some patients with MCD and PEL. Anti-IL-6R and anti-IL-6 antibodies have shown efficacy in the treatment of MCD.

2.3. PEL

PEL is a rare and aggressive non-Hodgkin's lymphoma that typically presents as a liquid malignancy in the body cavities of patients with HIV-AIDS [40]. PEL cells are always infected with KSHV and often are co-infected with EBV [2,40]. Although PEL cells are of B-cell lineage as they display immunoglobulin gene rearrangement, they express the surface markers CD45, CD38, CD71 and CD30, but lack expression of CD20, CD19, surface immunoglobulin, CD79a and other typical B cell surface markers. Despite the absence of surface CD20, there is evidence that rituximab (anti-CD20 antibody) can be an effective treatment for some patients with PEL (Figure 1) [41,42]. Morphologically, PEL cells appear plasmablastic, immunoblastic or anaplastic lymphoid cells.

KSHV does not immortalize B-lymphocytes in culture and success in adaptation of primary PEL cells to culture has been limited. However, a few PEL cell lines have been established from PEL patients, which have been critical to KSHV research as they maintain KSHV infection through passage in culture. KSHV is necessary for the survival of established PEL cell lines [43–45]. In PEL cells KSHV is maintained as an oligoclonal or monoclonal episome, and is mostly latent although a small proportion of cells can spontaneously undergo lytic replication associated with vIL-6 expression [46,47].

Viral replication can be induced experimentally in PEL cells with TPA treatment [48]. A characteristic feature of PEL is high level VEGF secretion; VEGF is critical to increased vascular permeability and production of body cavity effusions that is typical of PEL disease, and through these functions contributes to PEL disease progression [49,50]. Cytogenetic characterization of PEL cells has failed to detect common chromosomal aberrations, but *Myc* is generally amplified [40]. KSHV LANA, which maintains the viral genome during cell division, functionally inhibits the tumor-suppressor genes p53 and Rb [51].

3. KSHV-pirated Inflammatory Genes: vIL-6, vFLIP and vMIPs

vIL-6 (*ORF K2*) is expressed in MCD lesions and in PEL cells in conjunction with LANA (*ORF73*) and other KSHV genes, including PF-8 (*ORF59*) and *vIRFs* (*ORFs K9*, *K10/10.1*, *K10.5*, *K11* and *K11.1*) [6,27,33–35]. Patients with MCD and KICS have detectable vIL-6 in the circulation, and flares of MCD are associated with spikes in circulating levels of vIL-6 [6,31].

The amino acid sequence of vIL-6 exhibits approximately 25% of similarity to that of human IL-6 [48,52]. Consistent with this modest amino acid conservation, signaling by cellular and vIL-6 differ. Cellular IL-6 requires binding to the non-signaling IL-6R prior to engagement of the signaling chain gp130 [53]. Instead, vIL-6 directly ligates and activates gp130 signaling without a requirement for IL-6R binding [33,54]. Since the distribution of gp130 is much wider than that of IL-6R, it follows that vIL-6 may affect a wider range of cells than its cellular counterpart, which requires the alpha subunit of the receptor, IL-6R. vIL-6 is inefficiently secreted. Nonetheless, vIL-6 can also signal from the intracellular compartment through direct binding to intracellular gp130 [55,56].

An early study reported that subcutaneous inoculation of vIL-6-expressing fibroblasts in nude mice resulted in accelerated fibroblast growth and formation of tumors that were much larger and more vascularized than observed in controls injected with control fibroblasts; tissue levels of VEGF were much higher than in controls [50]. vIL-6 may play a similar growth-promoting, permeability-enhancing and pro-angiogenic role in KSHV-MCD, KICS and PEL, conditions in which vIL-6 is detected in the circulation [6]. The potential importance of vIL-6 in MCD is confirmed by studies of vIL-6 transgenic mice: H2K promoter-driven vIL-6 expression in hematopoietic cells caused high mortality in most of the founder mice; in the surviving mouse lines, splenomegaly, lymph node enlargement and other manifestations typical of MCD were observed [38].

3.1. vFLIP

KSHV-infected cells in KS lesions, the “KS cells” show a characteristic spindle cell shape. KS cells express latency-related genes, including LANA, vFLIP (*ORF 71*) [viral Fas-associated death domain (FADD) interleukin-1β-converting enzyme (FLICE) inhibitory protein] and kaposin (*ORF K12*), and lytic genes, including vGPCR (*ORF 74*) and vCyclin (*ORF72*) [57]. Intriguingly, KSHV vFLIP alone is sufficient to change the typical cobblestone morphology of endothelial cells into that of elongated, spindle-like cells [17–19]. vFLIP was originally identified as a viral homologue of cFLIP (cellular FLICE-like inhibitory protein), which inhibits Caspase 8 activity induced by death domain-containing receptors [58]. vFLIP activates the canonical NF-κB pathway (Figure 1), and the morphologic change into spindle-cells induced by vFLIP is dependent upon vFLIP activation of the NF-κB pathway.

Constitutive NF-κB activation leads to transcriptional regulation of NF-κB target genes, including increased expression of proinflammatory cytokines (*GM-CSF*, *IL-6* and *IL-1β*), chemokines (*Mip1α*, *Rantes*, *Mcp-2*, *Ip-10* and *I-tac*) and interferon-responsive genes, which are likely critical contributors to the prominent proinflammatory phenotype of KS [19,57]. Furthermore, persistent endothelial NF-κB activation by vFLIP induces expression of the NF-κB regulator A20/TNFAIP3, which represses vFLIP-induced NF-κB activation and augments IKK1 protein expression [59]. A20, a ubiquitin-editing enzyme, inhibits NF-κB activation by TNFα and vFLIP, albeit through distinct mechanisms [59]. When TNFα induces NF-κB activation, A20 ubiquitinates IKKγ promoting proteasome-dependent degradation and reducing downstream signaling [60]. Instead, when vFLIP-induces NF-κB activation, the de-ubiquitination activity of A20 is dispensable for NF-κB inhibition [59]. High-level expression of A20 in vFLIP-expressing cells and tissues suggests an important role of the NF-κB pathway in KS [59].

Another function of vFLIP is regulation of cell death by autophagy, a tightly regulated process of cell degradation leading to the removal of cytoplasmic cell components [61]. vFLIP inhibits autophagy and promotes cell survival. This pro-survival function is mediated by vFLIP binding to Atg3, preventing Atg3 binding to the ubiquitin-like protein LC3, which is critical for autophagosome biogenesis [61]. Rapamycin, an mTOR inhibitor with anti-tumor activity, is a potent inducer of autophagy. Despite vFLIP inhibiting Rapamycin-induced autophagy [61], Rapamycin stops growth and KSHV reactivation in PEL cells, which express vFLIP (Figure 1) [62–64]. In addition, Rapamycin reduces PEL and KS tumor progression in mice [63,65]. The anti-PEL activity has been attributed to Rapamycin inhibiting secretion of IL-10, an essential autocrine growth factor for PEL [62,63].

3.2. vMIPs

KSHV encodes the proteins vMIP-I (*ORF K6*), vMIP-II (*ORF K4*) that share extensive sequence identity (43% and 52%, respectively) to the cellular cytokine MIP1α (macrophage inflammatory protein/CCL3), and vMIPIII (ORF K4.1), which is more distantly related to MIP1α [66,67]. vMIP-1 binds to the CCR8 receptor, resulting in Ca^{2+} ion-dependent signaling [68]. vMIP-2 was shown to bind to CCR3 and CCR8 acting as an agonist and to also bind to a variety of other chemokine receptors acting as a broad spectrum antagonist [68–71]. vMIP-III preferentially binds and activates CCR4, acting as a chemoattractant for Th2-type memory T cells, which express this receptor in the skin [67,72]. The cutaneous location of KS has suggested an important role of vMIP-III as an attractant of pro-tumorigenic Th2-type cells that would favor KS progression rather than Th1-type cells, which usually exert anti-tumor activity [67]. Importantly, all three vMIPs were shown to exert pro-angiogenic activity in model systems, and it was proposed that they contribute to the pro-angiogenic phenotype of KS and MCD [67,70]. vMIPs are expressed in the lytic phase of KSHV infection, and have been detected in KS and MCD, which contain spontaneously reactivated cells: vMIP-I and vMIP-II were detected in tissues affected with MCD, whereas only vMIP-II was detected in KS tissues [73].

Currently, there are no preclinical models to assess vMIPs function *in vivo*, and the complexities of chemokine receptor/ligand interactions and redundancy of viral and cellular chemokines suggests that specific targeting these viral products may be difficult. An intriguing approach to overcome some of the complexities of the chemokine system has focused on the development of a therapeutic neutralizing antibody that inactivates multiple chemokines [74].

4. The NF-κB and p53 Pathways: Common Targets of KSHV Gene Products Relevant to KSHV Malignancies

Non-Hodgkin's lymphomas generally display constitutive activation of NF-κB due to defects in its homeostatic control [75]. KSHV constitutively activates NF-κB via physical interaction of vFLIP with IKKγ/NEMO within the I-κB kinase (IKK) complex (Figure 1) [76,77]. The molecular chaperone HSP90 binds to the vFLIP/IKK complex in PEL cells (Figure 1) [78]. Consistent with an essential function of NF-κB activity in sustaining PEL cell survival, inhibition of NF-κB results in PEL cell death [79,80]. In addition, inhibition of HSP90 or vFLIP kills KSHV-infected PEL cell lines, inducing apoptosis and autophagy [44,78,81,82]. The function of HSP90 is to maintain or promote the proper conformation of other "client" proteins, which include many oncogenes and KSHV LANA [83–85]. Inhibition of HSP90 causes "client" proteins to acquire abnormal conformation leading to their ubiquitination and proteasome degradation [84]. Thus, HSP90 inhibitors (Figure 1) are being developed as potential therapeutics in cancer, and some drugs have reached clinical testing in different cancer types, but not in PEL or other KSHV-related malignancies [86,87]. Since HSP90 regulates the stability of several IKK kinases and disruption of HSP90 blocks NF-κB activation [88,89], HSP90 inhibitors hold promise in the treatment of PEL.

Other than vIL-6 and vFLIP, there are several viral inflammatory factors encoded by the KSHV genome [48]. vIRF1 (*ORF K9*), which aligns in the middle of viral genome [11], inhibits type-I IFN signaling and expression of genes under IFN regulatory control [90]. vIRF3/LANA2, which is expressed in PEL but not in KS tissues, inhibits p53 transcriptional activity and pro-apoptotic function (Figure 1) [91]. An important non-inflammatory viral homologue of a cellular protein is vCyclin, which can overcome retinoblastoma (RB) tumor suppressor protein-mediated cell cycle arrest [92], and inhibit p53 expression and function [93]. Transgenic expression of vCyclin in mice under the control of VEGFR3, alters lymphatic vessel structure and causes lymphatic vessel dysfunction [94]. Other than viral homologues, LANA, which is expressed in all KHSV-infected cells, also interacts with p53 and inhibits its transcriptional activity (Figure 1) [51]. The oncoprotein MDM2 (Monocyte to Macrophage Differentiation Factor-2) binds to p53 and negatively regulates its stability and pro-apoptotic activity [95]. Thus, inhibitors of MDM2 would be expected to activate p53 and increase cell death, which would be a valuable approach to reducing tumor cell growth. Nutlins are selective inhibitors of the p53-MDM2 interaction, which cause p53-dependent apoptosis in cancer cells (Figure 1) [96]. Nutlin-3a, a small molecule inhibitor of the p53/MDM2 interaction, which promotes p53 reactivation, kills PEL cells in culture and has potent anti-tumor activity in mice bearing PEL tumors [97,98].

5. Future Directions

We have described KSHV-pirated genes and their functions in KSHV-related diseases. It is clear that viral homologues of cellular genes that have been captured during virus-host co-evolution play important roles in KSHV life cycle and pathogenicity. Therefore, a clear understanding of the function of these factors can provide precise therapeutic targets directed at the pathogen. Precision medicine is a current goal in cancer therapy, as it ideally targets important unique tumor targets while sparing the normal cell counterparts. Targeting KSHV-specific factors is in line with current efforts.

Acknowledgments

This work is supported by the intramural research program of CCR/NCI/NIH. We thank R. Yarchoan, T. Uldrick and M. Polizzotto for helpful discussions.

Author Contributions

S.S. and G.T. wrote this paper.

References and Notes

1. Chang, Y.; Cesarman, E.; Pessin, M.S.; Lee, F.; Culpepper, J.; Knowles, D.M.; Moore, P.S. Identification of herpesvirus-like DNA sequences in AIDS-associated Kaposi's sarcoma. *Science* **1994**, *266*, 1865–1869.
2. Cesarman, E.; Chang, Y.; Moore, P.S.; Said, J.W.; Knowles, D.M. Kaposi's sarcoma-associated herpesvirus-like DNA sequences in AIDS-related body-cavity-based lymphomas. *N. Engl. J. Med.* **1995**, *332*, 1186–1191.
3. Nador, R.G.; Cesarman, E.; Knowles, D.M.; Said, J.W. Herpes-like DNA sequences in a body-cavity-based lymphoma in an HIV-negative patient. *N. Engl. J. Med.* **1995**, *333*, 943.
4. Soulier, J.; Grollet, L.; Oksenhendler, E.; Cacoub, P.; Cazals-Hatem, D.; Babinet, P.; D'Agay, M.F.; Clauvel, J.P.; Raphael, M.; Degos, L.; *et al.* Kaposi's sarcoma-associated herpesvirus-like DNA sequences in multicentric Castleman's disease. *Blood* **1995**, *86*, 1276–1280.
5. Dupin, N.; Diss, T.L.; Kellam, P.; Tulliez, M.; Du, M.Q.; Sicard, D.; Weiss, R.A.; Isaacson, P.G.; Boshoff, C. HHV-8 is associated with a plasmablastic variant of Castleman disease that is linked to HHV-8-positive plasmablastic lymphoma. *Blood* **2000**, *95*, 1406–1412.
6. Uldrick, T.S.; Wang, V.; O'Mahony, D.; Aleman, K.; Wyvill, K.M.; Marshall, V.; Steinberg, S.M.; Pittaluga, S.; Maric, I.; Whitby, D.; *et al.* An interleukin-6-related systemic inflammatory syndrome in patients co-infected with Kaposi sarcoma-associated herpesvirus and HIV but without Multicentric Castleman disease. *Clin. Infect. Dis.* **2010**, *51*, 350–358.
7. Martin, J.N.; Ganem, D.E.; Osmond, D.H.; Page-Shafer, K.A.; Macrae, D.; Kedes, D.H. Sexual transmission and the natural history of human herpesvirus 8 infection. *N. Engl. J. Med.* **1998**, *338*, 948–954.
8. Dollard, S.C.; Butler, L.M.; Jones, A.M.; Mermin, J.H.; Chidzonga, M.; Chipato, T.; Shiboski, C.H.; Brander, C.; Mosam, A.; Kiepiela, P.; *et al.* Substantial regional differences in human herpesvirus 8 seroprevalence in sub-Saharan Africa: Insights on the origin of the "Kaposi's sarcoma belt". *Int. J. Cancer* **2010**, *127*, 2395–2401.
9. Niedt, G.W.; Prioleau, P.G. Kaposi's sarcoma occurring in a dermatome previously involved by herpes zoster. *J. Am. Acad. Dermatol.* **1988**, *18*, 448–451.
10. Achenbach, C.J.; Harrington, R.D.; Dhanireddy, S.; Crane, H.M.; Casper, C.; Kitahata, M.M. Paradoxical immune reconstitution inflammatory syndrome in HIV-infected patients treated with combination antiretroviral therapy after AIDS-defining opportunistic infection. *Clin. Infect. Dis.* **2012**, *54*, 424–433.

11. Russo, J.J.; Bohenzky, R.A.; Chien, M.C.; Chen, J.; Yan, M.; Maddalena, D.; Parry, J.P.; Peruzzi, D.; Edelman, I.S.; Chang, Y.; *et al.* Nucleotide sequence of the Kaposi sarcoma-associated herpesvirus (HHV8). *Proc. Natl. Acad. Sci. USA* **1996**, *93*, 14862–14867.
12. Browning, P.J.; Sechler, J.M.; Kaplan, M.; Washington, R.H.; Gendelman, R.; Yarchoan, R.; Ensoli, B.; Gallo, R.C. Identification and culture of Kaposi's sarcoma-like spindle cells from the peripheral blood of human immunodeficiency virus-1-infected individuals and normal controls. *Blood* **1994**, *84*, 2711–2720.
13. Yao, L.; Salvucci, O.; Cardones, A.R.; Hwang, S.T.; Aoki, Y.; De La Luz Sierra, M.; Sajewicz, A.; Pittaluga, S.; Yarchoan, R.; Tosato, G. Selective expression of stromal-derived factor-1 in the capillary vascular endothelium plays a role in Kaposi sarcoma pathogenesis. *Blood* **2003**, *102*, 3900–3905.
14. Della Bella, S.; Taddeo, A.; Calabro, M.L.; Brambilla, L.; Bellinvia, M.; Bergamo, E.; Clerici, M.; Villa, M.L. Peripheral blood endothelial progenitors as potential reservoirs of Kaposi's sarcoma-associated herpesvirus. *PLoS One* **2008**, *3*, e1520.
15. Duus, K.M.; Lentchitsky, V.; Wagenaar, T.; Grose, C.; Webster-Cyriaque, J. Wild-type Kaposi's sarcoma-associated herpesvirus isolated from the oropharynx of immune-competent individuals has tropism for cultured oral epithelial cells. *J. Virol.* **2004**, *78*, 4074–4084.
16. Carroll, P.A.; Brazeau, E.; Lagunoff, M. Kaposi's sarcoma-associated herpesvirus infection of blood endothelial cells induces lymphatic differentiation. *Virology* **2004**, *328*, 7–18.
17. Grossmann, C.; Podgrabinska, S.; Skobe, M.; Ganem, D. Activation of NF-kappaB by the latent vFLIP gene of Kaposi's sarcoma-associated herpesvirus is required for the spindle shape of virus-infected endothelial cells and contributes to their proinflammatory phenotype. *J. Virol.* **2006**, *80*, 7179–7185.
18. Matta, H.; Surabhi, R.M.; Zhao, J.; Punj, V.; Sun, Q.; Schamus, S.; Mazzacurati, L.; Chaudhary, P.M. Induction of spindle cell morphology in human vascular endothelial cells by human herpesvirus 8-encoded viral FLICE inhibitory protein K13. *Oncogene* **2007**, *26*, 1656–1660.
19. Sakakibara, S.; Pise-Masison, C.A.; Brady, J.N.; Tosato, G. Gene regulation and functional alterations induced by Kaposi's sarcoma-associated herpesvirus-encoded ORFK13/vFLIP in endothelial cells. *J. Virol.* **2009**, *83*, 2140–2153.
20. Hong, Y.K.; Foreman, K.; Shin, J.W.; Hirakawa, S.; Curry, C.L.; Sage, D.R.; Libermann, T.; Dezube, B.J.; Fingeroth, J.D.; Detmar, M. Lymphatic reprogramming of blood vascular endothelium by Kaposi sarcoma-associated herpesvirus. *Nat. Genet.* **2004**, *36*, 683–685.
21. Boshoff, C.; Whitby, D.; Hatziioannou, T.; Fisher, C.; van der Walt, J.; Hatzakis, A.; Weiss, R.; Schulz, T. Kaposi's-sarcoma-associated herpesvirus in HIV-negative Kaposi's sarcoma. *Lancet* **1995**, *345*, 1043–1044.
22. Morris, V.A.; Punjabi, A.S.; Wells, R.C.; Wittkopp, C.J.; Vart, R.; Lagunoff, M. The KSHV viral IL-6 homolog is sufficient to induce blood to lymphatic endothelial cell differentiation. *Virology* **2012**, *428*, 112–120.
23. Cheng, F.; Pekkonen, P.; Laurinavicius, S.; Sugiyama, N.; Henderson, S.; Gunther, T.; Rantanen, V.; Kaivanto, E.; Aavikko, M.; Sarek, G.; *et al.* KSHV-initiated notch activation leads to membrane-type-1 matrix metalloproteinase-dependent lymphatic endothelial-to-mesenchymal transition. *Cell Host Microbe* **2011**, *10*, 577–590.

24. Gasperini, P.; Espigol-Frigole, G.; McCormick, P.J.; Salvucci, O.; Maric, D.; Uldrick, T.S.; Polizzotto, M.N.; Yarchoan, R.; Tosato, G. Kaposi sarcoma herpesvirus promotes endothelial-to-mesenchymal transition through Notch-dependent signaling. *Cancer Res.* **2012**, *72*, 1157–1169.
25. Emuss, V.; Lagos, D.; Pizzey, A.; Gratrix, F.; Henderson, S.R.; Boshoff, C. KSHV manipulates Notch signaling by DLL4 and JAG1 to alter cell cycle genes in lymphatic endothelia. *PLoS Pathog.* **2009**, *5*, e1000616.
26. Oksenhendler, E.; Carcelain, G.; Aoki, Y.; Boulanger, E.; Maillard, A.; Clauvel, J.P.; Agbalika, F. High levels of human herpesvirus 8 viral load, human interleukin-6, interleukin-10, and C reactive protein correlate with exacerbation of multicentric castleman disease in HIV-infected patients. *Blood* **2000**, *96*, 2069–2073.
27. Polizzotto, M.N.; Uldrick, T.S.; Hu, D.; Yarchoan, R. Clinical manifestations of kaposi sarcoma herpesvirus lytic activation: Multicentric Castleman Disease (KSHV-MCD) and the KSHV inflammatory cytokine syndrome. *Front. Microbiol.* **2012**, *3*, 73.
28. Cronin, D.M.; Warnke, R.A. Castleman disease: An update on classification and the spectrum of associated lesions. *Adv. Anat. Pathol.* **2009**, *16*, 236–246.
29. Pluda, J.M.; Yarchoan, R.; Jaffe, E.S.; Feuerstein, I.M.; Solomon, D.; Steinberg, S.M.; Wyvill, K.M.; Raubitschek, A.; Katz, D.; Broder, S. Development of non-Hodgkin lymphoma in a cohort of patients with severe human immunodeficiency virus (HIV) infection on long-term antiretroviral therapy. *Ann. Intern. Med.* **1990**, *113*, 276–282.
30. Chadburn, A.; Hyjek, E.M.; Tam, W.; Liu, Y.; Rengifo, T.; Cesarman, E.; Knowles, D.M. Immunophenotypic analysis of the Kaposi sarcoma herpesvirus (KSHV; HHV-8)-infected B cells in HIV+ multicentric Castleman disease (MCD). *Histopathology* **2008**, *53*, 513–524.
31. Aoki, Y.; Tosato, G.; Fonville, T.W.; Pittaluga, S. Serum viral interleukin-6 in AIDS-related multicentric Castleman disease. *Blood* **2001**, *97*, 2526–2527.
32. Polizzotto, M.N.; Uldrick, T.S.; Wang, V.; Aleman, K.; Wyvill, K.M.; Marshall, V.; Pittaluga, S.; O'Mahony, D.; Whitby, D.; Tosato, G.; *et al.* Human and viral interleukin-6 and other cytokines in Kaposi sarcoma herpesvirus-associated multicentric Castleman disease. *Blood* **2013**, *122*, 4189–4198.
33. Aoki, Y.; Narazaki, M.; Kishimoto, T.; Tosato, G. Receptor engagement by viral interleukin-6 encoded by Kaposi sarcoma-associated herpesvirus. *Blood* **2001**, *98*, 3042–3049.
34. Du, M.Q.; Diss, T.C.; Liu, H.; Ye, H.; Hamoudi, R.A.; Cabecadas, J.; Dong, H.Y.; Harris, N.L.; Chan, J.K.; Rees, J.W.; *et al.* KSHV- and EBV-associated germinotropic lymphoproliferative disorder. *Blood* **2002**, *100*, 3415–3418.
35. Katano, H.; Sato, Y.; Kurata, T.; Mori, S.; Sata, T. Expression and localization of human herpesvirus 8-encoded proteins in primary effusion lymphoma, Kaposi's sarcoma, and multicentric Castleman's disease. *Virology* **2000**, *269*, 335–344.
36. Nishimoto, N.; Sasai, M.; Shima, Y.; Nakagawa, M.; Matsumoto, T.; Shirai, T.; Kishimoto, T.; Yoshizaki, K. Improvement in Castleman's disease by humanized anti-interleukin-6 receptor antibody therapy. *Blood* **2000**, *95*, 56–61.
37. Nishimoto, N.; Kanakura, Y.; Aozasa, K.; Johkoh, T.; Nakamura, M.; Nakano, S.; Nakano, N.;

Ikeda, Y.; Sasaki, T.; Nishioka, K.; *et al.* Humanized anti-interleukin-6 receptor antibody treatment of multicentric Castleman disease. *Blood* **2005**, *106*, 2627–2632.
38. Suthaus, J.; Stuhlmann-Laeisz, C.; Tompkins, V.S.; Rosean, T.R.; Klapper, W.; Tosato, G.; Janz, S.; Scheller, J.; Rose-John, S. HHV-8-encoded viral IL-6 collaborates with mouse IL-6 in the development of multicentric Castleman disease in mice. *Blood* **2012**, *119*, 5173–5181.
39. Marcelin, A.G.; Aaron, L.; Mateus, C.; Gyan, E.; Gorin, I.; Viard, J.P.; Calvez, V.; Dupin, N. Rituximab therapy for HIV-associated Castleman disease. *Blood* **2003**, *102*, 2786–2788.
40. Knowles, D.M.; Inghirami, G.; Ubriaco, A.; Dalla-Favera, R. Molecular genetic analysis of three AIDS-associated neoplasms of uncertain lineage demonstrates their B-cell derivation and the possible pathogenetic role of the Epstein-Barr virus. *Blood* **1989**, *73*, 792–799.
41. Perez, C.L.; Rudoy, S. Anti-CD20 monoclonal antibody treatment of human herpesvirus 8-associated, body cavity-based lymphoma with an unusual phenotype in a human immunodeficiency virus-negative patient. *Clin. Diagn. Lab. Immunol.* **2001**, *8*, 993–996.
42. Siddiqi, T.; Joyce, R.M. A case of HIV-negative primary effusion lymphoma treated with bortezomib, pegylated liposomal doxorubicin, and rituximab. *Clin. Lymphoma Myeloma* **2008**, *8*, 300–304.
43. Kliche, S.; Kremmer, E.; Hammerschmidt, W.; Koszinowski, U.; Haas, J. Persistent infection of Epstein-Barr virus-positive B lymphocytes by human herpesvirus 8. *J. Virol.* **1998**, *72*, 8143–8149.
44. Guasparri, I.; Keller, S.A.; Cesarman, E. KSHV vFLIP is essential for the survival of infected lymphoma cells. *J. Exp. Med.* **2004**, *199*, 993–1003.
45. Godfrey, A.; Anderson, J.; Papanastasiou, A.; Takeuchi, Y.; Boshoff, C. Inhibiting primary effusion lymphoma by lentiviral vectors encoding short hairpin RNA. *Blood* **2005**, *105*, 2510–2518.
46. Judde, J.G.; Lacoste, V.; Briere, J.; Kassa-Kelembho, E.; Clyti, E.; Couppie, P.; Buchrieser, C.; Tulliez, M.; Morvan, J.; Gessain, A. Monoclonality or oligoclonality of human herpesvirus 8 terminal repeat sequences in Kaposi's sarcoma and other diseases. *J. Natl. Cancer Inst.* **2000**, *92*, 729–736.
47. Speck, S.H.; Ganem, D. Viral latency and its regulation: Lessons from the gamma-herpesviruses. *Cell Host Microbe* **2010**, *8*, 100–115.
48. Moore, P.S.; Boshoff, C.; Weiss, R.A.; Chang, Y. Molecular mimicry of human cytokine and cytokine response pathway genes by KSHV. *Science* **1996**, *274*, 1739–1744.
49. Aoki, Y.; Tosato, G. Role of vascular endothelial growth factor/vascular permeability factor in the pathogenesis of Kaposi's sarcoma-associated herpesvirus-infected primary effusion lymphomas. *Blood* **1999**, *94*, 4247–4254.
50. Aoki, Y.; Jaffe, E.S.; Chang, Y.; Jones, K.; Teruya-Feldstein, J.; Moore, P.S.; Tosato, G. Angiogenesis and hematopoiesis induced by Kaposi's sarcoma-associated herpesvirus-encoded interleukin-6. *Blood* **1999**, *93*, 4034–4043.
51. Radkov, S.A.; Kellam, P.; Boshoff, C. The latent nuclear antigen of Kaposi sarcoma-associated herpesvirus targets the retinoblastoma-E2F pathway and with the oncogene Hras transforms primary rat cells. *Nat. Med.* **2000**, *6*, 1121–1127.
52. Nicholas, J.; Ruvolo, V.R.; Burns, W.H.; Sandford, G.; Wan, X.; Ciufo, D.; Hendrickson, S.B.; Guo, H.G.; Hayward, G.S.; Reitz, M.S. Kaposi's sarcoma-associated human herpesvirus-8

encodes homologues of macrophage inflammatory protein-1 and interleukin-6. *Nat. Med.* **1997**, *3*, 287–292.

53. Murakami, M.; Hibi, M.; Nakagawa, N.; Nakagawa, T.; Yasukawa, K.; Yamanishi, K.; Taga, T.; Kishimoto, T. IL-6-induced homodimerization of gp130 and associated activation of a tyrosine kinase. *Science* **1993**, *260*, 1808–1810.
54. Wan, X.; Wang, H.; Nicholas, J. Human herpesvirus 8 interleukin-6 (vIL-6) signals through gp130 but has structural and receptor-binding properties distinct from those of human IL-6. *J. Virol.* **1999**, *73*, 8268–8278.
55. Meads, M.B.; Medveczky, P.G. Kaposi's sarcoma-associated herpesvirus-encoded viral interleukin-6 is secreted and modified differently than human interleukin-6: Evidence for a unique autocrine signaling mechanism. *J. Biol. Chem.* **2004**, *279*, 51793–51803.
56. Chen, D.; Sandford, G.; Nicholas, J. Intracellular signaling mechanisms and activities of human herpesvirus 8 interleukin-6. *J. Virol.* **2009**, *83*, 722–733.
57. Ganem, D. KSHV infection and the pathogenesis of Kaposi's sarcoma. *Annu. Rev. Pathol.* **2006**, *1*, 273–296.
58. Chaudhary, P.M.; Jasmin, A.; Eby, M.T.; Hood, L. Modulation of the NF-kappa B pathway by virally encoded death effector domains-containing proteins. *Oncogene* **1999**, *18*, 5738–5746.
59. Sakakibara, S.; Espigol-Frigole, G.; Gasperini, P.; Uldrick, T.S.; Yarchoan, R.; Tosato, G. A20/TNFAIP3 inhibits NF-kappaB activation induced by the Kaposi's sarcoma-associated herpesvirus vFLIP oncoprotein. *Oncogene* **2013**, *32*, 1223–1232.
60. Shembade, N.; Ma, A.; Harhaj, E.W. Inhibition of NF-kappaB signaling by A20 through disruption of ubiquitin enzyme complexes. *Science* **2010**, *327*, 1135–1139.
61. Lee, J.S.; Li, Q.; Lee, J.Y.; Lee, S.H.; Jeong, J.H.; Lee, H.R.; Chang, H.; Zhou, F.C.; Gao, S.J.; Liang, C.; *et al.* FLIP-mediated autophagy regulation in cell death control. *Nat. Cell Biol.* **2009**, *11*, 1355–1362.
62. Sin, S.H.; Roy, D.; Wang, L.; Staudt, M.R.; Fakhari, F.D.; Patel, D.D.; Henry, D.; Harrington, W.J., Jr.; Damania, B.A.; Dittmer, D.P. Rapamycin is efficacious against primary effusion lymphoma (PEL) cell lines *in vivo* by inhibiting autocrine signaling. *Blood* **2007**, *109*, 2165–2173.
63. Gasperini, P.; Tosato, G. Targeting the mammalian target of Rapamycin to inhibit VEGF and cytokines for the treatment of primary effusion lymphoma. *Leukemia* **2009**, *23*, 1867–1874.
64. Nichols, L.A.; Adang, L.A.; Kedes, D.H. Rapamycin blocks production of KSHV/HHV8: Insights into the anti-tumor activity of an immunosuppressant drug. *PLoS One* **2011**, *6*, e14535.
65. Roy, D.; Sin, S.H.; Lucas, A.; Venkataramanan, R.; Wang, L.; Eason, A.; Chavakula, V.; Hilton, I.B.; Tamburro, K.M.; Damania, B.; *et al.* mTOR inhibitors block Kaposi sarcoma growth by inhibiting essential autocrine growth factors and tumor angiogenesis. *Cancer Res.* **2013**, *73*, 2235–2246.
66. Moore, P.S.; Gao, S.J.; Dominguez, G.; Cesarman, E.; Lungu, O.; Knowles, D.M.; Garber, R.; Pellett, P.E.; McGeoch, D.J.; Chang, Y. Primary characterization of a herpesvirus agent associated with Kaposi's sarcomae. *J. Virol.* **1996**, *70*, 549–558.
67. Stine, J.T.; Wood, C.; Hill, M.; Epp, A.; Raport, C.J.; Schweickart, V.L.; Endo, Y.; Sasaki, T.; Simmons, G.; Boshoff, C.; *et al.* KSHV-encoded CC chemokine vMIP-III is a CCR4 agonist,

stimulates angiogenesis, and selectively chemoattracts TH2 cells. *Blood* **2000**, *95*, 1151–1157.

68. Dairaghi, D.J.; Fan, R.A.; McMaster, B.E.; Hanley, M.R.; Schall, T.J. HHV8-encoded vMIP-I selectively engages chemokine receptor CCR8. Agonist and antagonist profiles of viral chemokines. *J. Biol. Chem.* **1999**, *274*, 21569–21574.
69. Kledal, T.N.; Rosenkilde, M.M.; Coulin, F.; Simmons, G.; Johnsen, A.H.; Alouani, S.; Power, C.A.; Luttichau, H.R.; Gerstoft, J.; Clapham, P.R.; *et al.* A broad-spectrum chemokine antagonist encoded by Kaposi's sarcoma-associated herpesvirus. *Science* **1997**, *277*, 1656–1659.
70. Boshoff, C.; Endo, Y.; Collins, P.D.; Takeuchi, Y.; Reeves, J.D.; Schweickart, V.L.; Siani, M.A.; Sasaki, T.; Williams, T.J.; Gray, P.W.; *et al.* Angiogenic and HIV-inhibitory functions of KSHV-encoded chemokines. *Science* **1997**, *278*, 290–294.
71. Sozzani, S.; Luini, W.; Bianchi, G.; Allavena, P.; Wells, T.N.; Napolitano, M.; Bernardini, G.; Vecchi, A.; D'Ambrosio, D.; Mazzeo, D.; *et al.* The viral chemokine macrophage inflammatory protein-II is a selective Th2 chemoattractant. *Blood* **1998**, *92*, 4036–4039.
72. Campbell, J.J.; Haraldsen, G.; Pan, J.; Rottman, J.; Qin, S.; Ponath, P.; Andrew, D.P.; Warnke, R.; Ruffing, N.; Kassam, N.; *et al.* The chemokine receptor CCR4 in vascular recognition by cutaneous but not intestinal memory T cells. *Nature* **1999**, *400*, 776–780.
73. Nakano, K.; Katano, H.; Tadagaki, K.; Sato, Y.; Ohsaki, E.; Mori, Y.; Yamanishi, K.; Ueda, K. Novel monoclonal antibodies for identification of multicentric Castleman's disease; Kaposi's sarcoma-associated herpesvirus-encoded vMIP-I and vMIP-II. *Virology* **2012**, *425*, 95–102.
74. Scalley-Kim, M.L.; Hess, B.W.; Kelly, R.L.; Krostag, A.R.; Lustig, K.H.; Marken, J.S.; Ovendale, P.J.; Posey, A.R.; Smolak, P.J.; Taylor, J.D.; *et al.* A novel highly potent therapeutic antibody neutralizes multiple human chemokines and mimics viral immune modulation. *PLoS One* **2012**, *7*, e43332.
75. Compagno, M.; Lim, W.K.; Grunn, A.; Nandula, S.V.; Brahmachary, M.; Shen, Q.; Bertoni, F.; Ponzoni, M.; Scandurra, M.; Califano, A.; *et al.* Mutations of multiple genes cause deregulation of NF-kappaB in diffuse large B-cell lymphoma. *Nature* **2009**, *459*, 717–721.
76. Field, N.P.; Gal-Oz, E.; Bonanno, G.A. Continuing bonds and adjustment at 5 years after the death of a spouse. *J. Consult. Clin. Psychol.* **2003**, *71*, 110–117.
77. Liu, L.; Eby, M.T.; Rathore, N.; Sinha, S.K.; Kumar, A.; Chaudhary, P.M. The human herpes virus 8-encoded viral FLICE inhibitory protein physically associates with and persistently activates the Ikappa B kinase complex. *J. Biol. Chem.* **2002**, *277*, 13745–13751.
78. Field, N.; Low, W.; Daniels, M.; Howell, S.; Daviet, L.; Boshoff, C.; Collins, M. KSHV vFLIP binds to IKK-gamma to activate IKK. *J. Cell Sci.* **2003**, *116*, 3721–3728.
79. Keller, S.A.; Schattner, E.J.; Cesarman, E. Inhibition of NF-kappaB induces apoptosis of KSHV-infected primary effusion lymphoma cells. *Blood* **2000**, *96*, 2537–2542.
80. Hussain, A.R.; Ahmed, S.O.; Ahmed, M.; Khan, O.S.; Al Abdulmohsen, S.; Platanias, L.C.; Al-Kuraya, K.S.; Uddin, S. Cross-talk between NFkB and the PI3-kinase/AKT pathway can be targeted in primary effusion lymphoma (PEL) cell lines for efficient apoptosis. *PLoS One* **2012**, *7*, e39945.
81. Gopalakrishnan, R.; Matta, H.; Chaudhary, P.M. A purine scaffold HSP90 inhibitor BIIB021 has selective activity against KSHV-associated primary effusion lymphoma and blocks vFLIP K13-induced NF-kappaB. *Clin. Cancer Res.* **2013**, *19*, 5016–5026.

82. Nayar, U.; Lu, P.; Goldstein, R.L.; Vider, J.; Ballon, G.; Rodina, A.; Taldone, T.; Erdjument-Bromage, H.; Chomet, M.; Blasberg, R.; *et al.* Targeting the Hsp90-associated viral oncoproteome in gammaherpesvirus-associated malignancies. *Blood* **2013**, *122*, 2837–2847.
83. Pearl, L.H.; Prodromou, C. Structure and mechanism of the Hsp90 molecular chaperone machinery. *Annu. Rev. Biochem.* **2006**, *75*, 271–294.
84. Messaoudi, S.; Peyrat, J.F.; Brion, J.D.; Alami, M. Recent advances in Hsp90 inhibitors as antitumor agents. *Anti-Cancer Agents Med. Chem.* **2008**, *8*, 761–782.
85. Chen, W.; Sin, S.H.; Wen, K.W.; Damania, B.; Dittmer, D.P. Hsp90 inhibitors are efficacious against Kaposi Sarcoma by enhancing the degradation of the essential viral gene LANA, of the viral co-receptor EphA2 as well as other client proteins. *PLoS Pathog.* **2012**, *8*, e1003048.
86. Biamonte, M.A.; Van de Water, R.; Arndt, J.W.; Scannevin, R.H.; Perret, D.; Lee, W.C. Heat shock protein 90: Inhibitors in clinical trials. *J. Med. Chem.* **2010**, *53*, 3–17.
87. Goloudina, A.R.; Demidov, O.N.; Garrido, C. Inhibition of HSP70: A challenging anti-cancer strategy. *Cancer Lett.* **2012**, *325*, 117–124.
88. Lewis, J.; Devin, A.; Miller, A.; Lin, Y.; Rodriguez, Y.; Neckers, L.; Liu, Z.G. Disruption of hsp90 function results in degradation of the death domain kinase, receptor-interacting protein (RIP), and blockage of tumor necrosis factor-induced nuclear factor-kappaB activation. *J. Biol. Chem.* **2000**, *275*, 10519–10526.
89. Thangjam, G.S.; Dimitropoulou, C.; Joshi, A.D.; Barabutis, N.; Shaw, M.C.; Kovalenkov, Y.; Wallace, C.M.; Fulton, D.J.; Patel, V.; Catravas, J.D. Novel mechanism of attenuation of LPS-induced NF-kappaB activation by the heat shock protein 90 inhibitor, 17-N-allylamino-17-demethoxygeldanamycin, in human lung microvascular endothelial cells. *Am. J. Respir. Cell Mol. Biol.* **2014**, *50*, 942–952.
90. Gao, S.J.; Boshoff, C.; Jayachandra, S.; Weiss, R.A.; Chang, Y.; Moore, P.S. KSHV ORF K9 (vIRF) is an oncogene which inhibits the interferon signaling pathway. *Oncogene* **1997**, *15*, 1979–1985.
91. Rivas, C.; Thlick, A.E.; Parravicini, C.; Moore, P.S.; Chang, Y. Kaposi's sarcoma-associated herpesvirus LANA2 is a B-cell-specific latent viral protein that inhibits p53. *J. Virol.* **2001**, *75*, 429–438.
92. Chang, Y.; Moore, P.S.; Talbot, S.J.; Boshoff, C.H.; Zarkowska, T.; Godden, K.; Paterson, H.; Weiss, R.A.; Mittnacht, S. Cyclin encoded by KS herpesvirus. *Nature* **1996**, *382*, 410.
93. Verschuren, E.W.; Hodgson, J.G.; Gray, J.W.; Kogan, S.; Jones, N.; Evan, G.I. The role of p53 in suppression of KSHV cyclin-induced lymphomagenesis. *Cancer Res.* **2004**, *64*, 581–589.
94. Sugaya, M.; Watanabe, T.; Yang, A.; Starost, M.F.; Kobayashi, H.; Atkins, A.M.; Borris, D.L.; Hanan, E.A.; Schimel, D.; Bryant, M.A.; *et al.* Lymphatic dysfunction in transgenic mice expressing KSHV k-cyclin under the control of the VEGFR-3 promoter. *Blood* **2005**, *105*, 2356–2363.
95. Freedman, D.A.; Wu, L.; Levine, A.J. Functions of the MDM2 oncoprotein. *Cell. Mol. Life Sci.* **1999**, *55*, 96–107.
96. Vassilev, L.T. Small-molecule antagonists of p53-MDM2 binding: Research tools and potential therapeutics. *Cell Cycle* **2004**, *3*, 419–421.

97. Sarek, G.; Kurki, S.; Enback, J.; Iotzova, G.; Haas, J.; Laakkonen, P.; Laiho, M.; Ojala, P.M. Reactivation of the p53 pathway as a treatment modality for KSHV-induced lymphomas. *J. Clin. Investig.* **2007**, *117*, 1019–1028.
98. Sarek, G.; Ojala, P.M. p53 reactivation kills KSHV lymphomas efficiently *in vitro* and *in vivo*: New hope for treating aggressive viral lymphomas. *Cell Cycle* **2007**, *6*, 2205–2209.

3

Multiple Regions of Kaposi's Sarcoma-Associated Herpesvirus ORF59 RNA are Required for its Expression Mediated by Viral ORF57 and Cellular RBM15

Maria Julia Massimelli [1,†], Vladimir Majerciak [1], Jeong-Gu Kang [1,‡], David J. Liewehr [2], Seth M. Steinberg [2] and Zhi-Ming Zheng [1,*]

[1] Tumor Virus RNA Biology Section, Gene Regulation and Chromosome Biology Laboratory, Center for Cancer Research, National Cancer Institute, National Institutes of Health, 1050 Boyles Street, Frederick, MD 21702, USA; E-Mails: julia.massimelli@uci.edu (M.J.M.); majerciv@mail.nih.gov (V.M.); jkang@jkbiopharma.com (J.-G.K.)

[2] Biostatistics & Data Management Section, Center for Cancer Research, National Cancer Institute, National Institutes of Health, Bethesda, MD 20892, USA; E-Mails: liewehrd@mail.nih.gov (D.J.L.); steinbes@mail.nih.gov (S.M.S.)

† Current address: Molecular Biology and Biochemistry Department, University of California Irvine, Irvine, CA 92697, USA.

‡ Current address: JK BioPharma Solutions, Inc., Rockville, MD 20850, USA.

* Author to whom correspondence should be addressed; E-Mail: zhengt@exchange.nih.gov;

Academic Editor: Eric O. Freed

Abstract: KSHV ORF57 (MTA) promotes RNA stability of ORF59, a viral DNA polymerase processivity factor. Here, we show that the integrity of both ORF59 RNA ends is necessary for ORF57-mediated ORF59 expression and deletion of both 5' and 3' regions, or one end region with a central region, of ORF59 RNA prevents ORF57-mediated translation of ORF59. The ORF59 sequence between nt 96633 and 96559 resembles other known MTA-responsive elements (MREs). ORF57 specifically binds to a stem-loop region from nt 96596–96572 of the MRE, which also binds cellular RBM15. Internal deletion of the MRE from ORF59 led to poor export, but accumulation of nuclear ORF59 RNA in the presence of ORF57 or RBM15. Despite of being translatable in the presence of ORF57, this deletion mutant exhibits translational defect in the presence of RBM15. Together, our results provide novel insight into the roles of ORF57 and RBM15 in ORF59 RNA accumulation and protein translation.

Keywords: KSHV; ORF59; ORF57; RNA accumulation; RNA export; RBM15

1. Introduction

Kaposi's sarcoma-associated herpesvirus is the causative agent of multiple human malignancies, including Kaposi's sarcoma (KS), primary effusion lymphomas (PELs), and multicentric Castleman's disease (MCD). Following primary infection, KSHV establishes a lifelong latency in which the KSHV episome is replicated by the host cell replication machinery. Upon reactivation to lytic cycle, the replication of KSHV genome depends on several viral proteins, including DNA polymerase (Pol-8, ORF9), processivity factor of DNA polymerase (PF-8, ORF59), helicase (HEL, ORF44), primase (PRI, ORF56), primase-associated factor (ORF40/41), single-strand DNA binding protein (SSB, ORF6), replication and transcription activator (ORF50 or RTA), and replication-associated protein (RAP, K8) [1–3]. ORF59 is a phosphoprotein that binds dsDNA and the DNA polymerase Pol-8, promoting the synthesis of full-length DNA by acting as a sliding clamp [4]. ORF59 protein is phosphorylated by KSHV viral Ser/Thr kinase (ORF36) and this step seems to be critical for ORF59 activity and viral DNA synthesis [5]. RTA activates ORF59 transcription [6–8] and recruits ORF59 protein to the origin of lytic replication (*ori*Lyt) [3,4,9,10]. RTA also activates other early/delayed-early/late genes necessary for the successful production of infectious viral particles [6,7]. Lytic replication accelerates host mRNA turnover [11], but promotes stability of viral RNAs to ensure successful virus lytic replication. KSHV ORF57 is a viral early, dimerable protein [12]. ORF57 promotes accumulation of numerous viral transcripts, thereby naming mRNA transcript accumulation (MTA) [13–16], and has many other functions [17]. ORF57 binds to a structured sequence motif in the RNA, which is named as an MTA-responsive element (MRE), in association with cellular proteins to promote viral RNA stability [18–20]. We showed that ORF57 binds to ORF59 RNA [13,18] and promotes ORF59 RNA accumulation by interacting with cellular proteins RBM15 and OTT3 [15]. Direct interactions of ORF57 with RBM15 and OTT3 interferes with RBM15 or OTT3 binding to ORF59 RNA and prevents RBM15/OTT3-mediated nuclear accumulation and hyperpolyadenylation of ORF59 RNA [15]. Here, we present the further evidence of RNA cis-elements in ORF59 in regulation of ORF59 RNA export and protein translation.

2. Materials and Methods

2.1. The Expression Vectors

pVM7 for FLAG-tagged ORF57, pVM68 for 3 × FLAG-tagged ORF57, pVM18 for FLAG-tagged ORF59 and RBM15-FLAG were described in previous reports [13,21,22]. The mutant (mt) ORF59-FLAG expression vectors constructed in this study are listed in Table S1. The oligos used for plasmid construction are listed in Table S2. Overlapped PCR for construction of truncation mutants was performed as described [23] and all of truncation mutations were constructed in-frame with a FLAG-tag to express a truncated ORF59-FLAG fusion.

2.2. Cells and Co-Transfection Assays

Human HEK293 cells were cultivated in DMEM supplemented with 10% fetal bovine serum (FBS). All co-transfection assays were carried out in HEK293 cells (5×10^5/well) plated in a six-well plate. Twenty four hrs after seeding the cells were co-transfected with 1 μg of individual ORF59 expression vector together with 0.2 μg of ORF57 (pVM7 or pVM68) or RBM15 vectors. The cells cotransfected with an empty pFLAG-CMV-5.1 vector were used as a negative control. All transfections were carried out with Lipofectamine 2000 (Invitrogen, Carlsbad, CA, USA) as recommended. Protein samples were obtained by direct cell lysis with 0.5 mL of 2× SDS protein loading solution (Quality Biological, Inc., Gathersburg, MD, USA) supplemented with 5% 2-mercaptoethanol and analyzed by Western blotting with following antibodies: rabbit polyclonal anti-ORF57 antibody [24] (1:3000), rabbit polyclonal anti-RBM15 antibody (Proteintech Group, Inc., Chicago, IL, USA), mouse monoclonal anti-ORF57 antibody (unpublished data, used at a dilution of 1:1000); monoclonal anti-FLAG M2 antibody (1:3000, Sigma, St. Louis, MO, USA, F1804); monoclonal anti-β-tubulin (1:3000, Sigma, T5201), together with corresponding peroxidase-conjugated secondary antibodies (1:10,000, Sigma). The signal on the Western blot was detected with SuperSignal West Pico Chemiluminiscence Substrate (Pierce, Rockford, IL, USA). Total cell RNA samples were prepared 24 or 48 h after transfection by the addition of 1 mL of TRIzol reagent (Life Technologies, Carlsbad, CA, USA). Extraction of cytoplasmic and nuclear total RNA and Northern blot analysis was performed as described before [13]. The expression level of gene of interest was determined by Northern blot analysis using the following γ-^{32}P-labeled antisense oligo probes: oVM11 for ORF57, oVM158 and oJM65 for ORF59, oZMZ270 for GAPDH and oST197 for U6 small nuclear RNA (see Table S2 for each oligo sequence). The autoradiograph was captured using a Molecular Dynamics PhosphorImager Storm 860 and analyzed with ImageQuant software (GE Healthcare Bio-Sciences, Pistataway, NJ, USA).

2.3. RNA-Protein Pulldown Assays

KSHV-infected, engineered BCBL1 cells, TREx BCBL1-vector and TREx BCBL1-RTA cells [25], were cultivated in RPMI 1640 supplemented with 10% FBS and hygromycin B (50 μg/mL). To induce the expression of KSHV lytic genes, TREx BCBL1-RTA or TREx BCBL1-vector (a negative control) cells were treated with 1 μg/mL of doxycycline (Dox) for 24 h. Total cell extract was prepared from ~5×10^6 of TREx BCBL1–RTA or –vector cells in RIPA buffer as described [19]. RNA-protein pull down assays were performed as described previously [18,26] using customized 5'-biotinylated RNA oligos oJM36, oJM37, oJM38, or oJM39 derived from ORF59 RNA (Table S2) and oNP42 or oNP41 derived from vIL6 (Table S2) [18]. The proteins associated with each RNA oligo were analyzed by Western blotting.

2.4. Quantitative RT-PCR and RNA Decay Analysis

ORF59 RNA decay was analyzed in HEK293 cells transiently transfected with an ORF59 expression vector in the presence or absence of ORF57 as described previously [15,19]. The following ORF59 TaqMan primers from IDT (Coralville, IA, USA) were used: ORF59 Probe, ORF59 primer 1 and ORF59 primer 2 (Table S2). The relative expression of ORF59 and non-linear regression analysis of the data were determined as described [15,19].

3. Results and Discussion

3.1. Mapping of the Regions in ORF59 RNA Responsible for ORF57-Mediated RNA Accumulation and Protein Translation

In an attempt to delineate ORF59 sequences responsible for regulation by ORF57, we constructed a series of progressive 5' to 3' and 3' to 5' truncations of ORF59 (Figure 1A) and evaluated the relevance of these sequences for accumulation of ORF59 RNA and translation of ORF59 protein in the presence or absence of ORF57. Progressive truncations from the 5'- or 3'-end of ORF59 were found to gradually increase the accumulation of total ORF59 RNA in the presence of ORF57 (Figure 1B,C), in particular from the 3'-end truncation (Figure 1C), but appeared to gradually decrease the translation of ORF59 protein (Figure 1D,E). Surprisingly, three expression vectors pJM33-35 with the progressive 5' to 3' ORF59 truncation and an expression vector pJM41 with the 3' to 5' ORF59 truncation exhibited no protein translation in the presence of ORF57 (Figure 1D, E) despite they were constructed in-frame with a C-terminal FLAG tag. As the expression vectors pJM32 and pJM40 could translate ORF59, we initially assumed that the sequence motifs from nt 96181 to 96070 and from nt 96353 to 96230 in ORF59 RNA are important for ORF59 translation in response to ORF57, although the truncation from nt 95984 to 96104 already displayed a marked reduction of ORF59 translation (compare pJM38 to pJM39) (Figure 1D, E). Further analysis of a construct pJM42, which has both deletion of 312 nts from the 5' end as seen in pJM30 and the deletion of 270 nts from the 3' end as seen in pJM37, also showed the lack of ORF59 translation in the presence of ORF57, indicating that two cis-elements, each at the 5' half or at the 3' half of ORF59 are important for ORF59 translation in the presence of ORF57. Two representative mutants, pJM33 with 670 nt truncation at the 5' end and pJM40 with 681 nt truncation at the 3' end, were analyzed for their RNA export by cytoplasmic (C) and nuclear (N) fractionation (Figure 1F). Mutant pJM33 expressed no detectable ORF59 protein (Figure 1D, Lane 7) while mutant pJM40 had ~60% reduction of ORF59 protein expression compared to the wild type (Figure 1E, Lane 6). Although the pJM33 did exhibit a relative lower C/N ratio (1.3) compared to that of pJM40 and wild type RNA (1.9 and 2.2, respectively) (Figure 1F), both ORF59 mutants were able to export RNA to the cytoplasm (Figure 1F, Lanes 3 and 6). Therefore, the reduced or absent ORF59 protein expression observed in these truncation mutants in the presence of ORF57 could not be resulted from this small change of C/N ratio.

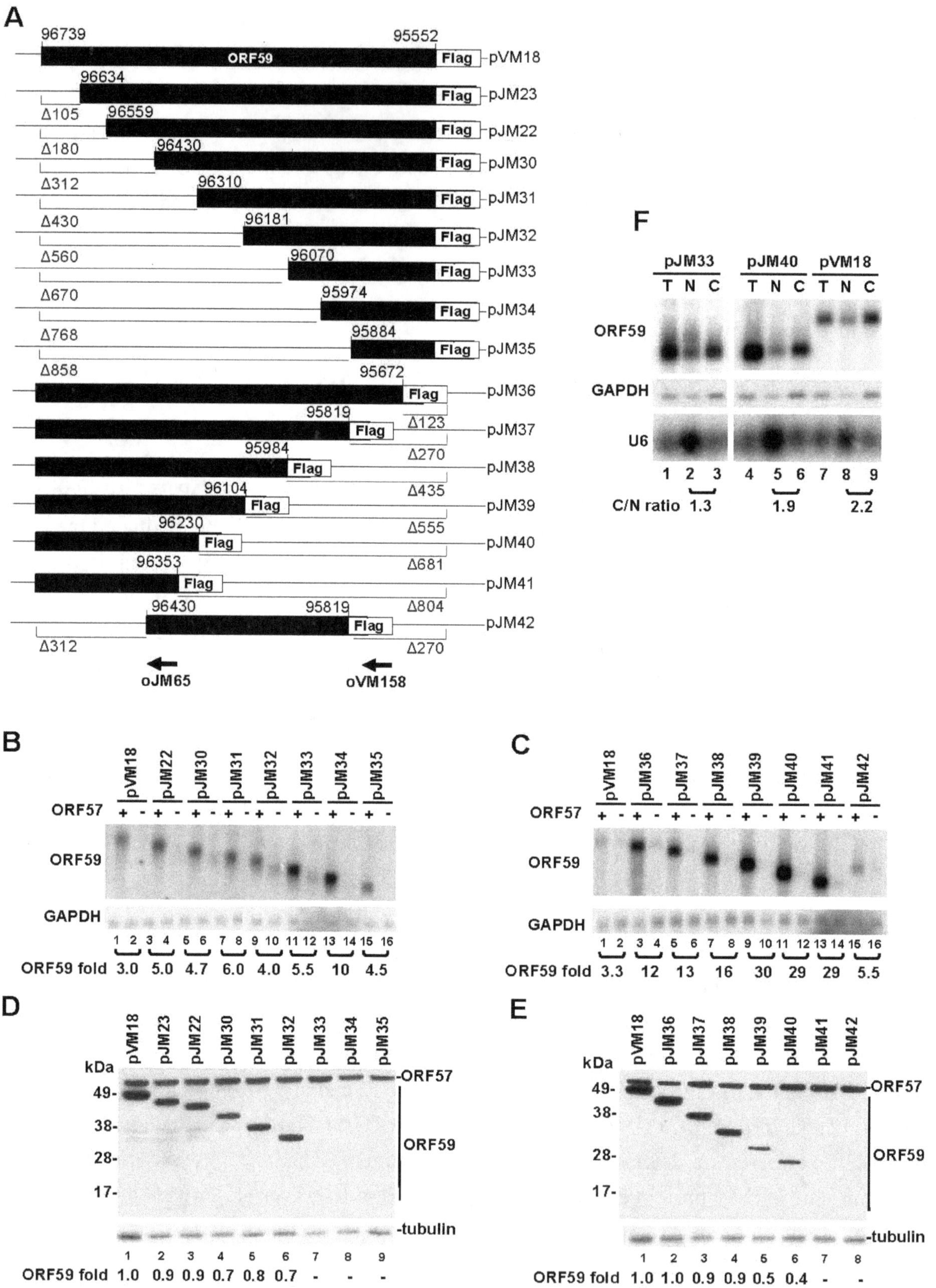

Figure 1. Mapping of ORF57-responsive elements in ORF59 RNA. (**A**) Diagrams of wt ORF59 and its progressive 5' and 3' deletion mutants fused in-frame with FLAG-tag on the C-terminus. The size (in nts) of each deletion (Δ) is indicated below the gray line. Numbers above each construct are nt positions in the KSHV genome (GenBank Ac. No. U75698.1).

Arrows in the corresponding ORF59 positions are antisense oligomers (oVM158 or oJM65) used for detection of ORF59 RNA in panel (B) and (C) by Northern blot. (**B–E**) The RNA and protein expression of ORF59 wt and its deletion mutants in the presence or absence of ORF57. HEK293 cells were transfected with ORF59-FLAG expression vectors in the presence (+) or absence (−) of ORF57 and total RNA and proteins were isolated 24 h after transfection. Northern blot analysis of ORF59 expression with a γ-^{32}P-labeled antisense oligo probe oVM158 or oJM65 for ORF59 RNA in panels B and C. The relative level of ORF59 in each sample was calculated after normalization to GAPDH, detected by a γ-^{32}P-labeled antisense oligo oZMZ270, used as a loading control. A relative ratio (fold) of ORF59 in the presence over the absence of ORF57 was determined according to the normalized levels in each sample pair. Protein expression profiles of wt ORF59 and its in-frame deletion mutants in the presence of ORF57 are shown in panels (D) and (E). Protein samples from HEK293 cells with each construct transfection as described above in the presence of ORF57 were analyzed by Western blot with anti-FLAG for both ORF59 (wild type and mutants) and ORF57 expression or anti-β-tubulin for sample loading. The relative protein level showed at the bottom of each Western blot panel was calculated by measurement of protein signal intensity after normalization to β-tubulin, with the wt ORF59 protein level being set to 1. (**F**) RNA expressed from both wt ORF59 and its deletion mutants is exportable. Total (T) and fractionated nuclear (N) and cytoplasmic (C) RNA were isolated from HEK293 cells at 24 h of transfection with a wt ORF59 (pVM18) or its two deletion mutants (a 5' deletion mt pJM33 and a 3' deletion mutant pJM40) in the presence of ORF57 and were analyzed by Northern blot as described above. GAPDH RNA served as a control for sample loading and nuclear U6 RNA served as a fractionation efficiency control. The relative level of ORF59 RNA in the cytoplasm *versus* the nucleus for each expression vector was calculated as a C/N ratio.

It has previously been shown that ORF57 is capable of enhancing translation of viral mRNAs by recruiting PYM, a cellular protein capable of facilitating the assembly of the 48S pre-initiation complex onto viral intronless mRNAs [27]. As all truncated mutants analyzed in Figure 1 show higher RNA accumulation in the presence of ORF57, we hypothesized that ORF57 promotion of ORF59 translation requires at least one RNA cis-element either at the 5'-end or at the 3'-ends of ORF59 mRNA to sustain its translational initiation activity for a certain size of ORF59 RNA, but lacking of both leads to the complete abolishment of its activity in initiation of ORF59 translation. Alternatively, a truncated ORF59 protein expressed from pJM33, pJM34, pJM35, pJM41, or pJM42 might be a labile protein because of lacking one (pJM33, pJM34, pJM35, and pJM41) or two (pJM42) dimerization domains on either ends of ORF59 protein [3]. Another uncharacterized domain encoded from nt 96181(pJM32)-96070 (pJM33) might be also required to stabilize the protein when one dimerization domain is missing from ORF59. In supporting this notion, we also observed that a remarkable reduction of the truncated ORF59 protein expression from pJM39 over that of pJM38, when further deletion was made from nt 95984 (pJM38) to 96104 (pJM39) (Figure 1E). Further deletion to nt 96353 (pJM41) makes this truncated protein undetectable. Deletion of the phosphorylation sites (Ser376, Ser378, and Ser379) on the C-terminus of ORF59 [5] does not appear to affect the protein stability.

3.2. The 5' MRE of ORF59 Interacts with ORF57 and RBM15 and Exhibits a Role in RNA Export and Stability

Previous anti-ORF57 CLIP assays [18] identified two regions within ORF59 RNA interacting with ORF57 (Figure 2A), one at the 5'-end and the other at the 3'-end of ORF59 RNA. We selected the region at the 5' end of ORF59 (nt 96633–96559) that shows a pronounced structure and sequence similarity to previously identified MTA-responsive elements (MREs) [19,20,28] (Figure 2B). We called this region as the ORF59 MRE. In order to provide direct biochemical evidence of the ORF59 MRE interaction with ORF57, we performed RNA-pull down assays using customized biotinylated RNA oligos covering the entire ORF59 MRE (oJM36, oJM37, oJM38, oJM39) as showed in Figure 2C. KSHV ORF57 binding to individual oligomer was tested in the cell extract prepared from KSHV-infected TREx BCBL1-RTA (R) cells expressing ORF57 after induction of RTA expression with DOX. The cell extract from TREx BCBL1-vector (V) cells without ORF57 expression were used for comparison. Western blot analysis of the pulled down proteins showed the binding of ORF57 protein to oJM39 RNA overlapping nt 96595–96572, a loop structure of ORF59 MRE. In fact, ORF57 binds to oJM39 with even higher affinity than that observed with oNP42, an ORF57-binding RNA oligo from KSHV vIL-6 MRE used as a positive control [18] (Figure 2D). All remaining RNA oligos from other parts of the ORF59 MRE showed no binding affinity to ORF57 similarly to a negative control RNA oligo oNP41. These data indicate the importance of loop sequences of the ORF59 MRE for ORF57 binding, similar to the MREs indentified in vIL-6 and PAN [18,19].

Our previous studies have shown that ORF57 affects ORF59 accumulation by interacting with RBM15, an RNA binding export cofactor [15,29]. Therefore, we next tested the interaction of ORF59 MRE with RBM15 using the same set of RNA oligos. We found that, similarly to ORF57, RBM15 binds oJM39 and oNP42 (a positive control), but only very weakly to other ORF59 MRE RNA oligos (Figure 2D). RBM15 binding to oJM39 and oNP42 was independent of ORF57 even in the cell extract (V) without ORF57 expression from the TREx BCBL1-vector cells. The reduced RBM15 interaction with oJM39 and oNP42 in the extract from TREx BCBL1-RTA cells (R) than in the TREx BCBL1-vector cell extract (V) (Figure 2D) was caused by the reduced expression of RBM15 in TREx BCBL1–RTA cells (compare the input RBM15 of R *vs.* V in Figure 2D).

To evaluate the relevance and functionality of the ORF59 5' MRE sequence, we analyzed the MRE truncation mutants pJM22 and pJM23 (Figure 1A) further and constructed a mutant with an internal deletion of the MRE sequence (pJM15) (Figure 3A). These mutants were evaluated for their response to ORF57 in co-transfection experiments in HEK293 cells. Figure 3B shows that all three mutants exhibited increased RNA accumulation, compared to wt ORF59 RNA expressed from pVM18 in the presence of ORF57 (Figure 3B). However, the background RNA expression of all three ORF59 mutants was different from pVM18 wt ORF59 RNA (Figure 3B, Lanes 5–8). All three mutants also exhibited decreased efficiency of RNA export, in particular the nuclear export of pJM15 ORF59 RNA (Figure 3C) in the presence of ORF57, with a lower C/N ratio over that of the wt ORF59 RNA (pVM18). It remains to know why the mutant ORF59 RNA expressed from pJM23 with a 5' MRE displayed a similar pattern to the one expressed from pJM15 without a MRE. In addition, we found that the mutant ORF59 RNA expressed from pJM22 with a larger 5' deletion than pJM15 was less stable than wt ORF59 expressed from pVM18 in the presence of ORF57, whereas the mutant ORF59 RNA expressed from pJM15 was less stable than wt ORF59 RNA expressed from pVM18 in the absence of ORF57, with a half-life shorter than wt ORF59 (Figure 3D, Table S3). Insertion of the MRE alone or together with its upstream 105 nts into the 5' GFP displayed no effect on GFP expression (data not shown), suggesting that the MRE identified in the ORF59 has to work with other uncharacterized element(s) in ORF59 RNA for its function.

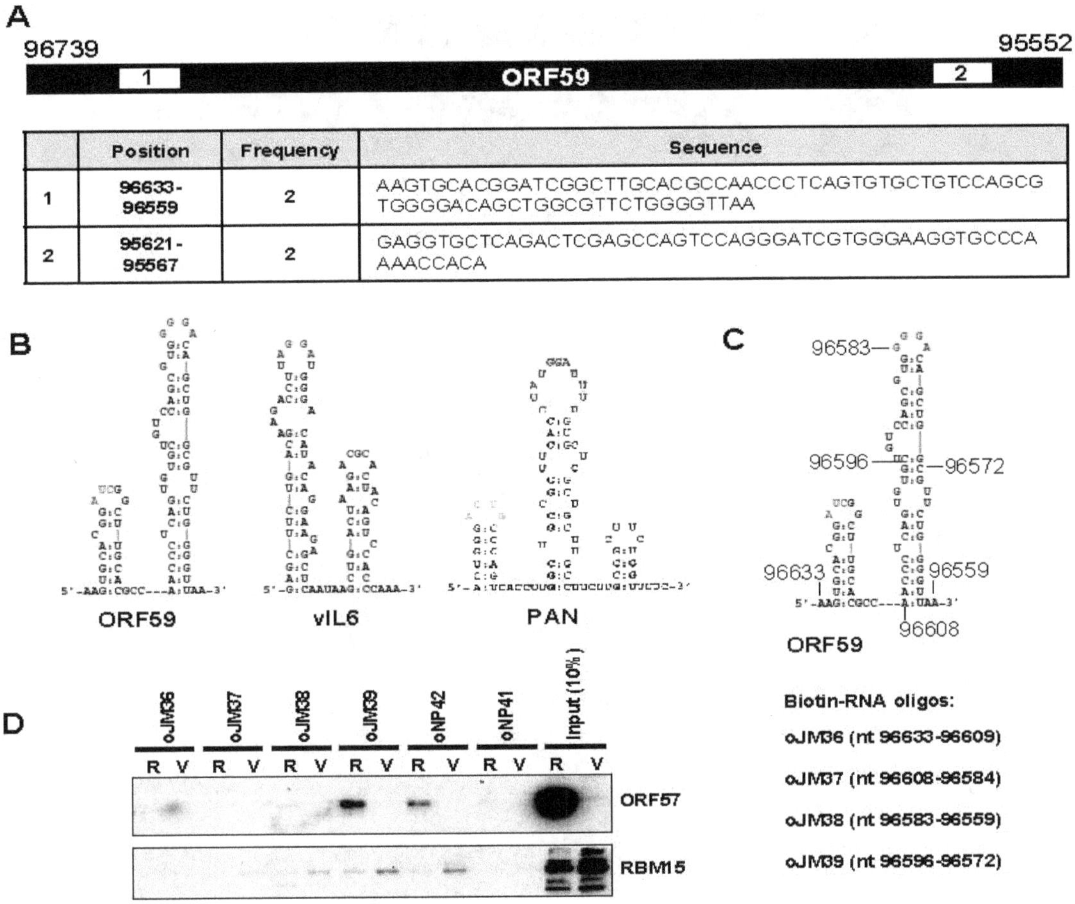

	Position	Frequency	Sequence
1	96633-96559	2	AAGTGCACGGATCGGCTTGCACGCCAACCCTCAGTGTGCTGTCCAGCG TGGGGACAGCTGGCGTTCTGGGGTTAA
2	95621-95567	2	GAGGTGCTCAGACTCGAGCCAGTCCAGGGATCGTGGGAAGGTGCCCA AAACCACA

Figure 2. A MRE in the 5' ORF59 interacts with ORF57 and RBM15. (**A**) Two ORF59 regions (numbered white boxes) identified in ORF57 CLIP assay [18]. Below is the table showing their positions, cloning frequency and nucleotide sequence. (**B**) Secondary structures of the identified RNA sequences in ORF59, vIL6 and PAN RNA by anti-ORF57 CLIP. The nucleotide sequences in the stem-loop in red color are the interacting site for ORF57 and its associated cellular proteins. (**C**) The ORF59 5' MRE structure with the nt positions of biotinylated RNA oligomers used in RNA pull-down assays. (**D**) KSHV ORF57 and cellular RBM15 preferentially bind to a stem-loop structure of the MRE. Cell lysates prepared from TREx BCBL1-RTA (R) or -vector (V) cells induced with Dox for 24 h were used for the RNA pulldown assays with an indicated RNA oligomer. The RNA oligo oNP42 derived from vIL6 RNA was used as a positive control and oNP41 oligo as negative control [18]. ORF57 or RBM15 associated with RNA oligos in the pulldowns was immunoblotted using an anti-ORF57 or anti-RBM15 antibody. The cell lysate (10%) prior to the pulldown served as an input control in Western blot.

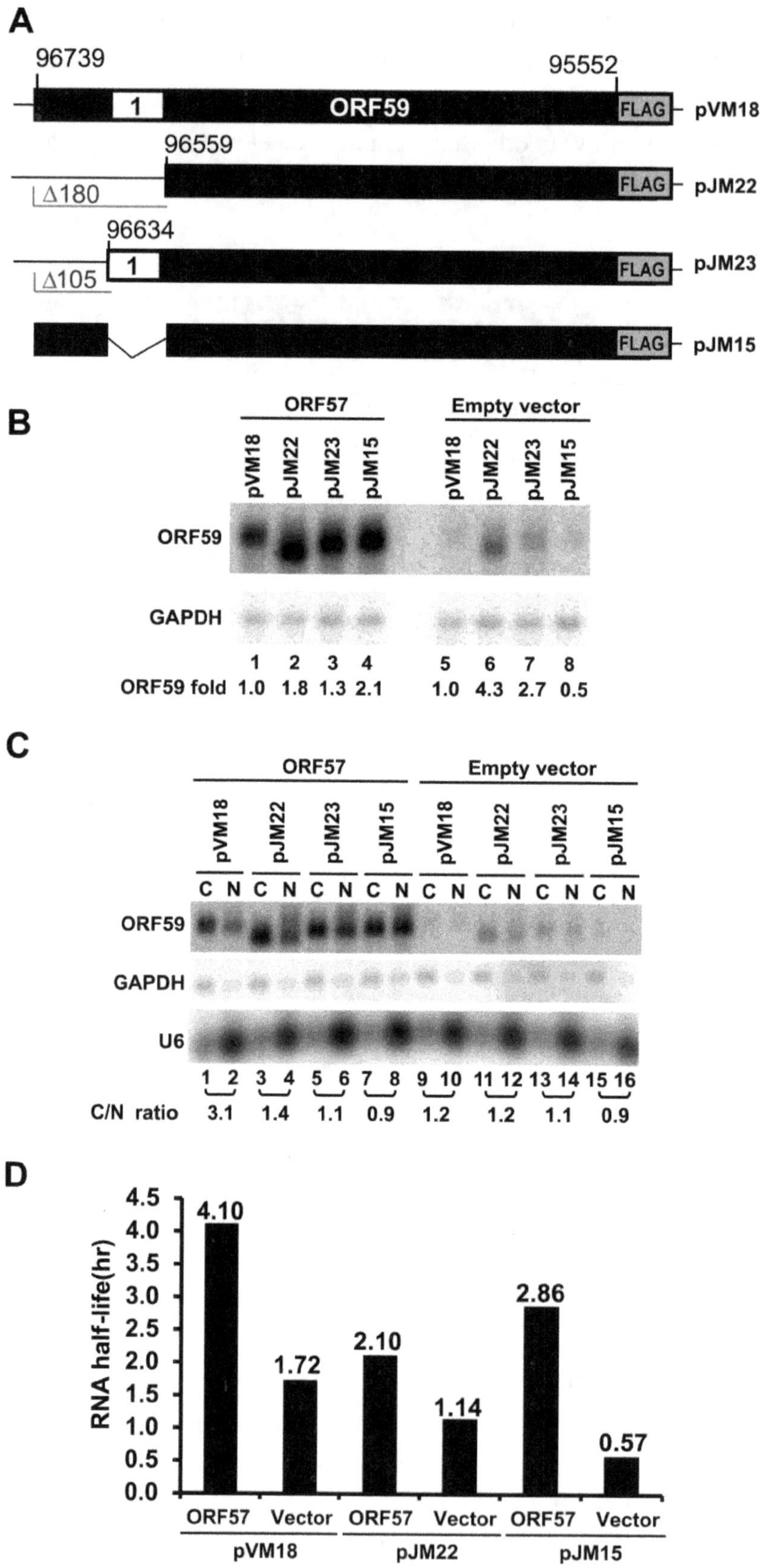

Figure 3. The role of 5' ORF59 MRE in ORF59 RNA export and half-life. (**A**) Diagrams of wt ORF59 and its 5' in-frame deletion mutants. See other details in Figure 1. (**B**) Expression of ORF59 RNA in HEK293 cells. Total RNA from HEK293 cells transfected with individual ORF59 construct in the presence or absence (empty vector) of ORF57-FLAG was examined

by Northern blot with a ^{32}P-labeled oligo probe (oVM158) for ORF59 RNA and oZMZ270 for GAPDH used as a loading control. A relative level of ORF59 RNA in each sample after normalization to the corresponding GAPDH level is shown at the bottom of Northern blots, with the level of wt ORF59 RNA being set to 1. (**C**) Examination of the 5' MRE in export of ORF59 RNA. Fractionated cytoplasmic (C) or nuclear (N) RNA from HEK293 cells cotransfected with wt ORF59 (pVM18) and its MRE deletion mutants (pJM22, pJM23 and pJM15) was examined by Northern blot as described in (**B**) for ORF59 RNA. GAPDH RNA and U6 RNA served as fractionation efficiency controls. C/N ratio was calculated based on relative level of ORF59 in the cytoplasm *versus* the nucleus. (**D**) Half-life of wt and mt ORF59 RNA in the presence of ORF57 or an empty vector. HEK293 were transfected with a vector expressing wt or mt ORF59 together with an ORF57 expression vector or an empty vector control. At 24 h of transfection, RNA transcription was stopped by addition of 10 μg/mL of actinomycin D and total RNA was extracted from each sample over the time (hr) for the remaining ORF59 and GAPDH (for normalization) RNA quantified by qRT-PCR. Results are presented as calculated half-life for each study group in three separate experimental repeats. Holm's method adjusted *p*-values for pair-wise comparisons between the corresponding estimated decay rate parameters as shown before [15,19] are presented in detail in Table S3.

3.3. ORF59 RNA without the 5' MRE is Translatable in the Presence of ORF57, but Poorly Translatable in the Presence of RBM15

KSHV ORF57 interacts with RNA export factor Aly/REF and NXF1 [20,30] and cofactors RBM15 and OTT3 by protein-protein interaction [15]. ORF57 affects the RNA-binding activity of RBM15 and prevents RBM15-mediated nuclear accumulation of hyperpolyadenylated ORF59 RNA [15]. Accordingly, we tested whether the ORF59 mutant in pJM15 carrying an internal deletion of the ORF59 MRE responds to RBM15 upon cotransfection of HEK293 cells (Figure 4A). As expected, ORF57 enhanced the expression of both wt ORF59 in pVM18 and mutant ORF59 in pJM15 both at their total RNA and protein levels. To our surprise, RBM15 cotransfection led to enhanced expression of both wt and mt RNAs, but only to increased protein expression of wt ORF59 in pVM18, not the mutant ORF59 in pJM15 (Figure 4A,B). Analysis of the relative C/N ratio in RNA fractionation experiments showed that ectopic RBM15 led to more nuclear RNA of both wt and mutant ORF59, indicating that RBM15 promotes the nuclear accumulation independently of the 5' MRE (Figure 4C). Thus, nuclear accumulation of the mt ORF59 RNA mediated by RBM15 would not be a plausible explanation for the lack of protein expression from the mt ORF59 RNA. Together with the data from Figure 3, our results suggest that the identified 5' MRE motif and its surrounding region in ORF59 RNA are more likely involved in the export of ORF59 RNA and are responsible for RBM15-mediated protein translation.

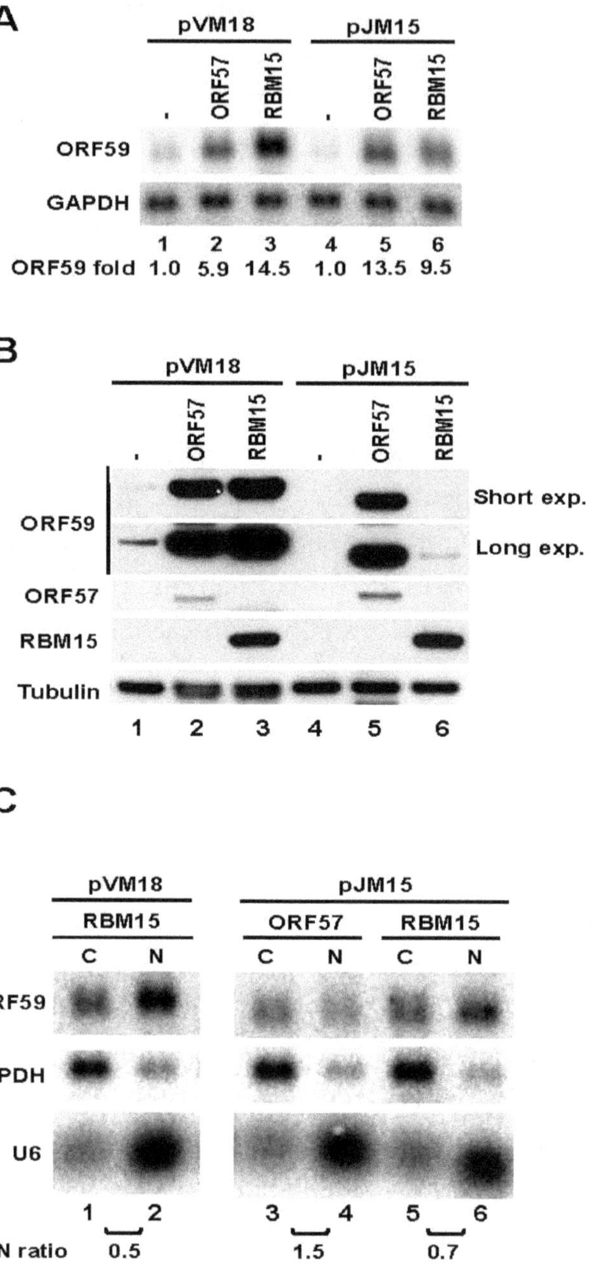

Figure 4. ORF57 but not RBM15 promotes protein translation from an ORF59 MRE-deletion mutant. (**A**) and (**B**) Total RNA from HEK293 cells transfected with each of the ORF59 constructs in the absence or presence of ORF57, or RBM15 was examined by Northern blot (**A**) as described in Figure 1B for the expression of ORF59 and GAPDH. Protein samples from the same set of HEK293 cells were examined by Western blot (**B**) using an anti-FLAG antibody to detect the expression of both ORF59 and ORF57 or an anti- β-tubulin antibody for sample loading. Two exposures (exp.), short and long, are shown for anti-FLAG ORF59 blot. (**C**) Fractionated cytoplasmic (C) or nuclear (N) RNA from HEK293 cells transfected with wt ORF59 (pVM18) or its MRE-deletion mutant (pJM15) in the presence of ectopic expression of RBM15 or ORF57 was examined by Northern blot as described in Figure 3C for ORF59 RNA expression and calculated as a C/N ratio for its cytoplasmic *versus* nuclear RNA levels. Both GAPDH and U6 RNAs were blotted for fractionation efficiency.

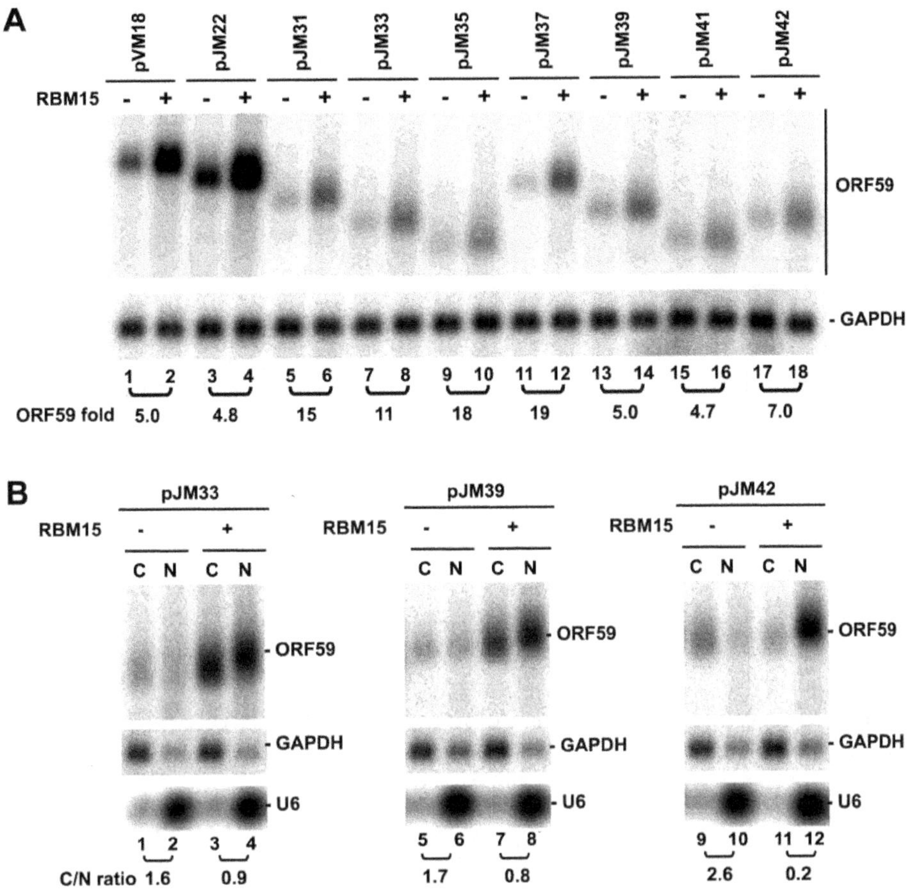

Figure 5. RBM15-mediated RNA accumulation of wt and mt ORF59 in the absence of ORF57. (**A**) Total RNA from HEK293 cells transfected with each of ORF59 constructs described in Figure 1A in the presence (+) or absence (−) of RBM15 were analyzed by Northern blot analysis as described in Figure 1B for ORF59 RNA expression using antisense oligo probe oVM158 for pVM18, pJM22, pJM31, pJM33 and pJM35 or antisense oligo probe oJM65 for pJM37, pJM39, pJM41 and pJM42. The same membrane was re-probed separately with GAPDH-specific probe for sample loading. (**B**) RBM15 promotes ORF59 expression and nuclear accumulation of the hyperpolyadenylated ORF59 RNA. Fractionated cytoplasmic (C) and nuclear (N) RNA from HEK293 cells cotransfected with each ORF59 construct in the presence (+) or absence (−) of RBM15 was examined by Northern blot for ORF59 RNA expression as a C/N ratio for its cytoplasmic *versus* nuclear RNA levels as described above. GAPDH RNA and U6 RNA served for fractionation efficiency.

We also analyzed RBM15's ability to accumulate RNA from other progressive 3'- and 5'-truncation mutants of ORF59. All of the mutants analyzed showed higher levels of ORF59 RNA expression in the presence of RBM15 (Figure 5A). In agreement with our previous study, RBM15-enhanced ORF59 RNA migrated slower and appeared with smear when compared with the RNA in the absence of RBM15 (Figure 5A). Nuclear and cytoplasmic fractionation analysis indicated that three selected ORF59 mutants exhibited RNA accumulation in the nuclear fraction in the presence of RBM15 and slower migration than its counterpart in the cytoplasmic fraction in electrophoresis (Figure 5B), indicating that the accumulated nuclear ORF59 RNA is of a larger size due to hyperpolyadenylation [15] regardless of where the truncation was made. Although the observed RNA hyperpolyadenylation is preventable in the presence of ORF57 as we reported for the wt ORF59 RNA [15], further experiments are needed to confirm the prediction.

4. Conclusions

In summary, this study has demonstrated that KSHV ORF59 contains multiple regions or motifs responsible for viral ORF57 and cellular RBM15. ORF57 promotes protein translation from ORF59 RNA and requires two terminal regions or one terminal region plus a central region of ORF59 for this function. The MRE identified by anti-ORF57 CLIP binds ORF57 protein and RBM15 and appears to be important for RNA export and stability of ORF59. However, this MRE motif is also important for RBM15-mediated protein translation of ORF59. Together, these studies provide further insight into how ORF57 promotes ORF59 expression by balanced interplay with RBM15.

Acknowledgments

This work was supported by the Intramural Research Program of the National Cancer Institute, Center for Cancer Research, NIH. We thank Barbara Felber at NCI for RBM15 plasmid and Jae Jung of the University of Southern California for inducible TREx BCBL1-RTA and -vector cell lines.

Author Contributions

MJM, VM and ZMZ conceived and designed the experiments. MJM, VM and JGK performed the experiments. DL and SS performed the biostatistics analysis. MJM, VM, ZMZ analyzed the results and wrote the manuscript.

References

1. Wu, F.Y.; Ahn, J.H.; Alcendor, D.J.; Jang, W.J.; Xiao, J.; Hayward, S.D.; Hayward, G.S. Origin-independent assembly of Kaposi's sarcoma-associated herpesvirus DNA replication compartments in transient cotransfection assays and association with the ORF-K8 protein and cellular PML. *J. Virol.* **2001**, *75*, 1487–1506.
2. AuCoin, D.P.; Colletti, K.S.; Cei, S.A.; Papouskova, I.; Tarrant, M.; Pari, G.S. Amplification of the Kaposi's sarcoma-associated herpesvirus/human herpesvirus 8 lytic origin of DNA replication is dependent upon a cis-acting AT-rich region and an ORF50 response element and the trans-acting factors ORF50 (K-Rta) and K8 (K-bZIP). *Virology* **2004**, *318*, 542–555.
3. Chen, X.; Lin, K.; Ricciardi, R.P. Human Kaposi's sarcoma herpesvirus processivity factor-8 functions as a dimer in DNA synthesis. *J. Biol. Chem.* **2004**, *279*, 28375–28386.
4. Chan, S.R.; Chandran, B. Characterization of human herpesvirus 8 ORF59 protein (PF-8) and mapping of the processivity and viral DNA polymerase-interacting domains. *J. Virol.* **2000**, *74*, 10920–10929.
5. McDowell, M.E.; Purushothaman, P.; Rossetto, C.C.; Pari, G.S.; Verma, S.C. Phosphorylation of Kaposi's sarcoma-associated herpesvirus processivity factor ORF59 by a viral kinase modulates its ability to associate with RTA and oriLyt. *J. Virol.* **2013**, *87*, 8038–8052.

6. Lukac, D.M.; Renne, R.; Kirshner, J.R.; Ganem, D. Reactivation of Kaposi's sarcoma-associated herpesvirus infection from latency by expression of the ORF 50 transactivator, a homolog of the EBV R protein. *Virology* **1998**, *252*, 304–312.
7. Gradoville, L.; Gerlach, J.; Grogan, E.; Shedd, D.; Nikiforow, S.; Metroka, C.; Miller, G. Kaposi's sarcoma-associated herpesvirus open reading frame 50/Rta protein activates the entire viral lytic cycle in the HH-B2 primary effusion lymphoma cell line. *J. Virol.* **2000**, *74*, 6207–6212.
8. Liu, Y.; Cao, Y.; Liang, D.; Gao, Y.; Xia, T.; Robertson, E.S.; Lan, K. Kaposi's sarcoma-associated herpesvirus RTA activates the processivity factor ORF59 through interaction with RBP-Jkappa and a cis-acting RTA responsive element. *Virology* **2008**, *380*, 264–275.
9. Lin, K.; Dai, C.Y.; Ricciardi, R.P. Cloning and functional analysis of Kaposi's sarcoma-associated herpesvirus DNA polymerase and its processivity factor. *J. Virol.* **1998**, *72*, 6228–6232.
10. Rossetto, C.C.; Susilarini, N.K.; Pari, G.S. Interaction of Kaposi's sarcoma-associated herpesvirus ORF59 with oriLyt is dependent on binding with K-Rta. *J. Virol.* **2011**, *85*, 3833–3841.
11. Glaunsinger, B.; Ganem, D. Lytic KSHV infection inhibits host gene expression by accelerating global mRNA turnover. *Mol. Cell* **2004**, *13*, 713–723.
12. Majerciak, V.; Pripuzova, N.; Chan, C.; Temkin, N.; Specht, S.I.; Zheng, Z.M. Stability of structured Kaposi sarcoma-associated herpesvirus ORF57 protein is regulated by protein phosphorylation and homodimerization. *J. Virol.* **2015**, doi:10.1128/jvi.03721-14.
13. Majerciak, V.; Yamanegi, K.; Nie, S.H.; Zheng, Z.M. Structural and functional analyses of Kaposi sarcoma-associated herpesvirus ORF57 nuclear localization signals in living cells. *J. Biol. Chem.* **2006**, *281*, 28365–28378.
14. Majerciak, V.; Zheng, Z.M. Kaposi's sarcoma-associated herpesvirus ORF57 in viral RNA processing. *Front. Biosci.* **2009**, *14*, 1516–1528.
15. Majerciak, V.; Uranishi, H.; Kruhlak, M.; Pilkington, G.R.; Massimelli, M.J.; Bear, J.; Pavlakis, G.N.; Felber, B.K.; Zheng, Z.M. Kaposi's sarcoma-associated herpesvirus ORF57 interacts with cellular RNA export cofactors RBM15 and OTT3 to promote expression of viral ORF59. *J. Virol.* **2011**, *85*, 1528–1540.
16. Kirshner, J.R.; Lukac, D.M.; Chang, J.; Ganem, D. Kaposi's sarcoma-associated herpesvirus open reading frame 57 encodes a posttranscriptional regulator with multiple distinct activities. *J. Virol.* **2000**, *74*, 3586–3597.
17. Majerciak, V.; Zheng, Z.M. KSHV ORF57, a protein of many faces. *Viruses* **2015**, in press.
18. Kang, J.G.; Pripuzova, N.; Majerciak, V.; Kruhlak, M.; Le, S.Y.; Zheng, Z.M. Kaposi's Sarcoma-Associated Herpesvirus ORF57 Promotes Escape of Viral and Human Interleukin-6 from MicroRNA-Mediated Suppression. *J. Virol.* **2011**, *85*, 2620–2630.
19. Massimelli, M.J.; Kang, J.G.; Majerciak, V.; Le, S.Y.; Liewehr, D.J.; Steinberg, S.M.; Zheng, Z.M. Stability of a Long Noncoding Viral RNA Depends on a 9-nt Core Element at the RNA 5' End to Interact with Viral ORF57 and Cellular PABPC1. *Int. J. Biol. Sci.* **2011**, *7*, 1145–1160.
20. Massimelli, M.J.; Majerciak, V.; Kruhlak, M.; Zheng, Z.M. Interplay between polyadenylate-binding protein 1 and Kaposi's sarcoma-associated herpesvirus ORF57 in accumulation of polyadenylated nuclear RNA, a viral long noncoding RNA. *J. Virol.* **2013**, *87*, 243–256.
21. Majerciak, V.; Yamanegi, K.; Zheng, Z.M. Gene structure and expression of Kaposi's sarcoma-associated herpesvirus ORF56, ORF57, ORF58, and ORF59. *J. Virol.* **2006**, *80*, 11968–11981.

22. Lindtner, S.; Zolotukhin, A.S.; Uranishi, H.; Bear, J.; Kulkarni, V.; Smulevitch, S.; Samiotaki, M.; Panayotou, G.; Felber, B.K.; Pavlakis, G.N. RNA-binding motif protein 15 binds to the RNA transport element RTE and provides a direct link to the NXF1 export pathway. *J. Biol. Chem.* **2006**, *281*, 36915–36928.
23. Zheng, Z.M.; Reid, E.S.; Baker, C.C. Utilization of the bovine papillomavirus type 1 late-stage-specific nucleotide 3605 3' splice site is modulated by a novel exonic bipartite regulator but not by an intronic purine-rich element. *J. Virol.* **2000**, *74*, 10612–10622.
24. Majerciak, V.; Pripuzova, N.; McCoy, J.P.; Gao, S.J.; Zheng, Z.M. Targeted disruption of Kaposi's sarcoma-associated herpesvirus ORF57 in the viral genome is detrimental for the expression of ORF59, K8alpha, and K8.1 and the production of infectious virus. *J. Virol.* **2007**, *81*, 1062–1071.
25. Nakamura, H.; Lu, M.; Gwack, Y.; Souvlis, J.; Zeichner, S.L.; Jung, J.U. Global changes in Kaposi's sarcoma-associated virus gene expression patterns following expression of a tetracycline-inducible Rta transactivator. *J. Virol.* **2003**, *77*, 4205–4220.
26. Jia, R.; Liu, X.; Tao, M.; Kruhlak, M.; Guo, M.; Meyers, C.; Baker, C.C.; Zheng, Z.M. Control of the papillomavirus early-to-late switch by differentially expressed SRp20. *J. Virol.* **2009**, *83*, 167–180.
27. Boyne, J.R.; Jackson, B.R.; Taylor, A.; Macnab, S.A.; Whitehouse, A. Kaposi's sarcoma-associated herpesvirus ORF57 protein interacts with PYM to enhance translation of viral intronless mRNAs. *EMBO J.* **2010**, *29*, 1851–1864.
28. Kang, J.G.; Majerciak, V.; Uldrick, T.S.; Wang, X.; Kruhlak, M.; Yarchoan, R.; Zheng, Z.M. Kaposi's sarcoma-associated herpesviral IL-6 and human IL-6 open reading frames contain miRNA binding sites and are subject to cellular miRNA regulation. *J. Pathol.* **2011**, *225*, 378–389.
29. Uranishi, H.; Zolotukhin, A.S.; Lindtner, S.; Warming, S.; Zhang, G.M.; Bear, J.; Copeland, N.G.; Jenkins, N.A.; Pavlakis, G.N.; Felber, B.K. The RNA binding motif protein 15B (RBM15B/OTT3) acts as co-factor of the nuclear export receptor NXF1. *J. Biol. Chem.* **2009**, *284*, 26106–26116.
30. Malik, P.; Blackbourn, D.J.; Clements, J.B. The evolutionarily conserved Kaposi's sarcoma-associated herpesvirus ORF57 protein interacts with REF protein and acts as an RNA export factor. *J. Biol. Chem.* **2004**, *279*, 33001–33011.

4

Seroprevalence of Human Herpesvirus 8 and Hepatitis C Virus among Drug Users in Shanghai, China

Tiejun Zhang [1,†], **Ying Liu** [1,2,†], **Yuyan Zhang** [2], **Jun Wang** [1], **Veenu Minhas** [3], **Charles Wood** [3] **and Na He** [1,*]

[1] Department of Epidemiology, School of Public Health, Fudan University, and The Key Laboratory of Public Health Safety of Ministry of Education, Shanghai 200032, China; E-Mails: tjzhang@shmu.edu.cn (T.Z.); hailingilei@126.com (Y.L.); 13211020065@fudan.edu.cn (J.W.)

[2] Putuo District Center for Disease Control and Prevention, Shanghai 200032, China; E-Mail: zhangyy1@shpt.gov.cn

[3] Nebraska Center for Virology, School of Biological Sciences, University of Nebraska-Lincoln, Lincoln, NE 68583, USA; E-Mails: vminhas2@unl.edu (V.M.); cwood@unl.edu (C.W.)

† These authors contributed equally to this work.

* Author to whom correspondence should be addressed; E-Mail: nhe@shmu.edu.cn;

Abstract: To elucidate and compare the seroprevalence of human herpesvirus 8 (HHV8) and hepatitis C virus (HCV) among Chinese drug users, a cross-sectional study of 441 participants, was conducted in Shanghai, China, from 2012 through 2013. Seventy-seven (17.5%) participants were found to be positive for HHV8 antibodies, while 271 (61.5%) participants were positive for HCV. No significant association between HHV8 seropositivity and drug use characteristics, sexual behaviors, HCV, or syphilis was observed. In contrast, a statistically significant association between HCV seropositivity and injected drug history (OR, 2.18, 95% CI 1.41–3.37) was detected, whereas no statistically significant association between HCV seropositivity and syphilis infection (OR, 7.56, 95% CI 0.94–60.57) were observed. Pairwise comparisons showed no significant differences between latent and lytic antibodies regarding HCV and HHV8 serostatus. The study demonstrated a moderate but elevated prevalence of HHV8 infection among drug users. The discordance between HHV8 and HCV infections suggests that blood borne transmission of HHV8 might not be the predominant mode of transmission in this population, which is in contrast to HCV.

Keywords: human herpesvirus 8; hepatitis C virus; drug user; prevalence; China

---

1. Introduction

Human herpesvirus 8 (HHV8), also known as Kaposi's Sarcoma associated herpesvirus (KSHV), was discovered in 1994 and is now known to be the etiologic agent for Kaposi's sarcoma (KS), primary effusion lymphoma (PEL), and multicentric Castleman's disease [1–4]. Since its initial discovery, the HHV8 epidemiology has been widely studied globally, showing significant differences in distribution among various geographical areas and various populations analyzed. A number of epidemiological studies have indicated that HHV8 seroprevalence vary considerably among countries and risk groups, but the routes of transmission have yet to be clearly defined [5]. Nonsexual and vertical transmission routes are believed to be of importance in endemic areas in a number of African countries, and a number of studies have demonstrated that saliva contact may be the major mode of transmission [6,7]. Sexual transmission has been shown to occur frequently among homosexual men in non-endemic areas, such as United States and Western Europe [8]. However, evidence of transmission via blood contact remains controversial [9–11]. The possibility that HHV8 can be transmitted by blood contact raises important public health concerns. Of particular concern are drug users, since the possibility of blood transmission may facilitate HHV8 transmission, and increase the prevalence of infection among this population. Evidence of blood borne transmission of HHV8 among injection-drug users has been suggested previously, but the data generated are from different population and could not exclude the possibility that the virus is transmitted via other modes of infection, such as via sexual contact and/or general behaviors accompanying the use of drugs rather than the drug use itself [12–14].

Drug users have been well documented to be at high risk of blood borne infections (e.g., hepatitis C virus) in China [15,16], but seroprevalence of HHV8 among drug users remained poorly documented, especially in China where there is a large population of intravenous drug users, and they are known to be at risk for Human immunodeficiency virus (HIV) and other associating co-infections. Prior studies among drug users have associated injection drug use practices with higher prevalence of HHV8 infection but with inconsistent results regarding the possibility of HHV8 transmission through drug use. Whether this route of transmission can efficiently occur among this population remains controversial [13,14,17,18]. Moreover, this route of transmission has not yet been evaluated in China, therefore, we recruited a group of drug users which has provided us with the unique opportunity to assess the possibility of HHV8 blood transmission through injection drug use practices. We hypothesized that if HHV8 can be transmitted through blood contact like hepatitis C virus (HCV), the seroprevalence of HHV8 should also be elevated like HCV in this population. Therefore, the prevalence of HHV8 and HCV, as well as several potential infectious agents were analyzed in parallel in this population.

2. Materials and Methods

2.1. Study Setting and Populations

This cross-sectional study was conducted in Putuo district, Shanghai, China, from 2012 through 2013. All individuals, ≥18 years of age who had a history of drug use in the past six months and resided in Shanghai, were eligible for enrollment. Subjects were recruited consecutively using convenience sampling from a drop-in center (DIC) serving drug users in Shanghai. Potential subjects were not required to disclose their names for participation. They received modest monetary compensation for time spent for the interview.

The purpose and methods of the study were explained to all the participants. Informed consent procedures were carried out individually, and written consent was obtained from all the participants before any procedures were performed. This study was approved by the Institutional Review Board of Fudan University, China.

2.2. Data Collection

After obtaining written consent, participants were interviewed by a trained health professional using a standard survey questionnaire covering demographics, drug use practices and sexual risk behaviors. Interviews were administered in-person in a private location (such as methadone clinics). Completed questionnaires were placed in a large black bag containing other completed questionnaires to reassure the participants about confidentiality of the provided information.

2.3. Sample Collection

Venous blood was collected by experienced nurses using sterilized needles, syringes and tubes, and transferred to the laboratory within 2 h after collection. Plasma samples were stored at −80 °C until serological testing. All specimens were coded by a unique identification number given to each study participant and were analyzed by two experienced technicians without any knowledge of the study participants identities.

2.4. Specimen Testing

HIV serology. All plasma samples were screened for HIV antibody using an enzyme-linked immunosorbent assay (ELISA; Vironostika HIV Uni-Form II plus O ELISA Kit, Biomerieux Shanghai Company Ltd., Shanghai, China), according to the manufacturer's instructions. All positive samples were further confirmed by Western blot assay (Genelabs Diagnostic, Singapore, Singapore).

HHV8 serology. Plasma samples were tested by immunofluoresence assay, as reported previously [19]. Briefly, two HHV8 serology tests were performed: first, BC-3 cells (HHV8 positive and Epstein-Barr virus negative B cell line, American Type Culture Collection, Manassas, VA, USA), stimulated by tetradecanoyl phorbol acetate (TPA) were fixed and permeabilized and used for monoclonal enhanced immunefluorescence assay. Second, *Spodoptera frugiperda* clone 9 expressing viral recombinant proteins, ORF73, ORF65, and ORF-K8.1, was used for testing. The procedure was similar to the BC-3 immunofluoresence assay. A sample was considered HHV8 seropositive only if it was positive at a standard serum dilution of 1:40 with both the BC-3 and *S. frugiperda* assay. Each slide was read independently by two experienced laboratory workers.

HBV and HCV serology. HBsAg was tested using an ELISA kit (Wantai Biotech Pharmacy Enterprise Co. Beijing, China). The test was performed following the procedures recommended by the manufacturer. Anti-HCV immunoglobulin G (IgG) antibody was tested to determine HCV infection status according to the manufacturer's protocol (Wantai Biomedical, Beijing, China). All the plasma samples were blindly assayed in duplicate.

Syphilis Serology. Syphilis was screened by using a rapid plasma reagent test (Span Diagnostics Ltd., India), and confirmed by the *Treponema pallidum* hemaglutination test (TPHA, Syphagen TPHA, Biokit, Barcelona, Spain).

All the above serological tests were performed by the same two experienced technicians, with duplicate negative, positive, and blank controls being tested in parallel.

2.5. Statistical Analysis

Original questionnaires and laboratory testing results were entered and managed in EpiData3.0, and then transferred to a SAS database for further analyses. Demographic characteristics and risk behaviors were analyzed using descriptive statistics, *i.e.*, mean, median, and interquartile range (IQR) for continuous variables, and proportions for categorical variables.

HHV8 seroprevalence was computed using the normal approximation to a binomial distribution, and tabulated by sociodemographic characteristics of study subjects, followed by Pearson's chi-squared tests to determine statistical significance. Initially, a univariate logistic regression analysis was conducted, followed by multivariate logistic regression analysis to explore associations between sexual behaviors and HHV8 seropositivity. Odds ratio (OR) and 95% confidence interval (95% CI) were used to determine whether a variable was associated with HHV8 infection. The nonparametric Mann-Whitney *U* test was used to assess the difference in the geometric mean titers (GMTs) of anti-HHV8 IgG between the HHV8 mono-infection and co-infection groups. A *p*-value less than or equal to 0.05 was considered to be statistically significant for all analyses. All statistical analyses were carried out using the SAS System for Windows (Cary, NC, USA), version 8.0.

3. Results

3.1. Socio-Demographic Characteristics of Participants

A total of 441 drug users were interviewed for this study, including 334 males and 107 females. The participants' characteristics are summarized in Table 1. Most participants (99.5%) resided in the study area, and 97.1% were of the Han ethnicity. The age of the study population ranged from 20 to 61 years. Male participants were significantly older than the female participants (46.05 ± 8.31 *vs.* 43.31 ± 8.35, $p = 0.003$). Approximately, 95.7% participants had an education level above high school. Female participants were more likely to have a steady sex partner compared to the male participants. There were no significant sociodemographic differences between male and female in terms of residency, ethnicity and education level.

Table 1. Sociodemographic characteristics of study participants.

	Male (n = 334) No. (%)	Female (n = 107) No. (%)	Total (n = 441) No. (%)
Residency (p = 1.000)			
Local	332 (99.4)	107 (100.0)	439 (99.5)
Non-local	2 (0.6)	0 (0.0)	2 (0.5)
Ethnicity (p = 0.526)			
Han	325 (97.3)	103 (96.3)	428 (97.1)
Minority	9 (2.7)	4 (3.7)	13 (2.9)
Age (years) (p = 0.005)			
≤40	80 (24.0)	39 (36.4)	119 (27.0)
41–50	138 (41.3)	47 (43.9)	185 (42.0)
≥51	116 (34.7)	21 (19.6)	137 (31.1)
Education (p = 0.107)			
Primary or lower	18 (5.4)	1 (0.9)	19 (4.3)
Junior high	199 (59.6)	71 (66.4)	270 (61.2)
Senior high or college	117 (35.0)	35 (32.7)	152 (34.5)
Steady partner (p = 0.012)			
No	187 (56.0)	45 (42.1)	232 (52.6)
Yes	147 (44.0)	62 (57.9)	209 (47.4)
HHV8-Ab (p = 0.842)			
No	275 (82.3)	89 (83.2)	364 (82.5)
Yes	59 (17.7)	18 (16.8)	77 (17.5)
HCV-Ab (p = 0.689)			
No	127 (38.0)	43 (40.2)	170 (38.5)
Yes	207 (62.0)	64 (59.8)	271 (61.5)
HIV-Ab (p = 0.427)			
No	333 (99.7)	106 (99.1)	439 (99.5)
Yes	1 (0.3)	1 (0.9)	2 (0.5)
Syphilis (p = 0.174)			
No	327 (97.9)	102 (95.3)	429 (97.3)
Yes	7 (2.1)	5 (4.7)	12 (2.7)

The majority (67.1%) of the participants had a history of injection drug use, and used mainly heroin and/or cocaine. Among them, 3.7% reported ever sharing syringes. Meanwhile, about 7.3% of participants reported commercial sex behaviors, including four female respondents. In this study, 59.4% participants reported never using condom in commercial sex contact.

3.2. Seroprevalence of HIV, HHV8, HCV, and Syphilis

Of all the 441 participants, 77 (17.5%) were HHV8 seropositive. The majority (61.5%) of the study participants enrolled were HCV positive, while the HIV prevalence was extremely low with only two cases being HIV positive. Given this low frequency of HIV, it was not considered for further analysis. As shown in Table 2, among the 77 HHV8 positive individuals, 44 (57.1%) were coinfected with HCV, and one case coinfected with HCV and syphilis concurrently. With the 364 HHV8 negative individuals, 214 (58.8%) were infected only with HCV, 10 (2.8%) were dually infected with HCV and syphilis, and two were dually with HCV and HIV.

Table 2. Summary of coinfections by human herpesvirus 8 (HHV8), hepatitis C virus (HCV), Human immunodeficiency virus (HIV) and syphilis among study participants.

Co-infections	HHV8-Uninfected ($N_1 = 364$)		HHV8-Infected ($N_2 = 77$)	
	No.	Prevalence (%)	No.	Prevalence (%)
None	137	37.6	32	41.6
Single pathogen				
HIV	0	0.0	0	0.0
Syphilis	1	0.3	0	0.0
HCV	214	58.8	44	57.1
Dual pathogens				
HIV + HCV	2	0.5	0	0.0
HIV + Syphilis	0	0.0	0	0.0
HCV + Syphilis	10	2.8	1	1.3
Total	364	100.0	77	100.0

3.3. Correlates of HHV8 and HCV Seropositivity

With regard to HHV8 infection, the univariate analysis showed that few variables were associated with HHV8 positive status among study participants. No significant association was detected between HHV8 infection and any sociodemographic characteristics, drug use or sex behaviors. Moreover, no statistically significant association between HHV8 infection and HCV or syphilis were found either (data not shown). The lack of association between HHV8 seropositivity and potential variables remained in both male and female subgroups, when separated analyses were performed. No significant association was observed between HHV8 seropositivity and history of commercial sex, after adjusting for sociodemographic characteristics by using multiple logistic regression analysis. Although those who had commercial sex contact were more likely to be HHV8 positive (OR, 6.05; 95% CI 0.80–45.67), the association did not achieve significance (Table 3).

With regard to HCV infection, the univariate analysis indicated that education level, ever injected drugs, ever had commercial sex, and never use condom for commercial sex were associated with HCV infection. Moreover those who ever shared syringes were all found to be positive for HCV. Multivariate analysis, adjusting for potential confounder, indicated ever injected drug history were independently associated with HHV8 infection among participants (OR, 2.18, 95% CI 1.41–3.37). Meanwhile syphilis infection seemed to be associated with HHV8 seropositivity, but this association was not significant (OR 7.56, 95% CI 0.94–60.57). No additional risk factors were identified when men and women were analyzed as an independent group.

Table 3. Correlates of HHV8 seropositivity among study participants.

Risk Factors	No. HHV8 Infection/No. Tested (%)	aORs (95%CI) *	*p*-values
Ever injected drugs			
No	21/145 (14.5)	1.00	
Yes	56/296 (18.9)	1.50 (0.83–2.69)	0.181
Ever sharing syringe			
No	75/430 (17.4)	1.00	
Yes	2/11 (18.2)	1.06 (0.20–5.46)	0.949
Steady partner			
No	41/232 (17.7)	1.00	
Yes	36/209 (17.2)	0.97 (0.55–1.72)	0.922
Ever had sex in the past month			
No	53/288 (18.4)	1.00	
Yes	24/153 (15.7)	0.85 (0.42–1.71)	0.650
Condom use in the last sex intercourse			
Never or no sex	69/380 (18.2)	1.00	
Yes	8/61 (13.1)	0.57 (0.21–1.56)	0.273
Ever had commercial sex			
No	69/409 (16.9)	1.00	
Yes	8/32 (25.0)	6.05 (0.80–45.67)	0.081
Frequency of condom use in commercial sex			
Always or no sex	71/414 (17.1)	1.00	
Sometimes	3/8 (37.5)	1.59 (0.16–6.12)	0.120
Never	3/19 (15.8)	0.72 (0.06–8.32)	0.790
HIV			
No	77/439 (17.5)	-	
Yes	0/2 (0.0)	-	-
HCV			
No	32/170 (18.8)	1.00	
Yes	45/271 (16.6)	0.82 (0.48–1.41)	0.476
Syphilis			
No	76/429 (17.7)	1.00	
Yes	1/12 (8.3)	0.50 (0.06–4.09)	0.519

* aORs: adjusted Odds ratios and 95% CI: 95% confidence interval.

3.4. HHV8 Antibody Titers by Different Characteristics

Since HHV8 and HCV coinfection is common in the study population, geometric mean titers (GMT) of antibodies to lytic and latent antigens of HHV8 were further compared according to their coinfection status. Figure 1 presents the distribution of HHV8 lytic and latent antibodies. The GMTs for HHV8 latent antibody were 480 (95% CI 362.5–597.5) and 480 (95% CI 351.0–609.0) in participants infected with HHV8 only and HHV8/HCV coinfection, respectively. While the GMTs for HHV8 lytic antibody were 405.3 (95% CI 290.0–520.7) and 622.2 (95% CI 387.8–856.6) for HHV8 mono infection and HHV8/HCV coinfection, respectively. Overall, the pairwise comparison indicated that neither lytic nor latent antibodies significantly differed across the two groups (for lytic antibody, Mann-Whitney $U = 158.0$, $p = 0.214$; for latent antibody Mann-Whitney $U = 318.0$, $p = 0.695$). Similarly, no significant differences for either lytic or latent antibody were observed with each group (for HHV8 group, Mann-Whitney $U = 161.5$, $p = 0.437$; for HHV8/HCV group Mann-Whitney U = 310.5, $p = 0.326$) (Figure 1).

Figure 1. Anti-HHV8 IgG antibody titer among patients with HHV8 monoinfection *versus* patients with HHV8/HCV confection.

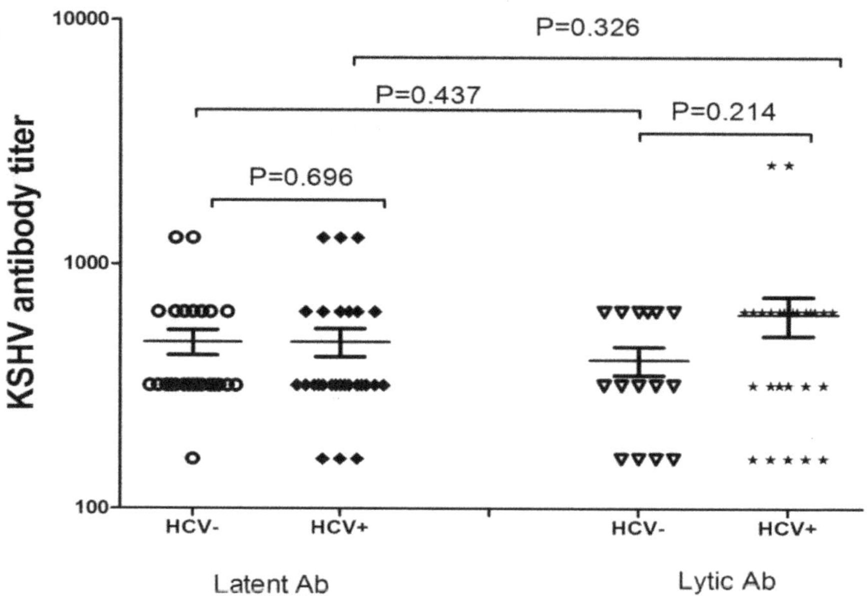

4. Discussion

The present study aimed to address the status of HHV8 infection amongst a group of drug users from mainland China. In the present study, a moderate seroprevalence (17.5%) of HHV8 was identified, which is relatively lower than that (32.7%) among men who have sex with men (MSM), but higher than the general population in China [20]. Meanwhile, their HCV infection status was evaluated in parallel, and as expected a high HCV seroprevalence (61.5%) was observed among the study participants. This high seroprevalence of HCV among drug users is in accordance with findings from previous studies in China [15,16].

Of importance, the results of this study suggest that HHV8 seropositivity was not directly associated with injection drug use behaviors, or with HCV and syphilis infections in the participants. This is consistent with two previous reports one among a cohort of Netherlands drug users, and the other from a minority population in China [14,21]. Interestingly, although sexual behaviors were found to be associated with HHV8 seropositivity [18,22,23], the association was not well established in the present study. Specifically, there was no significant association between commercial sex and HHV8 seropositivity. Furthermore the seropositivity of HHV8 was not associated with syphilis, a typical sexually transmitted disease either, in the present study. This discrepancy could possibly be explained by the fact that the majority of the participants were heterosexual individual, and the possibility of

HHV8 transmission via heterosexual contacts is still controversial. In this regard, more extensive studies are needed to address the association of sexual contact and HHV8 transmission in this population.

As expected, a high HCV prevalence was detected and injection drug use behaviors were found to be independent risk factors for HCV seropositivity. Most participants who have injected drugs and shared needles were seropositive for HCV, demonstrating that the blood-to-blood contact is highly efficient in transmitting HCV. Therefore, if the needle sharing and blood-to-blood contact is a major risk for HHV8 transmission, the observed prevalence of HHV8 should have been higher, and similar epidemiologic profiles should have been observed for HHV8 and HCV. The obvious difference in the prevalence of HHV8 and HCV infection in the study population, however, suggested that these two viruses should use distinct routes for efficient viral transmission. Interestingly, only a small proportion of the study participants self-reported needle sharing experience, possibly due to the effectiveness of the needle exchange program in reducing blood borne infectious disease transmission in China. Previous studies have also shown that indirect drug sharing and drug preparation practices, such as splitting drugs prepared by a user with subsequent transfer of the prepared drug from one syringe to a second syringe for another user, sharing cotton, filters, cooker, water, and water containers, are associated with HCV transmission [24–27]. These observations suggest that HHV8 is not transmitted efficiently through needle sharing or other behaviors among drug users. Therefore, it is likely that HCV, but not HHV8, could be efficiently transmitted through contaminated utensils during drug use practices. Taken together, findings from the present study, along with our previous study conducted in a community with high risks of blood borne infections particularly HCV, demonstrate that HHV8 and HCV do not share the same transmission routes [28]. This divergent characteristic between HHV8 and HCV further confirms that blood borne transmission is not a predominant transmission route for HHV8 but the possibility of blood borne transmission could not be completely ruled out.

This study is also subjected to certain limitations. First, the subjects in this study were recruited using a convenience sampling approach and the findings may not be representative of the whole drug-using population in China. Second, like most studies on sexual and drug use behavior, this study is potentially subjected to socially desirable responding or reporting bias. Third, only asymptomatically infected HHV8 individuals were recruited. Therefore, it is difficult to delineate the potential relationships between host lytic and latent antibody response to HHV8 antigens with KS risk. Nonetheless, in the present study, the parallel analysis of HHV8 and HCV serving as control for blood borne transmission, our findings could provide important information to a better understanding of the HHV8 epidemiology in China.

In conclusion, HHV8 seroprevalence is relatively low in the drug user and mirrors the low KS disease burden in this population in China. Injection drug use and needle sharing practice were not found to be a risk factor for HHV8 transmission among the drug users in the current study. In contrast, HCV was highly prevalent among this population and confirms that injection drug use behaviors are independent risk factors of HCV infection. Given the importance of this population in HIV prevention, more extensive study regarding HHV8 transmission and the effects of infection in this risk group is warranted.

Acknowledgments

This study was supported by Doctoral Fund of Ministry of Education of China (Grant No. 20120071120050), the Fundamental Research Funds for the Central Universities (10FX058 and 20520133104), the Shanghai Municipal Bureau of Health (grant No. XBR2011043 to NH, grant No. 08GWQ058), the Natural Science Foundation of China (grant No. 81161120407), and the United States National Institutes of Health Fogarty International Center (grant No. D43 TW001492, RO1 CA75903 and P30 GM103509 to CW). TZ was a Fogarty Fellow.

Author Contributions

T.Z. designed the research. T.Z. and Y.L. drafted the Manuscript. Y.L. and Y.Z. assisted with recruitment of participants, collection of samples and data collection. J.W. performed the experiments. T.Z. and V.M. performed data analysis. N.H. and C.W. were responsible for overseeing the study and initiation of collaborations. All authors contributed to study design, final interpretation of results and assisted in writing the manuscript.

References

1. Cesarman, E.; Chang, Y.; Moore, P.S.; Said, J.W.; Knowles, D.M. Kaposi's sarcoma-associated herpesvirus-like DNA sequences in AIDS-related body-cavity-based lymphomas. *N. Engl. J. Med.* **1995**, *332*, 1186–1191.
2. Chang, Y.; Cesarman, E.; Pessin, M.S.; Lee, F.; Culpepper, J.; Knowles, D.M.; Moore, P.S. Identification of herpesvirus-like DNA sequences in AIDS-associated Kaposi's sarcoma. *Science* **1994**, *266*, 1865–1869.
3. Soulier, J.; Grollet, L.; Oksenhendler, E.; Cacoub, P.; Cazals-Hatem, D.; Babinet, P.; d'Agay, M.F.; Clauvel, J.P.; Raphael, M.; Degos, L.; *et al.* Kaposi's sarcoma-associated herpesvirus-like DNA sequences in multicentric Castleman's disease. *Blood* **1995**, *86*, 1276–1280.
4. Whitby, D.; Howard, M.R.; Tenant-Flowers, M.; Brink, N.S.; Copas, A.; Boshoff, C.; Hatzioannou, T.; Suggett, F.E.; Aldam, D.M.; Denton, A.S.; *et al.* Detection of Kaposi sarcoma associated herpesvirus in peripheral blood of HIV-infected individuals and progression to Kaposi's sarcoma. *Lancet* **1995**, *346*, 799–802.
5. Bagni, R.; Whitby, D. Kaposi's sarcoma-associated herpesvirus transmission and primary infection. *Curr. Opin. HIV AIDS* **2009**, *4*, 22–26.
6. Butler, L.M.; Dorsey, G.; Hladik, W.; Rosenthal, P.J.; Brander, C.; Neilands, T.B.; Mbisa, G.; Whitby, D.; Kiepiela, P.; Mosam, A.; *et al.* Kaposi sarcoma-associated herpesvirus (KSHV) seroprevalence in population-based samples of African children: Evidence for at least 2 patterns of KSHV transmission. *J. Infect. Dis.* **2009**, *200*, 430–438.
7. Dedicoat, M.; Newton, R.; Alkharsah, K.R.; Sheldon, J.; Szabados, I.; Ndlovu, B.; Page, T.; Casabonne, D.; Gilks, C.F.; Cassol, S.A.; *et al.* Mother-to-child transmission of human herpesvirus-8 in South Africa. *J. Infect. Dis.* **2004**, *190*, 1068–1075.
8. Nawar, E.; Mbulaiteye, S.M.; Gallant, J.E.; Wohl, D.A.; Ardini, M.; Hendershot, T.; Goedert, J.J.;

Rabkin, C.S. Risk factors for Kaposi's sarcoma among HHV-8 seropositive homosexual men with AIDS. *Int. J. Cancer* **2005**, *115*, 296–300.

9. Cannon, M.J.; Operskalski, E.A.; Mosley, J.W.; Radford, K.; Dollard, S.C. Lack of evidence for human herpesvirus-8 transmission via blood transfusion in a historical US cohort. *J. Infect. Dis.* **2009**, *199*, 1592–1598.
10. Hladik, W.; Dollard, S.C.; Mermin, J.; Fowlkes, A.L.; Downing, R.; Amin, M.M.; Banage, F.; Nzaro, E.; Kataaha, P.; Dondero, T.J.; *et al.* Transmission of human herpesvirus 8 by blood transfusion. *N. Engl. J. Med.* **2006**, *355*, 1331–1338.
11. Vamvakas, E.C. Is human herpesvirus-8 transmitted by transfusion? *Transfus. Med. Rev.* **2010**, *24*, 1–14.
12. Goedert, J.J.; Charurat, M.; Blattner, W.A.; Hershow, R.C.; Pitt, J.; Diaz, C.; Mofenson, L.M.; Green, K.; Minkoff, H.; Paul, M.E.; *et al.* Risk factors for Kaposi's sarcoma-associated herpesvirus infection among HIV-1-infected pregnant women in the USA. *AIDS* **2003**, *17*, 425–433.
13. Atkinson, J.; Edlin, B.R.; Engels, E.A.; Kral, A.H.; Seal, K.; Gamache, C.J.; Whitby, D.; O'Brien, T.R. Seroprevalence of human herpesvirus 8 among injection drug users in San Francisco. *J. Infect. Dis.* **2003**, *187*, 974–981.
14. Renwick, N.; Dukers, N.H.; Weverling, G.J.; Sheldon, J.A.; Schulz, T.F.; Prins, M.; Coutinho, R.A.; Goudsmit, J. Risk factors for human herpesvirus 8 infection in a cohort of drug users in The Netherlands, 1985–1996. *J. Infect. Dis.* **2002**, *185*, 1808–1812.
15. Tao, Y.L.; Tang, Y.F.; Qiu, J.P.; Cai, X.F.; Shen, X.T.; Wang, Y.X.; Zhao, X.T. Prevalence of hepatitis C infection among intravenous drug users in Shanghai. *World J. Gastroenterol.* **2013**, *19*, 5320–5325.
16. Xia, Y.H.; Chen, W.; Tucker, J.D.; Wang, C.; Ling, L. HIV and hepatitis C virus test uptake at methadone clinics in Southern China: Opportunities for expanding detection of bloodborne infections. *BMC Public Health* **2013**, *13*, 899.
17. Zavitsanou, A.; Malliori, M.; Sypsa, V.; Petrodaskalaki, M.; Psichogiou, M.; Rokka, C.; Giannopoulos, A.; Kalapothaki, V.; Whitby, D.; Hatzakis, A. Seroepidemiology of human herpesvirus 8 (HHV-8) infection in injecting drug users. *Epidemiol. Infect.* **2010**, *138*, 403–408.
18. Bernstein, K.T.; Jacobson, L.P.; Jenkins, F.J.; Vlahov, D.; Armenian, H.K. Factors associated with human herpesvirus type 8 infection in an injecting drug user cohort. *Sex. Transm. Dis.* **2003**, *30*, 199–204.
19. Minhas, V.; Crosby, L.N.; Crabtree, K.L.; Phiri, S.; M'Soka, T.J.; Kankasa, C.; Harrington, W.J.; Mitchell, C.D.; Wood, C. Development of an immunofluorescence assay using recombinant proteins expressed in insect cells to screen and confirm presence of human herpesvirus 8-specific antibodies. *Clin. Vaccine Immunol.* **2008**, *15*, 1259–1264.
20. Zhang, T.; Lin, H.; Minhas, V.; Zhu, W.; Wood, C.; He, N. Prevalence and correlates of Kaposi's sarcoma-associated herpesvirus infection in a sample of men who have sex with men in Eastern China. *Epidemiol. Infect.* **2013**, *141*, 1823–1830.
21. Yang, P.; Guo, S.; Tan, X.; Yang, L.; Fu, B.; Wang, L.; Zhang, J.; Zhang, G. Seroepidemiology of Kaposi's sarcoma–associated herpesvirus in Uigur male drug users from a place in Xinjiang. *J. Shihezi Univ.* **2010**, *28*, 68–71.
22. Martro, E.; Esteve, A.; Schulz, T.F.; Sheldon, J.; Gambus, G.; Munoz, R.; Whitby, D.; Casabona, J.

Risk factors for human Herpesvirus 8 infection and AIDS-associated Kaposi's sarcoma among men who have sex with men in a European multicentre study. *Int. J. Cancer* **2007**, *120*, 1129–1135.

23. Eltom, M.A.; Mbulaiteye, S.M.; Dada, A.J.; Whitby, D.; Biggar, R.J. Transmission of human herpesvirus 8 by sexual activity among adults in Lagos, Nigeria. *AIDS* **2002**, *16*, 2473–2478.
24. Thiede, H.; Hagan, H.; Campbell, J.V.; Strathdee, S.A.; Bailey, S.L.; Hudson, S.M.; Kapadia, F.; Garfein, R.S. Prevalence and correlates of indirect sharing practices among young adult injection drug users in five U.S. cities. *Drug Alcohol. Depend.* **2007**, *91*, S39–S47.
25. Doerrbecker, J.; Behrendt, P.; Mateu-Gelabert, P.; Ciesek, S.; Riebesehl, N.; Wilhelm, C.; Steinmann, J.; Pietschmann, T.; Steinmann, E. Transmission of hepatitis C virus among people who inject drugs: Viral stability and association with drug preparation equipment. *J. Infect. Dis.* **2013**, *207*, 281–287.
26. Pouget, E.R.; Hagan, H.; des Jarlais, D.C. Meta-analysis of hepatitis C seroconversion in relation to shared syringes and drug preparation equipment. *Addiction* **2012**, *107*, 1057–1065.
27. Thorpe, L.E.; Ouellet, L.J.; Hershow, R.; Bailey, S.L.; Williams, I.T.; Williamson, J.; Monterroso, E.R.; Garfein, R.S. Risk of hepatitis C virus infection among young adult injection drug users who share injection equipment. *Am. J. Epidemiol.* **2002**, *155*, 645–653.
28. Zhang, T.; He, N.; Ding, Y.; Crabtree, K.; Minhas, V.; Wood, C. Prevalence of human herpesvirus 8 and hepatitis C virus in a rural community with a high risk for blood-borne infections in central China. *Clin. Microbiol. Infect.* **2011**, *17*, 395–401.

5

KSHV miRNAs Decrease Expression of Lytic Genes in Latently Infected PEL and Endothelial Cells by Targeting Host Transcription Factors

Karlie Plaisance-Bonstaff [1], Hong Seok Choi [1], Tyler Beals [1], Brian J. Krueger [1], Isaac W. Boss [1], Lauren A. Gay [1], Irina Haecker [1], Jianhong Hu [1] and Rolf Renne [1,2,3,*]

1 Department of Molecular Genetics and Microbiology, University of Florida, Gainesville, FL 32610, USA; E-Mails: karliebonstaff@gmail.com (K.P.-B.); hschoi@ufl.edu (H.S.C.); tbeals@ufl.edu (T.B.); brian.krueger@duke.edu (B.J.K.); iboss2022@gmail.com (I.W.B.); lagay@ufl.edu (L.A.G.); irina25@me.com (I.H.); Jianhong.Hu@bcm.edu (J.H.)

2 UF Health Cancer Center, University of Florida, Gainesville, FL 32610, USA

3 UF Institute of Genetics, University of Florida, Gainesville, FL 32610, USA

* Author to whom correspondence should be addressed; E-Mail: rrenne@ufl.edu;

External Editor: Zhi-Ming Zheng

Abstract: Kaposi's sarcoma-associated herpesvirus (KSHV) microRNAs are encoded in the latency-associated region. Knockdown of KSHV miR-K12-3 and miR-K12-11 increased expression of lytic genes in BC-3 cells, and increased virus production from latently infected BCBL-1 cells. Furthermore, iSLK cells infected with miR-K12-3 and miR-K12-11 deletion mutant viruses displayed increased spontaneous reactivation and were more sensitive to inducers of reactivation than cells infected with wild type KSHV. Predicted binding sites for miR-K12-3 and miR-K12-11 were found in the 3'UTRs of the cellular transcription factors MYB, Ets-1, and C/EBPα, which activate RTA, the KSHV replication and transcription activator. Targeting of MYB by miR-K12-11 was confirmed by cloning the MYB 3'UTR downstream from the luciferase reporter. Knockdown of miR-K12-11 resulted in increased levels of MYB transcript, and knockdown of miR-K12-3 increased both C/EBPα and Ets-1 transcripts. Thus, miR-K12-11 and miR-K12-3 contribute to maintenance of latency by decreasing RTA expression indirectly, presumably via down-regulation of MYB, C/EBPα and Ets-1, and possibly other host transcription factors.

Keywords: KSHV; miRNA; lytic reactivation; latency

1. Introduction

Kaposi's sarcoma-associated herpesvirus (KSHV), also known as Human Herpesvirus 8 (HHV-8), is a DNA tumor virus that infects endothelial cells *in vivo* and is the etiological agent of Kaposi's sarcoma (KS) [1]. KSHV has been linked to two B cell lymphoproliferative disorders, primary effusion lymphoma (PEL) and a subset of multicentric Castleman's disease (MCD) [2,3]. As with all herpesviruses, KSHV has both lytic and latent modes of replication. During lytic replication and reactivation, genome-wide expression occurs in a temporally regulated cascade of immediate early, early, and late genes, which results in lysis of the host cell and release of progeny virus. During latency most tumor cells express only a limited number of genes, the majority residing in the latency-associated region, which encodes the latency-associated nuclear antigen (LANA), v-FLIP, v-cyclin, kaposin, and 12 miRNA genes. A subset of cells also express vIRF3, vIL-6, and K1 during latency [4–12].

MicroRNAs (miRNAs) are non-coding RNAs 19–23 nucleotides in length that regulate gene expression post-transcriptionally by targeting 3' untranslated regions (UTRs) of messenger RNAs (for review see [13]). Since the discovery of highly expressed KSHV-encoded miRNAs in all KSHV-associated tumors [14,15], several cellular targets of KSHV-encoded miRNAs have been identified. Roles for KSHV miRNAs include promotion of angiogenesis, cell cycle regulation, inhibition of apoptosis and recently transformation [16] (for reviews see [17–19]). Early after the discovery of herpesvirus-encoded miRNAs it was hypothesized that these novel viral post-transcriptional regulators may promote latency by targeting lytic genes [20]. Indeed, one EBV microRNA, miR-BART2, is encoded antisense to BALF5, the EBV DNA polymerase, and targeting and cleavage of the BALF5 mRNA has been experimentally confirmed [21,22]. For KSHV, elegant work from the Ganem lab utilizing miRNA mimic- and antagomir-based screens provided evidence that KSHV miRNAs can modulate the latent/lytic transition through direct targeting of RTA by miR-K12-9 * [23]. RTA, the product of the immediate early lytic gene ORF50, is the KSHV replication and transcriptional activator that is expressed first during reactivation and initiates the cascade of lytic gene expression by activating several early lytic gene promoters [24,25]. Direct miRNA targeting of RTA was also demonstrated for miR-K12-5 and miR-K12-7 based on *in vitro* luciferase assays [26,27], consistent with *in silico* prediction of targets within the RTA 3'UTR. Hence, direct targeting of RTA by miRNAs may act as a gate keeper of latency by preventing reactivation.

Conceptually, miRNAs can also contribute to latency by targeting host factors that normally tip the balance toward reactivation. We present evidence here that KSHV miRNAs contribute to the maintenance of latency by targeting cellular transcription factors. Our data demonstrate that miR-K12-11 and miR-K12-3 prevent lytic reactivation by modulating expression of the transcription factors MYB, C/EBPα and Ets-1, which have previously been reported as activators of the RTA promoter [28–30].

2. Materials and Methods

2.1. Cell Lines

BC-3-G cells containing a PAN-GFP expression cassette were provided by Ren Sun (UCLA) [28]. BC-3-G, BC-3 and BCBL-1 primary effusion lymphoma (PEL) cells are latently infected with KSHV, and were cultured in RMPI supplemented with 10% FBS, 1% penicillin/streptomycin, and 1% sodium pyruvate. Human embryonic kidney 293 and 293T cells were cultured in DMEM with 10% FBS and 1% P/S. iSLK cells were kindly provided by Don Ganem (UCSF) [31] and were cultured under the same conditions as 293 cells. TIVE (Telomerase Immortalized Vein Endothelial) cells were cultured as previously described [32].

2.2. Antagomir Derepression Assays and Quantitative Reverse Transcription-PCR (RT-qPCR) Analysis

For inhibition of KSHV miRNAs, 2'OMe RNA antagomirs were used as previously described [33]. PEL cells (1×10^6) were transfected with 50–400 nM of antagomir using TransIT-TKO transfection reagent (Mirus) as described [34]. At 48 h post transfection (hpt), cells were harvested using RNA-Bee (Tel-Test) according to the manufacturer's instructions. 1 µg of DNase treated RNA was reverse transcribed using SuperScript III (Invitrogen, Grand Island, NY, USA) according to the manufacturer's instructions. Quantitative PCR (qPCR) analysis was carried out using an ABI StepOne Plus system along with ABI Fast SYBR reagent (Applied Biosystems, Carlsbad, CA, USA). Expression of all genes was normalized to β-actin expression, and Student's t-tests were performed to determine statistical significance compared to the mock control. The sequences of PCR primers can be found in Table 1.

Table 1. Primers used in this study.

Gene	Primer Sequence 5'-3'	Reference
MYB 3'UTR	FWD- GATGGAGGAGCAGATGACATC REV- AGGTAAAATAAGGGCAC	
Ets-1 3'UTR	FWD- CGTGTTGGTTGGACTCTGAA REV- TCTCCAGCAAAATGATGTGC	
C/EBPα 3'UTR	FWD-CTTGTGCCTTGGAAATGCAAACTCACC REV- AAGAAGAGAACCAAGCCGTCCTTC	
MYB	FWD- TCAGGAAACTTCTTCTGCTCACA REV- AGGTTCCCAGGTACTGCT	
Ets-1	FWD- AAGGGAGATCGAAGGAGGAA REV- TCTGCTATAGGAACTGCAGGAG	
C/EBPα	FWD- TGTATACCCCTGGTGGGAGA REV- TCATAACTCCGGTCCCTCTG	
RTA	FWD- CACAAAAATGGCGCAAGATGA REV- TGGTAGAGTTGGGCCTTCAGTT	[35]
ORF59	FWD- TTAGAAGTGGAAGGTGTGCC REV- TCCTGGAGTCCGGTATAGAATC	[36]
ORF19	FWD- GGCGAAAAAGTCAGCGGTGGT REV- CGGCGCGTCTTCCCTAAAGA	[37]
LANA N-Terminus	FWD- GCGCCCTTAACGAGAGGAAGTT REV- TTCCTTCGCGGTTGTAGATG	
β-actin	FWD- CATGTACGTTGCTATCCAGGC REV- CTCCTTAATGTCACGCACGAT	Primerbank ID 4501885a1

2.3. Virus Isolation and Quantitation

Virus particles were harvested from PEL cells 6 days post transfection (dpt) with antagomirs. Cells were pelleted at 1100 RPM for 5 min and media supernatant was passed through a 0.45 μM filter. Virus particles were pelleted by ultra-centrifugation using a Beckman SW-40 rotor at 100,000 g for 1 h on a 25% sucrose cushion. Virus pellets were resuspended in 1% of the original medium volume using serum-free RPMI. DNA was extracted from 25 μL of virus stocks using DNAzol (Molecular Research Center, Inc., Cincinnati, OH, USA) and resuspended in 25 μL of ddH_2O. Viral genome copy number was determined by qPCR assay using a standard curve based on serially diluted LANA expression plasmid.

2.4. Recombineering of miRNA Deletion Mutants in KSHV BAC16

KSHV BAC16 [38,39] was kindly provided by the Jung lab (USC). A modified version of the protocol by Tischer *et al.* [40] was used to generate mutant bacmids. BAC DNA was isolated from bacteria using the Large-Construct Kit (Qiagen) according to the manufacturer's recommendations. 293T cells were transfected with 2 μg of DNA using TransIT-293 reagent (Mirus) according to the manufacturer's instructions. Cells were selected with 100 μg/mL hygromycin B and expanded for 10–15 days. Transfected cells, which express GFP, were monitored under a fluorescence microscope. Once the expanded cell population was 100% GFP positive, cells were induced with 20 pg/mL of TPA and 1 mM valproic acid and co-cultured with iSLK cells [41]. Infected iSLK cells were selected using 1 μg/mL puromycin, 250 μg/mL G418, and 1.2 mg/mL hygromycin B. Stable KSHVΔmiRNA BAC16 iSLK cells were induced using 1 μg/mL doxycycline (DOX) and 1 mM sodium butyrate. Virus was collected and quantified at 4 days as described above.

2.5. Reporter Construction and Luciferase Assays

MYB, C/EBPα, and Ets-1 full length 3'UTR sequences were cloned downstream of the luciferase gene in pGL3 promoter (Promega) using pCRII-TOPO (Invitrogen) for MYB and GeneArt Seamless Cloning (Invitrogen) for C/EBPα and Ets-1. Primers for 3'UTR cloning can be found in Table 1. 293 cells were transfected using TransIT-293 reagent (Mirus) in 24-well cell culture dishes according to the manufacturer's protocol. Each transfection reaction contained 2 ng pCMV-Renilla (Promega) control vector, 20 ng pGL3 promoter-3'UTR reporter construct and 0, 400 or 800 ng pcDNA3.1 miRNA expression vector complemented with 800, 400, or 0 ng of empty pcDNA3.1 vector as filler to reach 800 ng total pcDNA3.1 in each transfection [42]. Cells were harvested 72 h post transfection and luciferase activity was quantified using the Dual Luciferase Reporter kit (Promega) according to the manufacturer's protocol. Each lysate was assayed for firefly luciferase activity using a FLUOstar OPTIMA reader (BMG Labtech, Cary, NC, USA). Firefly luciferase activity for each sample was normalized to Renilla expression and samples were compared to the miRNA mock transfection control. Transfection assays were performed in triplicate and repeated at least 3 times. Standard deviation was calculated for triplicates and displayed as error bars in the figures. Significance of the repression of the reporter construct relative to the 0 ng miRNA expression vector was tested by one-tailed, unpaired t-test.

3. Results

3.1. Screening for Effects of KSHV miRNA Knockdown on Reactivation in BC-3-G Cells

In order to determine which KSHV-encoded miRNAs may affect lytic reactivation, miRNA knockdown studies were performed in BC-3-G [28], a PEL-derived indicator cell line. BC-3-G cells contain a GFP cassette under the control of the KSHV PAN promoter, which is highly transactivated by RTA. Upon lytic reactivation, RTA is the first protein expressed, activates the PAN promoter and as a result switches on GFP expression [28].

To validate efficiency and specificity of 2'OMe antagomirs against KSHV miRNAs, knock-down was assessed by stem-loop TaqMan qPCR which measures the amount of mature miRNA. These experiments were done in BCBL-1 cells with antagomirs against miR-K12-3 and miR-K12-11. Transfection of antagomirs against miR-K12-3 yielded reduced levels of miR-K12-3, with the greatest effect at the highest concentration (400 nM) of antagomir (Figure 1A). Antagomirs against miR-K12-3 did not reduce the level of miR-K12-11. Similarly, miR-K12-11 antagomirs gave progressively reduced levels of miR-K12-11 with increasing antagomir concentration, but had no effect on the level of mature miR-K12-3 (Figure 1A). When antagomirs against miR-K12-3 and miR-K12-11 were co-transfected, strong inhibition of both miRs was observed.

BC-3-G cells were transfected with antagomirs for each of the 12 KSHV-encoded miRNAs. Cells were monitored for GFP expression by fluorescence microscopy at 72 h after transfection. Mock transfected control cells displayed a small number of GFP-expressing cells arising from spontaneous reactivation. Screening was performed with inhibition of all 12 KSHV-encoded miRNAs individually or in combinations. The greatest increase in GFP expression was observed when either miR-K12-3 or miR-K12-3 in combination with miR-K12-11 were knocked down (data not shown). MiR-K12-3 is the most abundant miRNA within RISC complexes in BC-3 cells but is only moderately expressed in BCBL-1 cells; miRK-12-11 is moderately expressed in both PEL lines. [43,44]. Interestingly, miR-K12-11 is a mimic of hsa-miR-155 and has been shown to play an important role in B cell proliferation [33–45]. To take into account miRNA expression differences between PEL lines, we tested the effects of antagomirs against miR K12-3 and miR-K12-11 in both BCBL-1 and BC-3 cells.

3.2. Lytic Gene Expression and Virus Production Increase upon miR-K12-3 and miR-K12-11 Knockdown in PEL Cells

Based on the preliminary results indicating that knockdown of miR-K12-3 and miR-K12-11 in BC-3-G cells increased RTA expression, we tested for effects on virus lytic gene expression downstream from RTA and on virus production. BC-3 cells, the parent of BC-3-G cells, were transfected with miR-K12-3 antagomir, or a combination of miR-K12-3 and miR-K12-11. RT-qPCR was performed at 48 h to monitor expression of RTA (immediate early gene), ORF59 (early) and ORF19 (late glycoprotein). Significant increases in the levels of RTA, ORF59 and ORF19 transcripts were observed upon addition of 400 nM antagomir against miR-K12-3 (Figure 1B–D). Moreover, expression of these lytic genes was induced more efficiently when both miR-K12-3 and miR-K12-11 were knocked down simultaneously (200 nM miR-K12-3 antagomir plus 200 nM miR-K12-11 antagomir (Figure 1B–D).

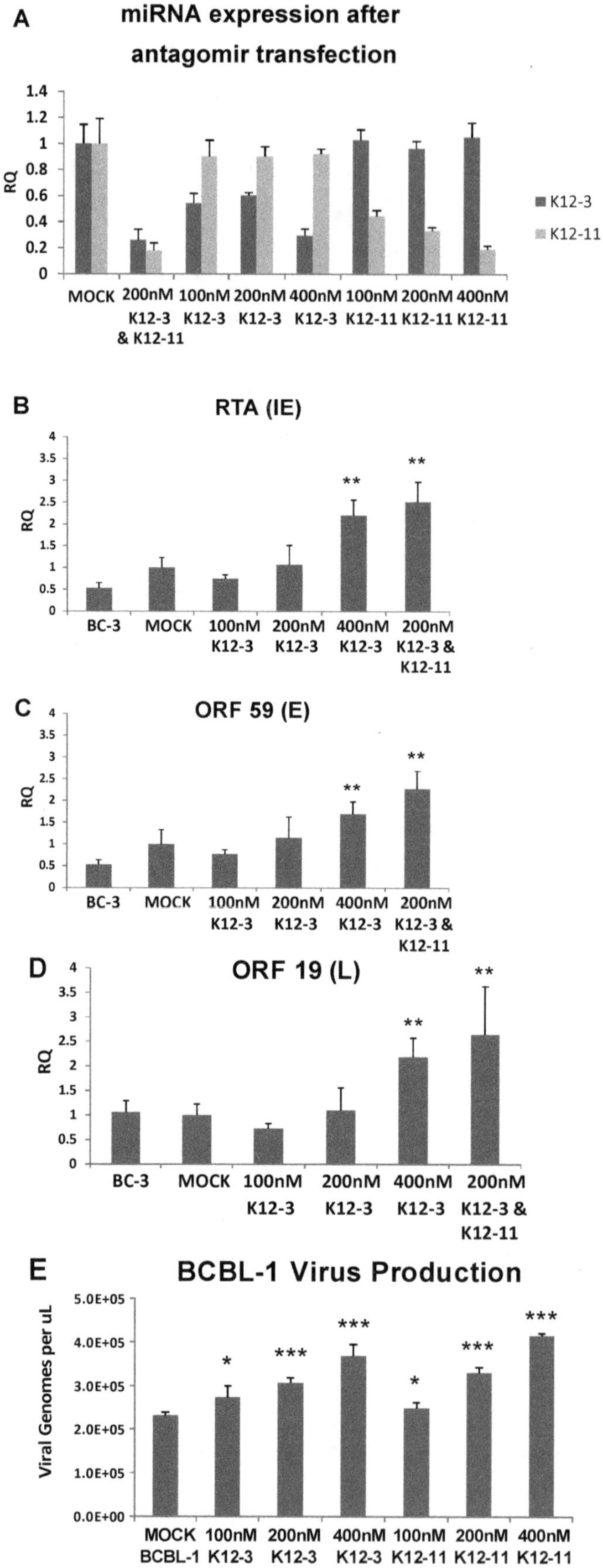

Figure 1. Lytic gene expression and virus production following miR-K12-3 and miR-K12-11 knockdown in PEL cells. (**A**) Confirmation of miRNA knockdown after antagomir transfection. BCBL-1 cells were transfected with the indicated amounts of miR-K12-3 or

miR-K12-11 antagomir or a combination of both. RNA was harvested at 48 h after transfection and TaqMan miRNA-RT and qPCR was performed using primers and probes specific to miR-K12-3 and miR-K12-11. Primers and a probe specific to RNU66 were used as a loading control and all samples were normalized to mock transfected controls; (**B–D**) BC-3 cells were transfected with the indicated amounts of miRNA-specific antagomir. Total RNA was collected at 48 h post-transfection and reverse transcribed; qPCR was performed using primers specific for RTA (**B**); ORF59 (**C**); and ORF19 (**D**); All samples were normalized to β-actin expression and compared to gene expression in the mock transfected control; (**E**) BCBL-1 cells were transfected with the indicated amounts of miRNA-specific antagomir. Cell media harboring progeny virus was collected at 6 days post-transfection, and the genome DNA concentration determined by qPCR using plasmid DNA as standard. $p < 0.05$ (*), $p < 0.01$ (**), $p < 0.001$ (***) compared to gene expression in the mock transfected control.

We asked if the increase in lytic gene expression translates into production of progeny virus. Latently infected BCBL-1 cells were used for these experiments. Cell-free virus was isolated from BCBL-1 supernatants 6 days post antagomir transfection, viral DNA was extracted, and viral genome copy number determined by qPCR. Figure 1E shows that virus production increased in a dose dependent manner with 100, 200 and 400 nM antagomir when either miR-K12-3 or miR-K12-11 was knocked down. These results indicate that knockdown of miR-K12-3 alone or of miR-K12-3 and miR-K12-11 together increases expression of lytic genes belonging to the immediate early, early and late classes. Knockdown of miR-K12-3 or of miR-K12-11 alone was sufficient to increase production of progeny virus.

3.3. Generation of KSHV miRNA Deletion Mutants and Latently Infected iSLK Cells

In order to investigate the role of miR-K12-3 and miR-K12-11 in the context of viral infection, we generated two miRNA deletion mutants using the KSHV bacmid BAC16, which was derived from the PEL cell line JSC-1 [38,39]. 20 to 25 bp regions were deleted from one arm of each pre-miRNA, destroying pre-miRNA hairpin formation without affecting neighboring miRNA expression. A modified version of the protocol of Tischer *et al.* was used to create markerless microRNA deletions within BAC16 [40,46]. Recombinant bacmids were validated by PCR and pulsed field gel electrophoresis to monitor for intact terminal repeats. Detailed experimental procedures to recover recombinant bacmids in iSLK cells will be reported elsewhere [47]. miRNA deletion mutant and wild-type (WT) bacmids were transfected into 293T cells. After selection, BAC16-containing 293T cells were induced with TPA and valproic acid, and co-cultured with iSLK cells, which harbor a doxycycline-inducible RTA gene and produce high levels of progeny virions [31,41]. WT BAC16-, KSHVΔmiRK-12-3-, or KSHVΔmiRK-12-11-containing iSLK cells produce high titer progeny virus after doxycycline induction (up to 1.14×10^7 genome copies/mL). As a final quality control, episomal DNAs isolated from latently infected iSLK cells containing miRNA deletion mutants or WT BAC16 were analyzed by Illumina-based genome-wide sequencing, which confirmed the presence of the appropriate deletion without detecting any mutations outside the miRNA region.

To test whether deletion of miR-K12-3 or miR-K12-11 had an effect on the maintenance of latency in iSLK cells, we analyzed spontaneous lytic gene expression during latent infection, and reactivation upon induction with sodium butyrate at a sub-optimal concentration. The expression levels of LANA determined by RT-qPCR in wild type- and mutant-infected iSLK cells were very similar (Figure 2A). However, iSLK cells latently infected with KSHVΔmiR-K12-11 displayed higher expression levels of RTA and ORF19 than iSLK cells infected with either WT virus or KSHVΔmiR-K12-3 (Figure 2A). Upon sub-optimal induction with 2 mM sodium butyrate, both KSHVΔmiR-K12-3- and KSHVΔmiR-K12-11-infected iSLK cells displayed ≥4-fold higher levels of RTA expression compared to KSHV wild type infected iSLK cells, and ≥6-fold higher levels of ORF19 expression (Figure 2B). Recently it has been shown that SLK cells, which are commonly used for KS-derived endothelial tumor biology studies, are not of endothelial origin, but are of epithelial cell origin instead [48]. To test for lytic gene expression levels in endothelial cells infected with KSHV miRNA-deletion mutants, we repeated experiments in telomerase-immortalized vascular endothelial (TIVE) cells [32]. Sub-optimal induction of infected TIVE cells gave a significant increase in expression of RTA and ORF19 lytic genes on infection with either KSHVΔmiR-K12-3 or KSHVΔmiR-K12-11 compared with WT (Figure 2C), as was seen for iSLK cells (Figure 2B). These results suggest that expression of miR-K12-3 and miR-K12-11 both contribute to the maintenance of latency in endothelial cells, and are consistent with the antagomir inhibition experiments in PEL cells which measured expression of RTA, ORF59, ORF19 and virus production during latent infection (Figure 1).

3.4. In Silico Seed Sequence Prediction for miR-K12-3 and miR-K12-11 Identified MYB, C/EBPα, and Ets-1 as Potential Targets

Scanning the 3'UTR of RTA revealed no seed sequence matches for miR-K12-3 or miR-K12-11. We therefore focused on transcription factors known to activate the viral RTA promoter [28–30], and searched for seed sequence matches of at least 6 nts [49]. The 3'UTR of the proto-oncogene MYB contained two potential 8 base binding sites for miR-K12-11 (Figure 3A). MYB is a transcriptional activator with a central role in hematopoiesis, and cooperates with transcription factors including members of the CEBP and Ets families. Targets of the MYB transcription factor include genes involved in development, cell survival, proliferation, and homeostasis [50]. C/EBPα is a member of the CCAAT/Enhancer-Binding Protein family, which are bZIP nuclear transcription factors. The 3'UTR of C/EBPα has two potential sites for miR-K12-3, each with a 7 base match (Figure 3B). C/EBPα is an important transcription factor involved in controlling tissue-specific gene expression in myeloid tissues and growth arrest [51].

Interestingly, we found 3 putative binding sites for miR-K12-3 and 3 for miR-K12-11 within the 3'UTR of Ets-1 (Figure 3C) ranging from 6 to 9 base matches. Ets-1 is a member of the Ets family of transcription factors, which are downstream effectors of the Ras-MAPK signaling cascades [52]. The Ets family binds to a specific DNA consensus sequence and activates genes involved in cell proliferation, differentiation, and survival [53].

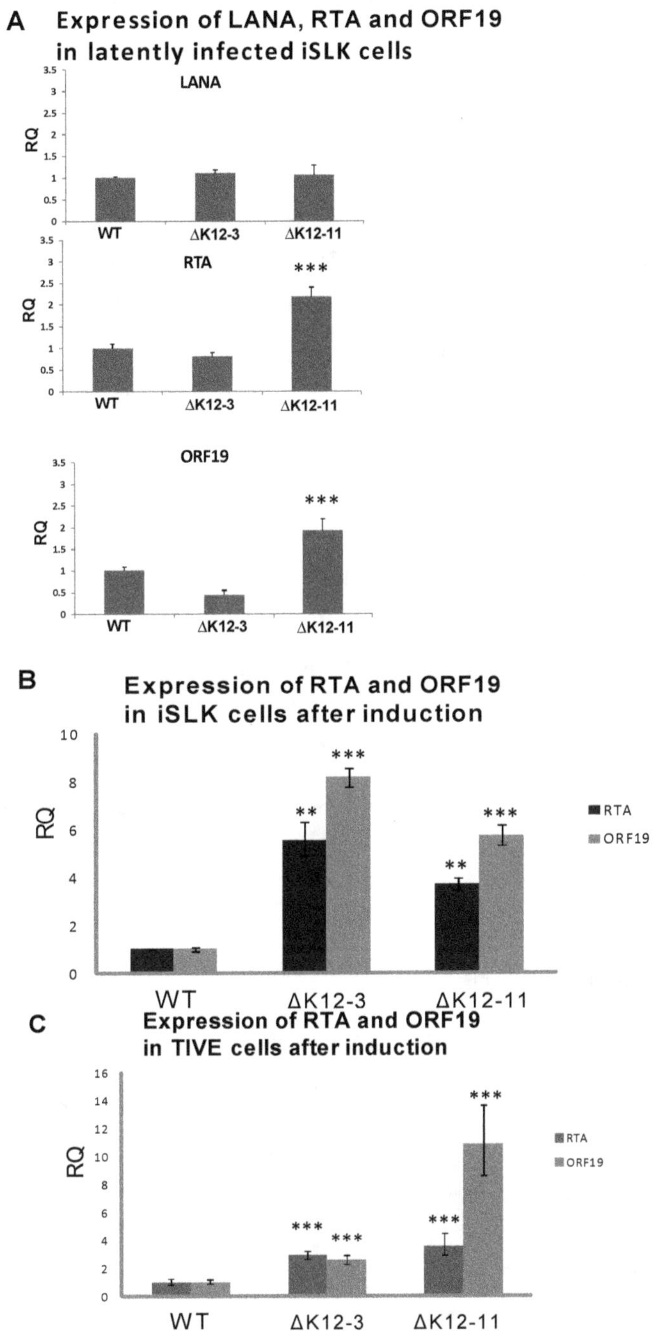

Figure 2. Expression of LANA and lytic genes in cells infected with KSHV miRNA deletion mutants. (**A**) iSLK cells latently infected with WT, ΔmiR-K12-3, or ΔmiR-K12-11 BAC16-derived KSHV were harvested for RNA and RT-qPCR was performed for LANA, RTA, and ORF19. All samples were normalized to β-actin expression; (**B**) Expression of RTA and ORF19 genes following induction of iSLK cells infected with WT, ΔmiR-K12-3, or ΔmiR-K12-11 KSHV mutants. Total RNA was harvested at 72 h post induction with 2 mM sodium butyrate and RT-qPCR performed for RTA and ORF19. All samples were normalized to β-actin and LANA expression. Samples were compared to the WT control. $p < 0.05$ (*), $p < 0.01$ (**), $p < 0.001$ (***); (**C**) Expression of RTA and ORF19 genes following induction of TIVE cells infected with WT, ΔmiR-K12-3, or ΔmiR-K12-11 KSHV mutants.

The 3'UTRs of MYB, Ets-1, and C/EBPα were separately cloned downstream of the firefly luciferase reporter, and assays performed to test directly for miR-K12-3 and/or miR-K12-11 targeting of the resulting transcripts. The luciferase reporter plasmid DNAs were transfected with either a miR-K12-3 or miR-K12-11 expression vector into 293 cells. Transfection with increasing amounts of miR-K12-11 expression vector gave progressively reduced luciferase activity from the luciferase-MYB 3'UTR construct (Figure 3D), indicating that miR-K12-11 targets the MYB 3'UTR. Despite the presence of two predicted binding sites in the C/EBPα 3'UTR for miR-K12-3, miR-K12-3 expression alone did not cause significant reduction in luciferase expression from the C/EBPα 3'UTR construct (Figure 3E). The pcDNA3.1/cluster expression plasmid, which expresses 10 KSHV miRNAs [42], was also tested for effects on the C/EBPα 3'UTR, and significantly reduced luciferase expression (Figure 3E). Bioinformatic prediction revealed that the C/EBPα 3'UTR contains additional seed sequence matches for miR-K12-8 and -9 (data not shown). The 3.6 kb long 3'UTR of Ets-1 has a number of putative binding sites for KSHV miRNAs (miR-K12-1, -6, and -7) in addition to the sites for miR-K12-3 and miR-K12-11. The luciferase-Ets-1 3'UTR construct was cotransfected with plasmids expressing miR-K12-3 and miR-K12-11 alone and in combination, and with pcDNA3.1/cluster. Expression of miR-K12-3 alone had no significant effect, but miR-K12-11 alone and miR-K12-11 in combination with miR-K12-3 led to significant repression (Figure 3F). Expression of the 10 cluster miRNAs also gave reduced luciferase expression. These results suggest that Ets-1 is targeted by multiple KSHV miRNAs. In summary, these *in vitro* data show that all three transcription factors can be targeted by KSHV miRNAs.

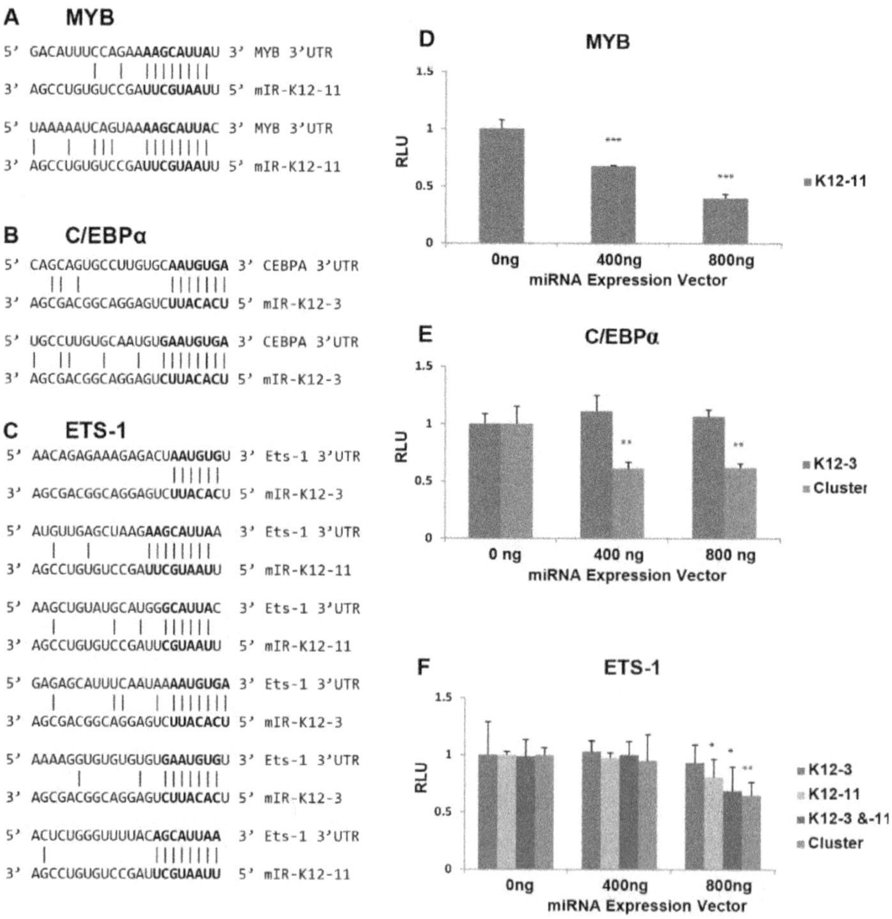

Figure 3. Effect of miRNAs on luciferase expression from 3'UTR reporter constructs. (**A–C**) Potential binding sites for miR-K12-3 and mIR-K12-11 in the 3'UTRs of the cellular MYB (**A**); C/EBPα (**B**) and Ets-1 (**C**) genes; (**D–F**) Dual luciferase reporter assays to assess targeting of the MYB (**D**); C/EBPα (**E**); and Ets-1 (**F**) 3'UTRs by KSHV miRNAs. 293 cells were transfected with pGL3-derived reporter plasmids carrying the MYB, C/EBPα or Ets-1 3'UTR downstream of the firefly luciferase together with the indicated amount of miRNA vectors expressing miR-K12-11 and/or miR-K12-3 or the 10 cluster miRNAs. Cells were harvested at 72 h post-transfection and the quantity of firefly luciferase measured by luminescence. Expression values were normalized to renilla luciferase expression from the pCMV-Renilla vector.

3.5. MYB, C/EBPα and Ets-1 Expression Is Increased upon miRNA Knockdown in PEL Cells

To investigate whether MYB, C/EBPα, and Ets-1 are regulated by KSHV miRNAs in latently infected PEL cells, BCBL-1 and BC-3 cells were transfected with miR-K12-11 or miR-K12-3 antagomirs. RT-qPCR performed 48 h after antagomir transfection revealed increased levels of MYB transcript upon miR-K12-11 knockdown with dose dependence (Figure 4A). MYB transcript was increased 1.5-fold in BCBL-1 cells compared to mock-transfected control in the presence of 200 nM miR-K12-11 antagomir. For C/EBPα, transcript levels were increased approximately two-fold in BCBL-1 cells on addition of miR-K12-3 antagomir (Figure 4B). In BC-3 cells, where miR-K12-3 is the most highly expressed KSHV miRNA [44,54], knockdown of miR-K12-3 gave increases of up to 2-fold in Ets-1 transcript levels in a dose-dependent fashion (Figure 4C). Addition of 200 nM miR-K12-3 and 200 nM miR-K12-11 antagomirs together gave a higher Ets-1 level than 200 nM miR-K12-3 antagomir alone.

These data suggest that expression of KSHV miR-K12-11 in latent infection normally reduces the level of MYB transcript, and miR-K12-3 behaves similarly against C/EBPα and Ets-1. Our results point to a mechanism in which miR-K12-3 and miR-K12-11 down-regulate RTA expression by modulating cellular transcription factors in cells of both endothelial and lymphoid origin.

4. Discussion

We used BC-3-G cells containing GFP driven by the RTA-responsive PAN promoter to ask whether antagomir-based inhibition of KSHV miRNAs activates RTA. Inhibition of single or multiple miRNAs revealed that in this experimental system only antagomirs against miR-K12-3 and mir-K12-11 led to significantly increased RTA expression levels. We note that miR-K12-3 and miR-K12-11 expression levels are higher in BC-3 cells compared to BCBL-1 cells used in previous screens [43,44]. Antagomir inhibition of miR-K12-7 and miR-K12-5 did not activate RTA expression in BC-3-G cells, as was observed in 293 cells infected with Bac36-derived KSHV [26,27]. Additionally, miR-K12-9 * is not expressed at all in BC-3 cells due to the presence of a highly polymorphic miR-K12-9 pre-miRNA [55,56], and therefore RTA inhibition by miR-K12-9 * [57] would not be expected in this cell line. We note that recent ribonomics reports on KSHV miRNA targetomes using PAR-CLIP and HITS-CLIP [43,44] did not reveal direct targeting of RTA by KSHV miRNAs in two different PEL cell lines including BC-3 during latency. In contrast, targeting of host BCLAF1 by mIR-K12-5 has been reported to sensitize cells for reactivation [23], and BCLAF1 was validated as a viral miRNA target in BCBL-1 cells by PAR-CLIP and HITS-CLIP [43,44].

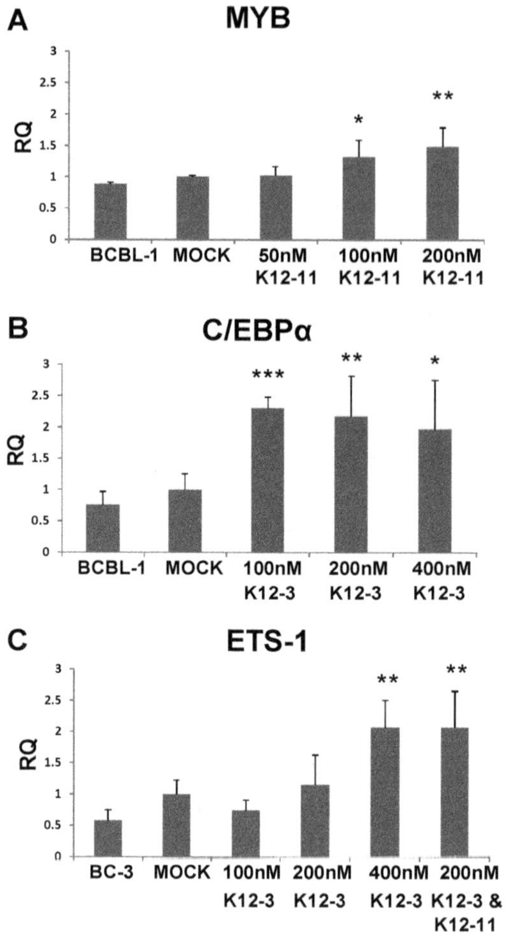

Figure 4. MYB, C/EBPα and Ets-1 expression after miR-K12-11 and miR-K12-3 knockdown in PEL cells. BCBL-1 and BC-3 cells were transfected with the indicated amounts of miRNA-specific antagomir. Total RNA was collected at 48 h, reverse transcribed, and MYB (**A**); C/EBPα (**B**); and Ets-1 (**C**) transcript levels determined by qPCR. All samples were normalized to β-actin expression and compared to gene expression in the mock transfected control. $p < 0.05$ (*), $p < 0.01$ (**), $p < 0.001$ (***).

Since neither the 3'UTR of RTA nor its coding region contains any mir-K12-3 or miR-K12-11 seed sequence matches, mir-K12-3 and miR-K12-11 apparently regulate RTA indirectly by targeting cellular genes that positively regulate RTA. A number of cellular transcription factors have been identified to activate the RTA promoter including MYB, Ets-1, and C/EBPα [28–30]. Lacoste and colleagues demonstrated that MYB transactivates the RTA promoter in the absence of any KSHV protein expression and furthermore that v-FLIP and v-GPCR induction of NF-κB leads to downregulation of MYB expression in PEL cells [29]. Our data indicate that miR-K12-11 directly targets the MYB 3'UTR (Figure 3) and that knockdown of miR-K12-11 increases MYB transcript level (Figure 4). Thus, miR-K12-11 fine tunes MYB expression post-transcriptionally in conjunction with virally encoded proteins that regulate MYB through NF-κB signaling. Interestingly, the NF-κB pathway or more specifically the IκBα super repressor, was shown to be regulated by miR-K12-1, thereby also inhibiting lytic growth [35].

Wang and colleagues observed that C/EBPα activates RTA directly and mapped three C/EBPα binding sites within the RTA promoter [30]. In addition, C/EBPα functions during lytic reactivation by interacting with RTA to bind and activate the K8 promoter, an early replication-associated protein. K8 then interacts with and stabilizes C/EBPα leading to upregulation of C/EBPα, in turn leading to further activation of RTA and K8 gene expression. It was also shown that C/EBPα activates additional early lytic genes PAN and ORF57 [30]. Even though miR-K12-3 does not appear to target C/EBPα directly by itself in 293 cells (Figure 3E), antagomir knockdown of miR-K12-3 in BCBL-1 cells does increase C/EBPα mRNA (Figure 4). Therefore, miR-K12-3 contributing to downregulation of C/EBPα represents targeting a central regulatory node to negatively modulate several immediate early and early genes.

Ets-1 was originally identified as an activator of RTA by utilizing BC-3-G cells in a genome-wide screen to identify cellular proteins and pathways that reactivate KSHV [28]. The Raf/MEK/ERK pathway was demonstrated to mediate KSHV reactivation and Ets-1, which is a downstream effector of this pathway, directly activates the RTA promoter. Ets-1 contains five putative KSHV miRNA binding sites in addition to a total of six seed sequence matches for miR-K12-3 and miR-K12-11. KSHV miRNA targeting of Ets-1 presumably contributes to maintenance of latency by inhibiting Raf/MEK/ERK induced activation of RTA. Ets-1 targeting by miRK-12-3 and -11 might be cell type specific since KSHV v-FLIP upregulates Ets-1 in a NF-κB-dependent fashion and moreover this is required for KSHV-dependent induction of lymphatic reprogramming of endothelial cells [58]. Hence, fine tuning of Ets-1 may be important to balance differentiation and reactivation. We recently demonstrated Ets-1 targeting in TIVE cells engineered to express miR-K12-11 in the absence of KSHV [59].

A common feature of these three transcription factors is their involvement in signaling pathways as part of cellular stress responses. The RTA promoter of γ-herpesviruses contains an array of transcription factor binding sites thereby sensing environmental changes and linking cell stress including innate and adapted immune responses against other pathogens to lytic reactivation [60,61]. Our data suggest a model where viral miRNAs contribute to the regulation of the latent to lytic transition by post-transcriptionally modulating multiple signaling pathways. We believe that KSHV miRNAs fine-tune the regulation of MYB, C/EBPα, and Ets-1, which increases the signaling threshold that has to be overcome before the lytic cascade can be initiated. Some of these cellular targets are regulated by other latency-associated genes, as in the case of MYB and NF-κB in lymphoid cells [29] and Ets-1 and NF-κB in endothelial cells [58,59]. Rather than acting like an on/off switch, viral miRNAs serve as gatekeepers of latency by manipulating multiple host cellular pathways that when activated would otherwise lead to RTA expression. In contrast to directly regulating RTA, targeting multiple cellular genes allows for greater flexibility in regulation of latency in different tissues and cell types where both the viral miRNAs and their cognate targets are expressed differentially [14,15,32,62]. For example, Ets-1 levels are higher in endothelial cells while MYB is a master regulator of hematopoietic cells.

We propose that the regulation of host transcription factors by KSHV miRNAs contributes to the ability to quickly respond to environmental stimuli, which can overcome the intricate balance between miRNA copy number and their cognate targets [63,64]. A similar miRNA-based "spring-loaded" regulatory model was first proposed for OCT and SOX transcription factors that regulate pluripotency during differentiation [65]. In conclusion, our results suggest that miR-K12-3 and miR-K12-11 contribute to regulation of latency in cells of endothelial and lymphoid origin by targeting the host cellular transcription factors MYB, C/EBPα and Ets-1, thereby indirectly manipulating RTA expression.

Acknowledgments

This work was supported by grants from National Institutes of Health (CA88763, CA 119917 and RC2 CA148407) to RR. KP-B was supported by T32 CA009126 and IWB was supported by T32AI060527 and T32AI007110. We thank Jae Jung, Don Ganem, and J. Myoung for their generosity in providing BAC16 and iSLK cells, and Ren Sun for kindly providing BC-3-G cells. Finally, we thank Peter Turner for carefully editing the manuscript.

Author Contributions

R.R. designed the experiments; K.P.-H., H.S.C., T.B., I.W.B., and L.A.G. performed the experiments. B.J.K. constructed the KSHV recombinants and optimized virus recovery in iSLK cells. I.H. provided bioinformatic support. K.P.-H., R.R., and H.S.C. wrote the manuscript.

References and Notes

1. Chang, Y.; Cesarman, E.; Pessin, M.S.; Lee, F.; Culpepper, J.; Knowles, D.M.; Moore, P.S. Identification of herpesvirus-like DNA sequences in AIDS-associated Kaposi's sarcoma (see comments). *Science* **1994**, *266*, 1865–1869.
2. Cesarman, E.; Chang, Y.; Moore, P.S.; Said, J.W.; Knowles, D.M. Kaposi's sarcoma-associated herpesvirus-like DNA sequences in AIDS-related body-cavity-based lymphomas. *N. Engl. J. Med.* **1995**, *332*, 1186–1191.
3. Soulier, J.; Grollet, L.; Oksenhendler, E.; Cacoub, P.; Cazals-Hatem, D.; Babinet, P.; d'Agay, M.F.; Clauvel, J.P.; Raphael, M.; Degos, L.; *et al.* Kaposi's sarcoma-associated herpesvirus-like DNA sequences in multicentric Castleman's disease. *Blood* **1995**, *86*, 1276–1280.
4. Dittmer, D.; Lagunoff, M.; Renne, R.; Staskus, K.; Haase, A.; Ganem, D. A cluster of latently expressed genes in Kaposi's sarcoma-associated herpesvirus. *J. Virol.* **1998**, *72*, 8309–8315.
5. Sarid, R.; Flore, O.; Bohenzky, R.A.; Chang, Y.; Moore, P.S. Transcription mapping of the Kaposi's sarcoma-associated herpesvirus (human herpesvirus 8) genome in a body cavity-based lymphoma cell line (BC-1). *J. Virol.* **1998**, *72*, 1005–1012.
6. Talbot, S.J.; Weiss, R.A.; Kellam, P.; Boshoff, C. Transcriptional analysis of human herpesvirus-8 open reading frames 71, 72, 73, K14, and 74 in a primary effusion lymphoma cell line. *Virology* **1999**, *257*, 84–94.
7. Rivas, C.; Thlick, A.E.; Parravicini, C.; Moore, P.S.; Chang, Y. Kaposi's sarcoma-associated herpesvirus LANA2 is a B-cell-specific latent viral protein that inhibits p53. *J. Virol.* **2001**, *75*, 429–438.

8. Chandriani, S.; Ganem, D. Array-based transcript profiling and limiting-dilution reverse transcription-PCR analysis identify additional latent genes in Kaposi's sarcoma-associated herpesvirus. *J. Virol.* **2010**, *84*, 5565–5573.
9. Pfeffer, S.; Sewer, A.; Lagos-Quintana, M.; Sheridan, R.; Sander, C.; Grasser, F.A.; van Dyk, L.F.; Ho, C.K.; Shuman, S.; Chien, M.; *et al*. Identification of microRNAs of the herpesvirus family. *Nat. Methods* **2005**, *2*, 269–276.
10. Cai, X.; Lu, S.; Zhang, Z.; Gonzalez, C.M.; Damania, B.; Cullen, B.R. Kaposi's sarcoma-associated herpesvirus expresses an array of viral microRNAs in latently infected cells. *Proc. Natl. Acad. Sci. USA* **2005**, *102*, 5570–5575.
11. Samols, M.A.; Hu, J.; Skalsky, R.L.; Renne, R. Cloning and identification of a microRNA cluster within the latency-associated region of Kaposi's sarcoma-associated herpesvirus. *J. Virol.* **2005**, *79*, 9301–9305.
12. Grundhoff, A.; Sullivan, C.S.; Ganem, D. A combined computational and microarray-based approach identifies novel microRNAs encoded by human gamma-herpesviruses. *RNA* **2006**, *12*, 733–750.
13. Bartel, D.P. MicroRNAs: Genomics, biogenesis, mechanism, and function. *Cell* **2004**, *116*, 281–297.
14. O'Hara, A.J.; Vahrson, W.; Dittmer, D.P. Gene alteration and precursor and mature microRNA transcription changes contribute to the miRNA signature of primary effusion lymphoma. *Blood* **2008**, *111*, 2347–2353.
15. O'Hara, A.J.; Chugh, P.; Wang, L.; Netto, E.M.; Luz, E.; Harrington, W.J.; Dezube, B.J.; Damania, B.; Dittmer, D.P. Pre-micro RNA signatures delineate stages of endothelial cell transformation in Kaposi sarcoma. *PLoS Pathog.* **2009**, *5*, e1000389.
16. Moody, R.; Zhu, Y.; Huang, Y.; Cui, X.; Jones, T.; Bedolla, R.; Lei, X.; Bai, Z.; Gao, S.J. KSHV microRNAs mediate cellular transformation and tumorigenesis by redundantly targeting cell growth and survival pathways. *PLoS Pathog.* **2013**, *9*, e1003857.
17. Plaisance-Bonstaff, K.; Renne, R. Viral miRNAs. *Methods Mol. Biol.* **2011**, *721*, 43–66.
18. Skalsky, R.L.; Cullen, B.R. Viruses, microRNAs, and host interactions. *Annu. Rev. Microbiol.* **2010**, *64*, 123–141.
19. Zhu, Y.; Haecker, I.; Yang, Y.; Gao, S.J.; Renne, R. gamma-Herpesvirus-encoded miRNAs and their roles in viral biology and pathogenesis. *Curr. Opin. Virol.* **2013**, *3*, 266–275.
20. Murphy, E.; Vanicek, J.; Robins, H.; Shenk, T.; Levine, A.J. Suppression of immediate-early viral gene expression by herpesvirus-coded microRNAs: Implications for latency. *Proc. Natl. Acad. Sci. USA* **2008**, *105*, 5453–5458.
21. Pfeffer, S.; Zavolan, M.; Grasser, F.A.; Chien, M.; Russo, J.J.; Ju, J.; John, B.; Enright, A.J.; Marks, D.; Sander, C.; Tuschl, T. Identification of virus-encoded microRNAs. *Science* **2004**, *304*, 734–736.
22. Barth, S.; Pfuhl, T.; Mamiani, A.; Ehses, C.; Roemer, K.; Kremmer, E.; Jaker, C.; Hock, J.; Meister, G.; Grasser, F.A. Epstein-Barr virus-encoded microRNA miR-BART2 down-regulates the viral DNA polymerase BALF5. *Nucleic Acids Res.* **2008**, *36*, 666–675.
23. Ziegelbauer, J.M.; Sullivan, C.S.; Ganem, D. Tandem array-based expression screens identify host mRNA targets of virus-encoded microRNAs. *Nat. Genet.* **2009**, *41*, 130–134.

24. Sun, R.; Lin, S.F.; Gradoville, L.; Yuan, Y.; Zhu, F.; Miller, G. A viral gene that activates lytic cycle expression of Kaposi's sarcoma- associated herpesvirus. *Proc. Natl. Acad. Sci. USA* **1998**, *95*, 10866–10871.
25. Lukac, D.M.; Renne, R.; Kirshner, J.R.; Ganem, D. Reactivation of Kaposi's sarcoma-associated herpesvirus infection from latency by expression of the ORF 50 transactivator, a homolog of the EBV R protein. *Virology* **1998**, *252*, 304–312.
26. Lu, F.; Stedman, W.; Yousef, M.; Renne, R.; Lieberman, P.M. Epigenetic regulation of Kaposi's sarcoma-associated herpesvirus latency by virus-encoded microRNAs that target Rta and the cellular Rbl2-DNMT pathway. *J. Virol.* **2010**, *84*, 2697–2706.
27. Lin, X.; Liang, D.; He, Z.; Deng, Q.; Robertson, E.S.; Lan, K. miR-K12–7-5p encoded by Kaposi's sarcoma-associated herpesvirus stabilizes the latent state by targeting viral ORF50/RTA. *PLoS One* **2011**, *6*, e16224.
28. Yu, F.; Harada, J.N.; Brown, H.J.; Deng, H.; Song, M.J.; Wu, T.T.; Kato-Stankiewicz, J.; Nelson, C.G.; Vieira, J.; Tamanoi, F.; *et al.* Systematic identification of cellular signals reactivating Kaposi sarcoma-associated herpesvirus. *PLoS Pathog.* **2007**, *3*, e44.
29. Lacoste, V.; Nicot, C.; Gessain, A.; Valensi, F.; Gabarre, J.; Matta, H.; Chaudhary, P.M.; Mahieux, R. In primary effusion lymphoma cells, MYB transcriptional repression is associated with v-FLIP expression during latent KSHV infection while both v-FLIP and v-GPCR become involved during the lytic cycle. *Br. J. Haematol.* **2007**, *138*, 487–501.
30. Wang, S.E.; Wu, F.Y.; Yu, Y.; Hayward, G.S. CCAAT/enhancer-binding protein-alpha is induced during the early stages of Kaposi's sarcoma-associated herpesvirus (KSHV) lytic cycle reactivation and together with the KSHV replication and transcription activator (RTA) cooperatively stimulates the viral RTA, MTA, and PAN promoters. *J. Virol.* **2003**, *77*, 9590–9612.
31. Myoung, J.; Ganem, D. Generation of a doxycycline-inducible KSHV producer cell line of endothelial origin: Maintenance of tight latency with efficient reactivation upon induction. *J. Virol. Methods* **2011**, *174*, 12–21.
32. An, F.Q.; Folarin, H.M.; Compitello, N.; Roth, J.; Gerson, S.L.; McCrae, K.R.; Fakhari, F.D.; Dittmer, D.P.; Renne, R. Long-term-infected telomerase-immortalized endothelial cells: A model for Kaposi's sarcoma-associated herpesvirus latency *in vitro* and *in vivo*. *J. Virol.* **2006**, *80*, 4833–4846.
33. Skalsky, R.L.; Samols, M.A.; Plaisance, K.B.; Boss, I.W.; Riva, A.; Lopez, M.C.; Baker, H.V.; Renne, R. Kaposi's sarcoma-associated herpesvirus encodes an ortholog of miR-155. *J. Virol.* **2007**, *81*, 12836–12845.
34. Boss, I.W.; Nadeau, P.E.; Abbott, J.R.; Yang, Y.; Mergia, A.; Renne, R. A KSHV encoded ortholog of miR-155 induces human splenic B-cell expansion in NOD/LtSz-scid IL2R{gamma}null mice. *J. Virol.* **2011**, *85*, 9877–9886.
35. Lei, X.; Bai, Z.; Ye, F.; Xie, J.; Kim, C.G.; Huang, Y.; Gao, S.J. Regulation of NF-kappaB inhibitor IkappaBalpha and viral replication by a KSHV microRNA. *Nat. Cell Biol.* **2010**, *12*, 193–199.
36. Majerciak, V.; Uranishi, H.; Kruhlak, M.; Pilkington, G.R.; Massimelli, M.J.; Bear, J.; Pavlakis, G.N.; Felber, B.K.; Zheng, Z.M. Kaposi's sarcoma-associated herpesvirus ORF57 interacts with cellular RNA export cofactors RBM15 and OTT3 to promote expression of viral ORF59. *J. Virol.* **2011**, *85*, 1528–1540.

37. Persson, L.M.; Wilson, A.C. Wide-scale use of Notch signaling factor CSL/RBP-Jkappa in RTA-mediated activation of Kaposi's sarcoma-associated herpesvirus lytic genes. *J. Virol.* **2010**, *84*, 1334–1347.
38. Brulois, K.; Toth, Z.; Wong, L.-Y.; Feng, P.; Gao, S.-J.; Ensser, A.; Jung, J.U. Kaposi's Sarcoma-Associated Herpesvirus K3 and K5 Ubiquitin E3 Ligases Have Stage-Specific Immune Evasion Roles during Lytic Replication. *J. Virol.* **2014**, *88*, 9335–9349.
39. Brulois, K.F.; Chang, H.; Lee, A.S.; Ensser, A.; Wong, L.Y.; Toth, Z.; Lee, S.H.; Lee, H.R.; Myoung, J.; Ganem, D.; *et al.* Construction and manipulation of a new Kaposi's sarcoma-associated herpesvirus bacterial artificial chromosome clone. *J. Virol.* **2012**, *86*, 9708–9720.
40. Tischer, B.K.; von Einem, J.; Kaufer, B.; Osterrieder, N. Two-step red-mediated recombination for versatile high-efficiency markerless DNA manipulation in Escherichia coli. *Biotechniques* **2006**, *40*, 191–197.
41. Myoung, J.; Ganem, D. Infection of lymphoblastoid cell lines by Kaposi's sarcoma-associated herpesvirus: Critical role of cell-associated virus. *J. Virol.* **2011**, *85*, 9767–9777.
42. Samols, M.A.; Hu, J.; Skalsky, R.L.; Maldonado, A.M.; Riva, A.; Lopez, M.C.; Baker, H.V.; Renne, R. Identification of cellular genes targeted by KSHV-encoded microRNAs. *PLoS Pathog.* **2007**, *3*, e65.
43. Gottwein, E.; Corcoran, D.L.; Mukherjee, N.; Skalsky, R.L.; Hafner, M.; Nusbaum, J.D.; Shamulailatpam, P.; Love, C.L.; Dave, S.S.; Tuschl, T.; *et al.* Viral microRNA targetome of KSHV-infected primary effusion lymphoma cell lines. *Cell Host Microbe* **2011**, *10*, 515–526.
44. Haecker, I.; Gay, L.A.; Yang, Y.; Hu, J.; Morse, A.M.; McIntyre, L.; Renne, R. Ago-HITS-CLIP Expands Understanding of Kaposi's Sarcoma-associated Herpesvirus miRNA Function in Primary Effusion Lymphomas. *PLoS Pathog.* **2012**, *8*, e1002884, doi:10.1371/journal.ppat.1002884.
45. Gottwein, E.; Mukherjee, N.; Sachse, C.; Frenzel, C.; Majoros, W.H.; Chi, J.T.; Braich, R.; Manoharan, M.; Soutschek, J.; Ohler, U.; *et al.* A viral microRNA functions as an orthologue of cellular miR-155. *Nature* **2007**, *450*, 1096–1099.
46. Krueger, B.; Plaisance, K.; Sangani, R.; Lanier, C.; Jain, V.; Hu, J.; Renne, R. A core laboratory for the generation of quality-controlled g-herpesvirus bacmids: Generation of KSHV miRNA mutants. *Infect. Agents Cancer* **2012**, *7*, 27.
47. Krueger, B.; Jain, V.; Plaisance-Bonstaff, K.; Renne, R. Generation of quality-controlled γ-herpesvirus bacmids to create KSHV miRNA mutants. University of Florida, Gainesville, FL, USA. Unpublished work, 2014.
48. Sturzl, M.; Gaus, D.; Dirks, W.G.; Ganem, D.; Jochmann, R. Kaposi's sarcoma-derived cell line SLK is not of endothelial origin, but is a contaminant from a known renal carcinoma cell line. *Int. J. Cancer* **2013**, *132*, 1954–1958.
49. Grimson, A.; Farh, K.K.; Johnston, W.K.; Garrett-Engele, P.; Lim, L.P.; Bartel, D.P. MicroRNA targeting specificity in mammals: Determinants beyond seed pairing. *Mol. Cell* **2007**, *27*, 91–105.
50. Ramsay, R.G.; Gonda, T.J. MYB function in normal and cancer cells. *Nat. Rev. Cancer* **2008**, *8*, 523–534.
51. Koschmieder, S.; Halmos, B.; Levantini, E.; Tenen, D.G. Dysregulation of the C/EBPalpha differentiation pathway in human cancer. *J. Clin. Oncol.* **2009**, *27*, 619–628.

52. Wasylyk, B.; Hagman, J.; Gutierrez-Hartmann, A. Ets transcription factors: Nuclear effectors of the Ras-MAP-kinase signaling pathway. *Trends Biochem. Sci.* **1998**, *23*, 213–216.
53. Dejana, E.; Taddei, A.; Randi, A.M. Foxs and Ets in the transcriptional regulation of endothelial cell differentiation and angiogenesis. *Biochim. Biophys. Acta* **2007**, *1775*, 298–312.
54. Gottwein, E.; Cai, X.; Cullen, B.R. A novel assay for viral microRNA function identifies a single nucleotide polymorphism that affects Drosha processing. *J. Virol.* **2006**, *80*, 5321–5326.
55. Marshall, V.; Parks, T.; Bagni, R.; Wang, C.D.; Samols, M.A.; Hu, J.; Wyvil, K.M.; Aleman, K.; Little, R.F.; Yarchoan, R.; *et al.* Conservation of virally encoded microRNAs in Kaposi sarcoma—Associated herpesvirus in primary effusion lymphoma cell lines and in patients with Kaposi sarcoma or multicentric Castleman disease. *J. Infect. Dis.* **2007**, *195*, 645–659.
56. Han, S.J.; Marshall, V.; Barsov, E.; Quinones, O.; Ray, A.; Labo, N.; Trivett, M.; Ott, D.; Renne, R.; Whitby, D. Kaposi's sarcoma-associated herpesvirus microRNA single-nucleotide polymorphisms identified in clinical samples can affect microRNA processing, level of expression, and silencing activity. *J. Virol.* **2013**, *87*, 12237–12248.
57. Bellare, P.; Ganem, D. Regulation of KSHV lytic switch protein expression by a virus-encoded microRNA: An evolutionary adaptation that fine-tunes lytic reactivation. *Cell Host Microbe* **2009**, *6*, 570–575.
58. Hung, T.; Wang, Y.; Lin, M.F.; Koegel, A.K.; Kotake, Y.; Grant, G.D.; Horlings, H.M.; Shah, N.; Umbricht, C.; Wang, P.; *et al.* Extensive and coordinated transcription of noncoding RNAs within cell-cycle promoters. *Nat. Genet.* **2011**, *43*, 621–629.
59. Yang, Y.; Boss, I.; McIntyre, L.; Renne, R. A systems biology approach identified different regulatory networks targeted by KSHV miR-K12-11 in B cells and endothelial cells. *BMC Genomics* **2014**, *15*, 668.
60. Reese, T.A.; Wakeman, B.S.; Choi, H.S.; Hufford, M.M.; Huang, S.C.; Zhang, X.; Buck, M.D.; Jezewski, A.; Kambal, A.; Liu, C.Y.; *et al.* Helminth infection reactivates latent gamma-herpesvirus via cytokine competition at a viral promoter. *Science* **2014**, *345*, 573–577.
61. Speck, S.H.; Ganem, D. Viral latency and its regulation: Lessons from the gamma-herpesviruses. *Cell Host Microbe* **2010**, *8*, 100–115.
62. An, F.Q.; Compitello, N.; Horwitz, E.; Sramkoski, M.; Knudsen, E.S.; Renne, R. The latency-associated nuclear antigen of Kaposi's sarcoma-associated herpesvirus modulates cellular gene expression and protects lymphoid cells from p16 INK4A-induced cell cycle arrest. *J. Biol. Chem.* **2005**, *280*, 3862–3874.
63. Baccarini, A.; Chauhan, H.; Gardner, T.J.; Jayaprakash, A.D.; Sachidanandam, R.; Brown, B.D. Kinetic analysis reveals the fate of a microRNA following target regulation in mammalian cells. *Curr. Biol.* **2011**, *21*, 369–376.
64. Mukherji, S.; Ebert, M.S.; Zheng, G.X.; Tsang, J.S.; Sharp, P.A.; van Oudenaarden, A. MicroRNAs can generate thresholds in target gene expression. *Nat. Genet.* **2011**, *43*, 854–859.
65. Marson, A.; Levine, S.S.; Cole, M.F.; Frampton, G.M.; Brambrink, T.; Johnstone, S.; Guenther, M.G.; Johnston, W.K.; Wernig, M.; Newman, J.; *et al.* Connecting microRNA genes to the core transcriptional regulatory circuitry of embryonic stem cells. *Cell* **2008**, *134*, 521–533.

6

KSHV Targeted Therapy: An Update on Inhibitors of Viral Lytic Replication

Natacha Coen, Sophie Duraffour, Robert Snoeck and Graciela Andrei *

Rega Institute for Medical Research, KU Leuven, B-3000 Leuven, Belgium;
E-Mails: natacha.coen@rega.kuleuven.be (N.C.); sophie.duraffour@rega.kuleuven.be (S.D.); robert.snoeck@rega.kuleuven.be (R.S.)

* Author to whom correspondence should be addressed; E-Mail: graciela.andrei@rega.kuleuven.be;

External Editor: Zhi-Ming Zheng

Abstract: Kaposi's sarcoma-associated herpesvirus (KSHV) is the causative agent of Kaposi's sarcoma, primary effusion lymphoma and multicentric Castleman's disease. Since the discovery of KSHV 20 years ago, there is still no standard treatment and the management of virus-associated malignancies remains toxic and incompletely efficacious. As the majority of tumor cells are latently infected with KSHV, currently marketed antivirals that target the virus lytic cycle have shown inconsistent results in clinic. Nevertheless, lytic replication plays a major role in disease progression and virus dissemination. Case reports and retrospective studies have pointed out the benefit of antiviral therapy in the treatment and prevention of KSHV-associated diseases. As a consequence, potent and selective antivirals are needed. This review focuses on the anti-KSHV activity, mode of action and current status of antiviral drugs targeting KSHV lytic cycle. Among these drugs, different subclasses of viral DNA polymerase inhibitors and compounds that do not target the viral DNA polymerase are being discussed. We also cover molecules that target cellular kinases, as well as the potential of new drug targets and animal models for antiviral testing.

Keywords: KSHV; antiviral; nucleoside analog; DNA polymerase inhibitors; lytic cycle; ganciclovir; cidofovir; foscarnet

---

1. Introduction

In 1872, Moritz Kaposi described a rare angiosarcoma that manifested mainly as skin lesions in elderly men [1]. More than a century passed between the first description of Kaposi's sarcoma (KS) and the discovery of its etiologic agent, Kaposi's sarcoma-associated herpesvirus (KSHV), by Chang and Moore in 1994 [2]. KS is a neoplasm derived from lymphatic endothelial cells infected with KSHV, composed of spindle-shaped cells and inflammatory mononuclear cells [3]. KS is grouped into four epidemiological forms: classic, endemic, iatrogenic and AIDS-related [4]. In addition, KSHV has been associated with two other diseases, primary effusion lymphoma (PEL) and multicentric Castleman's disease (MCD) [5]. PEL is a B-cell lymphoma that develops in pleural, pericardial or peritoneal cavity, while the B-cell lymphoproliferative disorder MCD is predominantly found in the lymph nodes and is characterized by vascular proliferation in the germinal centers [6,7].

2. Management of KSHV-Associated Diseases

There are no standard therapeutic guidelines for the management of KSHV-associated diseases, yet the main therapeutic options are discussed below for each disease. The treatment of choice for patients with KS depends on several parameters, such as the tumor location and variant of KS, rate of progression, distribution of the lesions, severity of the symptoms, and immune competence [8]. Therapeutic approaches for classic KS range from no treatment to surgical excision, local delivery of chemotherapeutic agents (such as bleomycin, vinblastine, vincristine and alitretinoin), and radiotherapy (Table 1) [9]. Management of iatrogenic KS often involves reduction or elimination of immunosuppressive therapy with or without local measures, whereas endemic KS is frequently responsive to systemic chemotherapy [10,11].

The current first-line systemic therapy for advanced, progressive acquired immunodeficiency syndrome (AIDS)-KS includes liposomal anthracyclines, such as daunorubicin and doxorubicin [12,13]. An essential component in the management of human immunodeficiency virus (HIV)-associated KS is the control of KS progression with highly active antiretroviral therapy (HAART), which leads to both immune reconstitution and control of HIV viremia [12,14,15]. Additional data from HIV cohorts also suggested that specific components of HAART might impact the incidence and resolution of KS [16]. Several randomized, placebo-controlled trials of high-dose zidovudine for the treatment of HIV demonstrated a 36% reduction in risk of developing KS compared with persons receiving placebo alone [17]. To date, there are no comprehensive studies conducted to evaluate whether HAART is able to inhibit KSHV viral production [18], but it has been shown that zidovudine is a substrate for KSHV thymidine kinase (TK) [19].

In addition, recent research has shown that HIV protease inhibitors, e.g., nelfinavir, have anti-angiogenic and anti-tumor properties [20]. Therefore, HAART combinations that contain HIV protease inhibitors may be superior for treatment of KS patients than those without [21]. Moreover, anti-herpetic agents, such as ganciclovir (GCV), were shown to reduce plasma viral load of KSHV and can prevent KS in KSHV-seropositive transplant recipients [9]. Additionally, target-based therapies, such as inhibition of angiogenesis, metalloproteinases, and cytokine signaling, may be an effective strategy to treat patients with KS that progresses despite chemotherapy and/or HAART [22].

PEL has usually been treated with chemotherapy (Table 1), but the prognosis is very poor in patients with a median survival of less than six months [23]. Individual case reports documented responses to antiviral therapy (GCV, foscarnet (PFA), intracavity cidofovir (CDV, HPMPC)), the proteosome inhibitor bortezomib, the immunosuppressive agent rapamycin, the monoclonal antibody rituximab (which targets the CD20 protein on the surface of B lymphocytes), and the antitumor antibiotic drug bleomycin [24–26].

In MCD patients, KSHV induces both human IL-6 and virus-encoded IL-6, and, therefore, treatment with tocilizumab, an anti-human IL-6 receptor antibody, has led to clinical responses in these patients [27]. Recently, siltuximab, a chimeric monoclonal antibody against IL-6, has been developed for the treatment of MCD patients showing promising results in a phase I clinical trial [28]. Rituximab therapy has been evaluated for the treatment of MCD and up to 70% of patients responded to the therapy (Table 1) [24,29,30]. In addition, antiviral therapy with GCV has been reported successful in MCD patients, since this disease is associated with active KSHV replication [31].

Table 1. Treatment modalities of KSHV-related diseases.

Treatment		KSHV-related Diseases
Intensification of HAART		AIDS-KS
Surgical excision		KS (single skin lesion)
Radiotherapy		KS
Immunotherapy	Reduction of immunosuppressive therapy	KS, PEL and MCD
	Anti-CD20 (Rituximab) Anti-human IL-6 receptor (Tocilizumab) Anti-IL6 chimeric monoclonal antibody (Siltuximab)	MCD
Chemotherapy	Liposomal anthracyclines	KS
	CHOP (cyclophosphamide, doxorubicin, vincristine, prednisone)	PEL and MCD
Antiviral drugs	(Val)ganciclovir, foscarnet	KS, PEL and MCD
	Intracavity cidofovir	PEL
Others	mTOR inhibitor (Rapamycin) Proteasome inhibitor (Bortezomib)	KS, PEL PEL
	Paclitaxel, anti-angiogenic agents, matrix metalloproteinase inhibitors	KS

3. Antiviral Therapy for the Treatment and Prevention of KSHV-Related Malignancies

Inhibition of KSHV lytic phase by antiviral drugs has not shown great efficacy for the treatment of KS, primarily due to the small amount of lytic KSHV present in KS tumors [32]. Though, the few cells showing lytic replication are known to play a central role in KS tumorigenesis [33]. However, a greater proportion of infected cells in PEL and MCD express lytic phase genes (up to 25% in MCD), as compared to KS, and, therefore, antiviral agents might be more effective in the treatment of MCD and PEL, than of KS [16,33,34].

The use of anti-herpes drugs in the protection against the development of AIDS-associated KS has been evaluated in a few studies. In 1996, analysis of data from 935 homosexual men with AIDS from the Multicenter AIDS Cohort Study showed that ACV did not appear to reduce the risk of KS [35]. In contrast, among men with cytomegalovirus (CMV) disease, GCV and PFA use were associated (although not significant) with a reduced risk of KS. An observational study has also suggested that GCV and PFA, but not acyclovir (ACV), may prevent the development of KS in HIV-infected patients [35]. Another study performed in the United Kingdom where a total of 3688 HIV patients were followed up for a median period of 4.2 years, during which time 598 patients developed KS, also indicated that GCV and PFA may have some activity in preventing the occurrence of KS, but that ACV had no benefit [36]. In a prospective, randomized, double-blind, placebo-controlled study including CMV-infected persons with advanced AIDS stage, prophylactic oral GCV significantly reduced the risk of CMV disease but not significant differences between the placebo and GCV groups were observed in the 12-month Kaplan-Meir estimates of KS (12% in the placebo group and 8% in the GCV group) [37]. The safety and efficacy of valganciclovir (VGCV, the oral prodrug of GCV) on HHV-8 replication in the oropharynx in HIV-seropositive and HIV-negative persons who were asymptomatically infected with HHV-8 was determined in a randomized, double-blind, placebo-controlled, crossover trial. VGCV administered orally once a day proved to be well-tolerated and significantly reduced the frequency and quantity of KSHV replication [38].

Regarding therapy of KS, a pilot study of CDV including seven patients with KS (five patients with AIDS-KS and two with classical KS) did not provide proof of principle for the treatment of KS with CDV [39]. Despite treatment with CDV (5 mg/kg/dose weekly for two weeks and then every other week) all patients had progression of their KS and there was no decrease in the virus load in peripheral blood mononuclear cells. Another report described the effects of CDV in two patients with AIDS-KS that received the same schedule of treatment for a period of 10 and 12 months [40]. An important regression of cutaneous KS lesions was observed after three months of treatment and reactivation of new KS lesions was not observed until six and 15 months after the end of the treatment. Treatment of classical KS with intralesional injections of CDV for five weeks gave no clinical, histological, immunohistological, or virological changes compared with saline-injected lesions [41].

KSHV gene expression was studied in CDV-treated and untreated PEL cells following induction to lytic replication with TPA (12-O-tetradecanoylphorbol-13-acetate) [42]. This study revealed that the expression of genes implicated in the pathogenesis of KS or KS-like tumors, such as vGPCR, vIL-6, viral interferon regulatory factor 1 (vIRF-1), and viral macrophage inflammatory protein II (vMIP-II), was not inhibited after treatment with CDV. This is likely also true for GCV or PFA, since they all block viral DNA replication and inhibiting DNA replication does not prevent expression of early genes implicated in viral pathogenesis. This might offer a rationale for the failure of CDV therapy in KSHV-related diseases.

Successful treatments of PEL with antiviral agents, alone or with adjunctive chemotherapy, immunotherapy or HAART, have been described to date with both GCV [43,44] and CDV [43,45,46]. However, the data on intravenous administration of CDV for treating PEL are limited and also controversial. Complete remission has been documented in two HIV-positive patients with PEL when CDV was combined with antiretroviral and interferon therapies, while another patient achieved only partial remission and died after six months [45,47]. The authors of this case study postulated that the concentrations of CDV achieved in the pleural fluid were not high enough when the drug was administered intravenously. On the contrary, several studies reported achieving partial or complete remission of PEL in HIV-negative patients with intracavity CDV after conventional chemotherapy failure [46,48].

Additionally, several MCD patients have been successfully treated with GCV [31,49] , whereas failures have been reported with CDV [50]. In addition, a pilot study conducted with high dose of zidovudine combined with VGCV in patients with symptomatic MCD demonstrated that 12 out of 14 patients had substantial clinical improvement. However, this study was not randomized and controlled trials would be needed to further evaluate the efficacy of zidovudine/VGCV and compare it with other approaches [51].

The impact of antiviral treatment on KS (*i.e.*, KSHV latently-infected cells) could be potentially improved by using combination therapy of antivirals together with lytic inducing agents (leading to KSHV reactivation) [33]. The approach of inducing lytic replication of γ-herpesviruses malignancies that employ lytic activation of viruses latently infecting tumors represent a novel strategy of anti-neoplastic therapy. This strategy, named lytic induction therapy, has been explored for KSHV, but in contrast to Epstein-Barr virus (EBV), has not yet been validated in clinical trials [52]. Agents that induce lytic replication include histone deacetylase inhibitors (HDAC) such as valproic acid, phorbol esters, calcium ionophores, and NF-kappaB inhibitors [53]. *In vitro*, valproic acid has been shown to significantly induce KSHV lytic cycle in PEL cells, resulting in apoptosis of the tumor cells [54]. However, a pilot clinical trial demonstrated that valproic acid was not able to induce KSHV lytic replication in patients with AIDS-associated KS on HAART [55]. Further researches are focusing on studying more potent lytic inducing agents, such as bortezomib, 5-azacytidine and vorinostat (suberanilohydroxamic acid), as well as on increasing the treatment regimens in patients with KS [26,56,57]. In addition to bortezomib, the non-tumor-promoting phorbol ester prostratin was also shown to induce immediate-early, early and late KSHV gene expression from two lymphoma cell lines *in vitro*, suggesting that both drugs could be used as therapeutic agents for KSHV-associated malignancies [58]. Recently, the combination of bortezomib with the HDAC inhibitor vorinostat was found to potently reactivate KSHV lytic replication and to induce PEL cell death, resulting in significantly prolonged survival of PEL-bearing mice [59]. These findings provide a strong rationale for using proteasome/HDAC inhibitor combination for the therapy of PEL.

4. Inhibitors of KSHV Lytic Replication under Investigation

Despite the fact that various antiviral agents inhibit KSHV replication *in vitro*, no drugs are currently licensed for the treatment of KSHV-related diseases. From the target perspective, candidate inhibitors for treatment of KSHV-related infections can be divided in two groups, (i) compounds that act on the viral DNA polymerase and (ii) drugs that do not interact with the viral DNA polymerase. The first group of compounds includes nucleoside analogs, acyclic nucleoside phosphonates (ANPs), pyrophosphate analogs and non-nucleoside inhibitors. Their anti-KSHV activities are summarized in Table 2. The second group of inhibitors targeting viral proteins comprises compounds isolated from plants.

4.1. DNA Polymerase Inhibitors: Nucleoside Analogs

Nucleoside analogs that are approved for the treatment of herpesvirus infections, other than KSHV and EBV, include ACV, penciclovir (PCV) and GCV and brivudin (BVDU) (Figure 1).

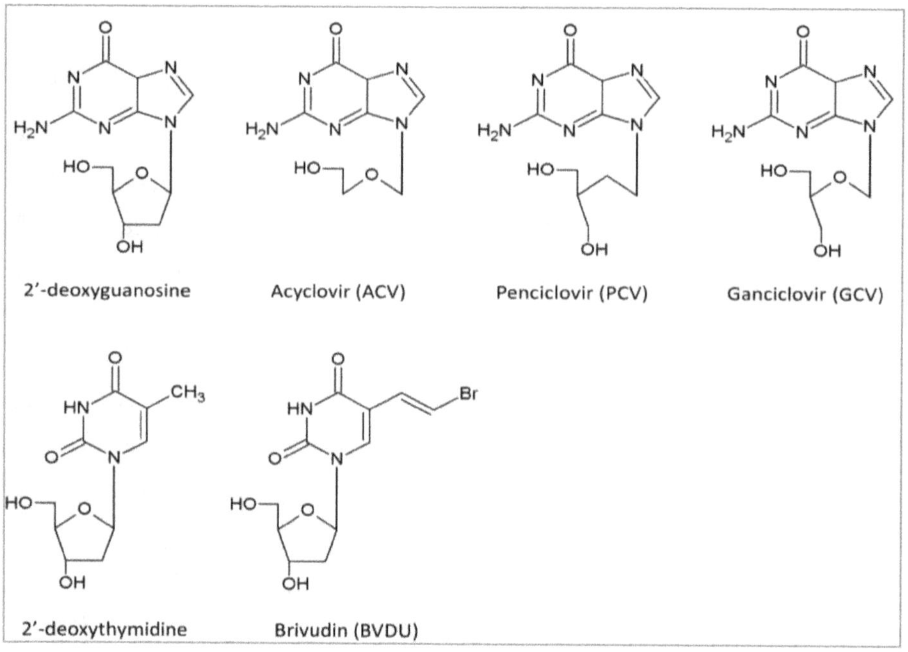

Figure 1. Structures of currently approved nucleoside analogs for herpesvirus infections. Acyclovir (ACV), penciclovir (PCV), ganciclovir (GCV) are derivatives of the natural nucleoside 2'-deoxyguanosine, whereas brivudin (BVDU) is an analog of the natural nucleoside 2'-deoxythymidine.

Nucleoside analogs in their active forms target and inhibit viral DNA polymerases by acting as competitive inhibitors of the natural dNTP substrates and/or by incorporation into the growing DNA chain where they can terminate DNA elongation. To become active, they require three intracellular phosphorylation steps to convert the nucleoside analogs into mono- (MP), di- (DP), and triphosphate (TP) forms (Figure 2). The first phosphorylation step is carried out by viral kinases, limiting this step to virus-infected cells [60], whereas the two subsequent phosphorylations are performed by cellular kinases [deoxyguanosine monophosphate (dGMP) and deoxynucleoside diphosphate (dNDP)] [61].

The viral TK of herpes simplex virus type 1 and type 2 (HSV-1 and HSV-2) and varicella-zoster virus (VZV) performs the initial phosphorylation of nucleoside analogs [62]. However, it has been demonstrated that KSHV TK has narrow substrate specificity since it recognizes pyrimidine derivatives (*i.e.*, BVDU) and not purine derivatives (*i.e.*, ACV and GCV) [63]. However, there is still a debate whether purine analogs could be phosphorylated by the KSHV TK to some degree [64]. Purine analogs are activated by the virus protein kinase (PK, ORF36) in KSHV-infected cells. KSHV PK is the homolog of the UL97 protein kinase encoded by human cytomegalovirus (HCMV), which is responsible for the conversion of GCV, and to a lesser extent of ACV, into their monophosphate forms in HCMV-infected cells [65].

BVDU is dependent on the virus-encoded TK and on its associated thymidylate kinase (dTMP) activity responsible for the first and second phosphorylations of BVDU and related analogs [66]. Previous studies have shown that KSHV and EBV TK also possesses thymidylate kinase activity [63,67]. The last phosphorylation step in the activation of BVDU is carried out by the cellular (d)NDP kinase.

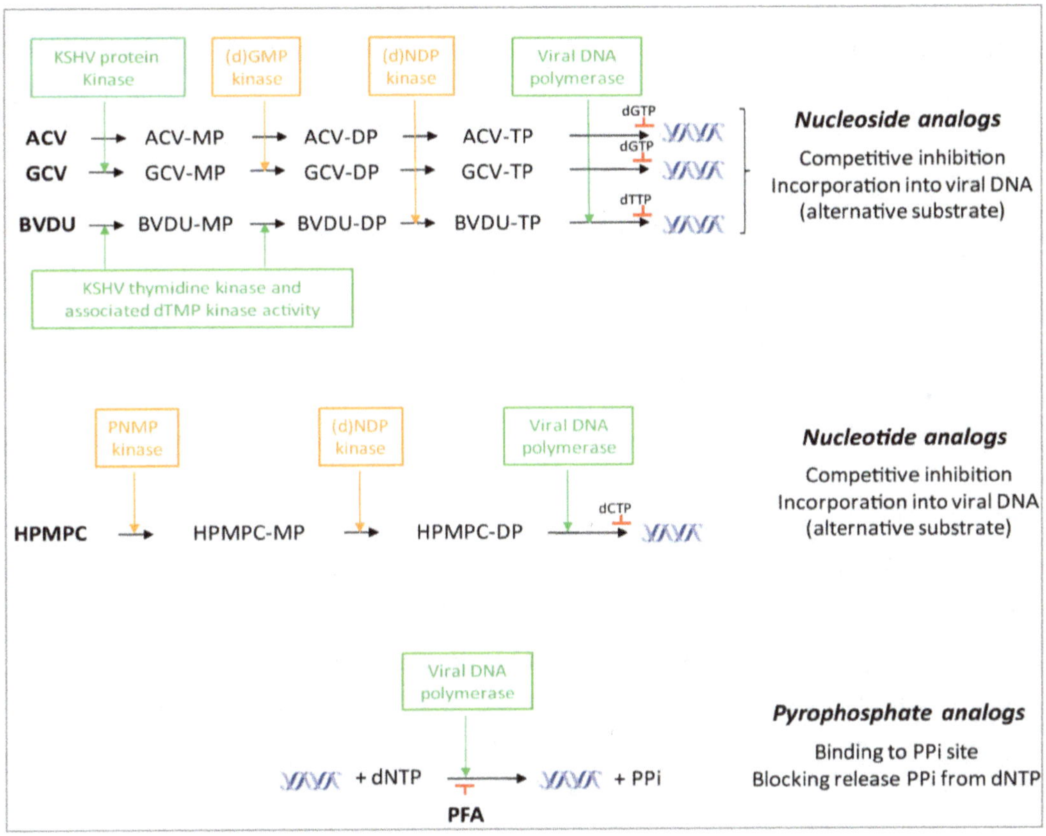

Figure 2. Mechanism of action of viral DNA polymerase inhibitors against KSHV replication. Nucleoside analogs require three phosphorylation steps to become active, being their conversions to the monophosphate (MP) forms carried out by the viral TK (BVDU) or PK (ACV and GCV). Further phosphorylation to the diphosphate (DP) is carried out by the viral TK for BVDU or cellular enzymes for ACV and GCV (*i.e.*, dGMP kinase). Conversion of these drugs to their triphosphate form (TP) by the nucleoside 5'-diphosphate (NDP) kinase results in inhibition of viral DNA polymerases because they act as competitive inhibitors of the natural substrate and/or as alternative substrates if incorporated into the growing DNA chain. ANPs, such as CDV, do not require activation by a virus-encoded enzyme to be active; instead, the two phosphorylations are done by cellular kinases (pyrimidine nucleoside monophosphate (PNMP) and 5'-diphosphate (NDP) kinase). ANP-DPs, recognized by the viral DNA polymerase, will then block DNA synthesis. PFA does not require modifications by viral or cellular kinases. PFA binds to the pyrophosphate (PPi) exchange site of the viral DNA polymerase and blocks the release of pyrophosphate from the terminal nucleoside triphosphate. As a consequence, 3'-5'-phosphodiester linkage necessary for viral DNA elongation is not possible (adapted from [62]).

Once activated, the nucleoside analogs in their triphosphate forms enter in competition with the natural substrates (dGTP or dTTP) for the viral DNA polymerase. They can inhibit the incorporation of natural dGTP or dTTP into the viral DNA in elongation or act as an alternate substrate and be incorporated into the DNA at its 3'-terminus [68]. Once ACV is incorporated, it cannot be excised by the DNA polymerase-associated 3'-5' exonuclease and it prevents further chain elongation because the

3'-hydroxyl group needed for DNA elongation is missing [69]. It has also been demonstrated that following incorporation of ACV-TP, the viral polymerase becomes trapped on the terminated DNA chain when the next deoxynucleoside triphosphate binds [69]. On the other hand, PCV, GCV and BVDU have a 3'-hydroxyl group on its acyclic side chain, allowing limited chain elongation when incorporated into the growing DNA strand [69,70].

ACV, marketed as Zovirax®, represents the first generation of effective anti-herpetic drugs with an excellent safety profile and potent activity against HSV and VZV infections [70]. Due to its limited oral bioavailability, the valine ester of ACV (VACV, Zelitrex® and Valtrex®) was developed and proved to be a safe and efficacious prodrug in patients with genital herpes, herpes *labialis*, and herpes zoster [62]. *In vitro*, ACV shows inhibitory activity against EBV, while several studies have consistently reported the weak anti-KSHV properties of ACV (Table 2) [71–76]. PCV (Denavir® and Vectavir®) and its orally available prodrug, famciclovir (Famvir®) are also indicated for the treatment of mucocutaneaous HSV-1 and HSV-2 infections, particularly recurrent herpes *labialis*, and have a spectrum of anti-KSHV activity similar to that of ACV [72,77,78]. Among several purine analogs, an ACV derivative, (1S',2R')-9-[[1',2'-bis(hydroxymethyl)cycloprop-1-yl]methyl]guanine (A-5021) was also shown to lack activity against KSHV, while this drugs was a potent inhibitor of HSV-1, HSV-2, VZV, HHV-6 and EBV replication [79]. Since KSHV does not show great sensitivity to ACV, its derivative H2G [R-9-[4-hydroxy-2-(hydroxymethyl)butyl]guanine] is also not inhibitory for KSHV replication [72].

GCV (Cymevene® and Cytovene®) was the first antiviral agent approved for the treatment of HCMV infections in immunocompromised patients and remains the first-line treatment of HCMV disease in transplant recipients [62,80]. In addition to HCMV, GCV has demonstrated efficacy against HSV, VZV, EBV, and KSHV replication [72–74,76,81]. Several reports have shown the efficacy of GCV against KSHV in the PEL cell line BCBL-1 for which the EC_{50} ranged from 1 μM to 10 μM (Table 2). However, major drawbacks of GCV are its significant bone marrow toxicity, its short half-life in tissue following oral administration (~5 h) and low bioavailability (~6% for GCV) [82]. Oral bioavailability is significantly improved with its L-valyl-ester derivative, namely VGCV (Valacyte®), to approximately 60% [62]. A structural analog of GCV, S2242 [(1,3-dihydroxy-2-propoxymethyl)purine], proved to be a more potent inhibitor of KSHV replication than GCV [72,83]. In contrast to GCV, S2242 is not phosphorylated by a herpesvirus kinase [84,85]. In fact, the cellular deoxyguanosine kinase is responsible for the first phosphorylation step, and subsequently the activation of S2242 [83]. However, the development of S2242 was halted due to toxicity [86].

Cyclopropavir, a methylenecyclopropane nucleoside resembling GCV, is under preclinical development for the treatment of HCMV-related infections. It is first phosphorylated by the HCMV PK and, under its active triphosphate form, the drug will further inhibit the viral DNA polymerase [87,88]. This compound has also shown good antiviral activity against KSHV replication *in vitro* for which the EC_{50} was 3.8 μM as measured by DNA hybridization assay [89]. The group of Prichard and colleagues reported that analogs of this class of compounds bearing 6-alkylthio substitutions had inhibitory efficacies comparable to cyclopropavir against KSHV replication *in vitro* (Table 2) [89].

The pyrimidine analog, BVDU (Zostex®, Brivirac®, Zerpex®), is a highly selective antiviral agent against HSV-1 and VZV replication and is indicated for the treatment of herpes *labialis* and herpes zoster [70] . BVDU has also shown activity against KSHV, albeit its anti-KSHV antiviral activity may be controversial as it varied from 0.9 μM to 24 μM in BCBL-1 cells depending on the report (Table 2) [72,73,90–92]. Still, the *in vitro* antiviral activity of BVDU could not be confirmed *in vivo* against murine γ-herpesvirus (MHV-68) replication in immunocompetent mice [90] and against MHV-68-induced mortality in immunocompromised mice [93]. However, BVDU administered orally at similar concentrations, is highly effective against HSV-1 and VZV infections in terms of reducing virus-induced mortality or diminishing virus titers in infected mice [94,95]. Thus, BVDU does not seem to be a suitable candidate drug for potential treatment of KSHV-related diseases.

Moreover, 2'-*exo*-methanocarbathymidine [(North)-methanocarbathymidine (N-MCT)], a thymidine analog, was identified as a potent drug with *in vitro* anti-KSHV activity. N-MCT blocked KSHV replication at EC_{50}s 5- to 10-fold lower than those of CDV and GCV, without notable cytotoxicity [96]. However, the *in vivo* antiviral efficacy was not investigated. Additionally, despite the higher *in vivo* (MHV-68 mouse model) antiviral efficacy of two thionucleoside derivatives, KAY-2-41 and KAH-39-149, as compared to HDVD, these molecules were active *in vitro* against EBV but not against KSHV [97].

While the EC_{50}s of zidovudine and stavudine have not been reported, these anti-HIV nucleoside reverse transcriptase inhibitors have been shown to be substrates of γ-herpesvirus TK, which efficiently converts them to their monophosphate forms [19,63].

Table 2. Anti-KSHV activity of viral DNA polymerase inhibitors.

Class	Subclass	Abbreviation	Drug Name	EC_{50} Range (µM) [a]	Stage of Development [d]	Refs.
Nucleoside analogs	Purine analogs	ACV	Acyclovir	26–138	Cohort study	[35,36,72–75,90]
		PCV	Penciclovir	43	*In vitro*	[72]
		A-5021	(1S,2R)-9-[[1,2-bis(hydroxymethyl) cycloprop-1yl]methyl]guanine	75	*In vitro*	[79]
		H2G	Omaciclovir	42	*In vitro*	[72]
		GCV	Ganciclovir	1.0–10	Randomized, controlled trial (with VGCV)	[31,38,72–75,90]
		S2242	2-Amino-7-[(1,3-dihydroxy-2-propoxy)-methyl]purine	0.1	*In vitro*	[72]
	Methylenecyclopropane nucleosides	CPV	Cyclopropavir	3.8 [b]	*In vitro*	[89]
			6-Alkoxy-substituted derivatives	1.8–3.5 [b]	*In vitro*	[89]
			6-Alkylthio-substituted derivatives	1.9–7.3 [b]	*In vitro*	[89]
	Pyrimidine analogs	AZT	Zidovudine		Randomized trial	[17]
		BVDU	Brivudine	0.9–24	*In vivo*	[72,73,90]
		N-MCT	(North)-methanocarbathymidine	0.08	*In vitro*	[96]
	L-dioxolane uracil analog	HDVD	1-[(2S,4S-2-(hydroxymethyl)-1,3-dioxolan-4-yl]5-vinylpyrimidine-2,4(1H,3H)-dione	0.09	*In vivo*	[90]
	Thiothymidine analogs	KAY-2-41	1-methyl substituted 4-thiothymidine	≥130	*In vivo*	[97]
		KAH-39-149	4-azido substituted 4-thiothymidine	>200	*In vivo*	[97]

Table 2. *Cont.*

Class	Subclass	Abbreviation	Drug Name	EC_{50} Range (μM) [a]	Stage of Development [d]	Refs.
cyclic nucleoside phosphonates	HPMP derivatives	HPMPC, CDV	Cidofovir	0.3–6.3	Pilot study	[39,46,72–75,90]
		CMX001	Brincidofovir	0.7	*In vitro*	[98]
		HPMP-5-azaC	1-(*S*)-[3-hydroxy-2-(phosphonomethoxy)-propyl]-5-azacytosine	0.7	*In vivo*	[98]
		HPMPA	(*S*)-9-[3-hydroxy-2-(phosphonomethoxy)-propyl]adenine	0.7	*In vitro*	[72,98]
		HPMPDAP	(*S*)-9-[3-hydroxy-2-(phosphonomethoxy)-propyl]-2,6-diaminopurine	0.9	*In vitro*	[98]
		HPMPO-DAPy	(*R*)-(2,4-diamino-3-hydroxy-6-[2-(phosphono-methoxy)propoxy])- pyrimidine	5.1	*In vitro*	[98]
	PME derivatives	PMEA	Adefovir	18–44	*In vitro*	[72,73,75,98]
		PMEDAP	(9-[2-(phosphonomethoxy)ethyl]-2,6-diamino-purine	16	*In vitro*	[98]
		PMEO-DAPy	2,4-diamino-6-[2-(phosphono-methoxy)ethoxy]-pyrimidine	12	*In vitro*	[98]
	PMP derivatives	PMPA	Tenofovir	>150	*In vitro*	ur unpublished data
Pyrophosphate analog		PFA	Foscarnet sodium	34–39	Cohort study	[35,36,74,75]
Non-nucleoside inhibitors	4-oxo-dihydroquinolines	83792, 529311, 68561, 570886		1.9–11.1 [c]	*In vitro*	[99]
	Pyrimidoquinoline analog	NSC 373989	(5-((3-(dimethylamino)propyl)amino) -3,10-dimethy-lpyrimido[4,5-b] quinoline-2,4(3H,-10H)-dione)	1.9	*In vitro*	[100]

[a] Concentration required to reduces KSHV DNA synthesis in TPA-stimulated BCBL-1 cells by 50% measured by real-time qPCR. The values are the range of the mean EC_{50s} of independent experiments as published in the original reports; [b] EC_{50} measured by flow cytometry; [c] EC_{50} measured by DNA hybridization assay. [d] Antiviral drug efficacy was evaluated *in vitro*, *in vivo* (MHV-68 mouse model) or in patients (cohort study, pilot study of randomized clinical trials). HPMP, 3-hydroxy-2-(phosphonomethoxy)propyl; PME, 2-(phosphonomethoxy)ethyl; PMP, 2-(phosphonomethoxy)propyl.

4.2. DNA Polymerase Inhibitors: Acyclic Nucleoside Phosphonates

The first ANP to be accredited with broad-spectrum antiviral activity against DNA viruses was (S)-9-(3-hydroxy-2-phosphonomethoxy-propyl)adenine or HPMPA [101]. In this nucleotide analog, the phosphate linkage (P-O-C) was replaced by the phosphonate (P-C-O) linkage, which was critical for the observed biological activity [102]. Subsequently, CDV was described as an antiviral agent active against HCMV and other DNA viruses [64]. In 1996, CDV was licensed for clinical use, under the trade name of Vistide®, for the treatment of HCMV retinitis in AIDS patients [68]. CDV is administered intravenously with the concomitant oral administration of probenecid, in order to block the drug uptake by the organic anion transporter in the proximal renal tubular cells that is responsible for the drug-related nephrotoxicity [102].

HPMPA and CDV (HPMPC) are ANPs classified as 'HPMP' (3-hydroxy-2-phosphonomethoxypropyl) derivatives, and differ from 'PME' (2-phosphonomethoxyethyl) derivatives, represented by adefovir (9-(2-phosphonomethoxyethyl) adenine, PMEA), according to their spectrum of antiviral activity (Figure 3) [62]. Adefovir was reported as an antiviral agent inhibiting *Herpesviridae*, *Hepadnaviridae*, and *Retroviridae* [68;101]. This drug was implemented for the treatment of chronic hepatitis B under the trade name of Hepsera®. PMPA or tenofovir is the representative of the 'PMP' (2-phosphonomethoxypropyl) derivatives of ANPs and has an antiviral spectrum restricted to *Hepadnaviridae* and *Retroviridae* [68]. The anti-HIV properties of tenofovir were first described in 1993 [101], and eight years later, the compound was licensed for clinical use for the treatment of HIV infections, under the trade name of Viread® [103]. Meanwhile, tenofovir in its oral prodrug form, tenofovir disoproxil fumarate, has become one of the cornerstones for anti-HIV therapy [104]. Recently, Hepsera® has been largely replaced by tenofovir, since the drug is approximately 30-times more potent against hepatitis B virus.

To accomplish their antiviral action, ANPs, must be first phosphorylated to their monophosphate form (MP) and, subsequently, to their diphosphate form (DP), which can be considered as the active metabolite that will finally interact with the viral DNA polymerase (Figure 2) [105]. These two phosphorylation steps are carried out by cellular enzymes. Of note, the intracellular phosphorylation of ANPs is thus independent of the herpesvirus-encoded TK or PK [68]. At the viral DNA polymerase level, the diphosphate form acts as a competitive inhibitor or alternate substrate with respect to the natural nucleoside, e.g., 2'-deoxycytidine-5'-triphosphate (dCTP) for CDV-DP, whereas the mechanism of action of adefovir is similar to that of CDV the integration of one molecule of adefovir at the 3'-end of the growing DNA chain terminates further chain elongation. On the other hand, CDV requires two consecutive ('tandem') incorporations to efficiently terminate DNA elongation by HCMV DNA polymerase [104].

As shown specifically for CDV, this compound offers a much longer antiviral response (several days) than nucleoside analogs, such as ACV, for which the antiviral response lasts for only a few hours [104,106] The prolonged antiviral action of CDV can be attributed to the long half-life of the CDV metabolites (CDV-MP, CDV-DP, and CDV phosphate-choline adduct) that are formed intracellular following uptake of CDV by the cells [107]. In particular the CDV phosphate-choline adduct serves as intracellular storage of CDV, since its intracellular half-life is 48 h [108].

A great deal of attention was given to the development of HPMPA analogs to improve its pharmacokinetic profile [109]. Promising anti-DNA virus effects were found for the 2,6-diaminopurine counterpart of HPMPA (*i.e.*, HPMPDAP) and PMEA (*i.e.*, PMEDAP) (Figure 3) [109]. The activity of HPMP-derivatives against DNA viruses is generally higher compared to their counterparts in the PME-series [110]. This was also seen against KSHV for which HPMP-derivatives were 10- to 100-fold more inhibitory than PME-derivatives (Table 2).

A second generation of ANPs has been described including the 6-[2-phosphono-methoxy)alkoxy]-2,4-diaminopyrimidines (DAPy). These compounds fall into two categories with as prototypes (*R*)-HPMPO-DAPy and PMEO-DAPy (Figure 3) [109]. The anti-DNA virus activity of (*R*)-HPMPO-DAPy is similar to that of CDV. Nevertheless, its inhibitory activity was five-fold less pronounced against KSHV than CDV (Table 2) [98].

A new class of promising antiviral compounds came with the discovery of ANPs bearing a triazine ring, especially 5-azacytosine as a base component [111]. The 5-azacytosine analog of CDV, *i.e.*, 1-(S)-[3-hydroxy-2-(phosphonomethoxy)propyl]-5-azacytosine (HPMP-5-azaC) (Figure 3), showed similar or higher anti-herpetic activity, including anti-KSHV activity, as compared to CDV. However, it remains to be investigated whether CDV derivatives, such as HPMP-5-azaC, would be more efficacious than CDV in the treatment of KSHV infections.

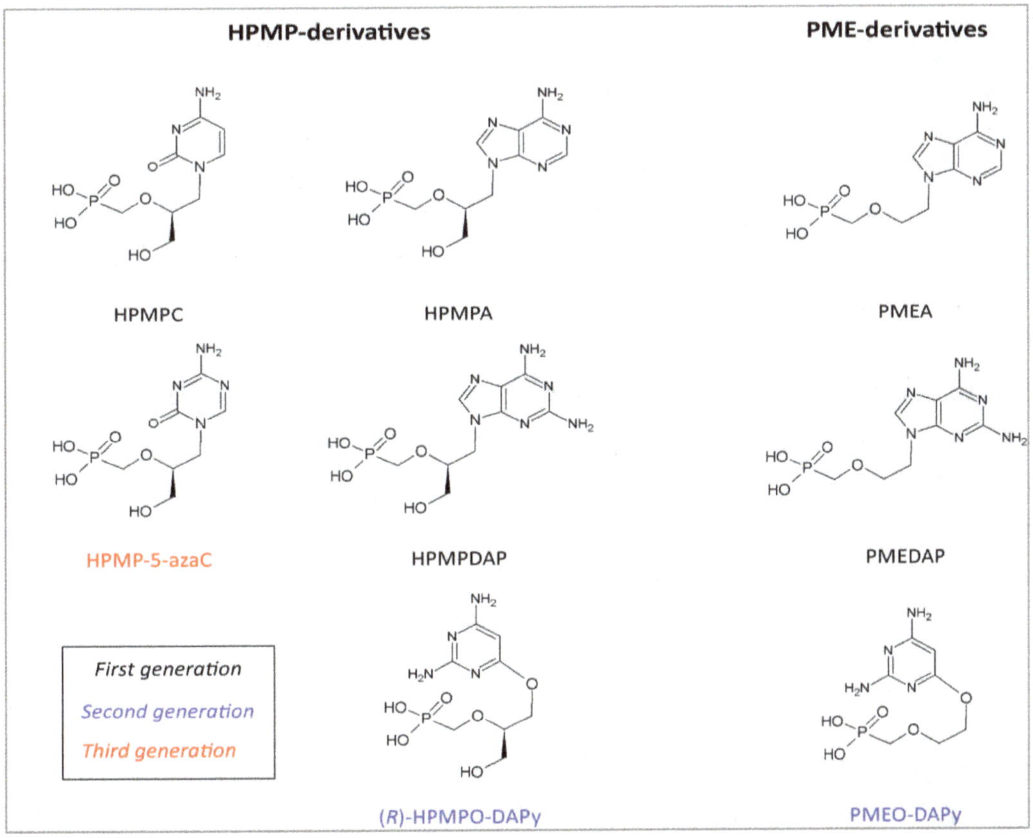

Figure 3. Structures of ANPs that exhibit anti-herpetic activity. Cidofovir (CDV, HPMPC), HPMPA and adefovir (PMEA) belong together with (*S*)-HPMPDAP and PMEDAP to the first generation of ANPs. (*R*)-HPMPO-DAPy and PMEO-DAPy belong to the second generation of ANPs, and HPMP-5-azaC is a molecule of the third generation of ANPs.

In addition to the dose-dependent nephrotoxicity of CDV, another major disadvantage that has restrained its use is its low oral bioavailability, due to the presence of the phosphonate group [61]. In order to achieve better oral bioavailability, the phosphonate group of the drug can be transformed to a phosphonic ester or amidate, which is enzymatically cleaved to the parent drug after passing the intestinal barrier, or inside the cells [112]. A considerable number of ANP prodrugs have been evaluated, but only a few of them passed preclinical studies. Additionally, lipophilic esters constitute an important class of phosphate and phosphonate prodrugs [112].

Particularly, the hexadecyloxypropyl (HDP) prodrug of CDV or CMX001 (brincidofovir) has shown promising results and is currently being developed as prophylactic and preemptive therapy of viral DNA infections [113]. A pharmacokinetic and safety study in humans reported that oral administration of CMX001 at different doses are well tolerated, with no dose-limiting toxicity, particularly, no nephrotoxicity or myelotoxicity, which are the dose-limiting toxicities of CDV or GCV, respectively [114]. In addition, CMX001 given orally at a dose of 100 mg to patients that received allogeneic hematopoietic-cell transplantation was shown to be well tolerated (*i.e.*, diarrhea was the only dose limiting adverse effect) and effective (*i.e.*, it reduced the incidence of HCMV events in these patients) [115]. In our studies, KSHV was similarly sensitive to CMX001 or CDV in BCBL-1 cells with EC_{50} values of 0.7 μM and 1.3 μM, respectively (Table 2) [98].

4.3. DNA Polymerase Inhibitors: Pyrophosphate Analogs

Phosphonoacetic acid and phosphonoformic acid are pyrophosphate analogs and non-competitive inhibitors of viral DNA polymerases by binding to the pyrophosphate-binding site of the enzyme. Hence, these compounds block the release of the pyrophosphate from the terminal nucleoside triphosphate added onto the growing DNA chain (Figure 2) [116]. PFA, the trisodium salt of phosphonoformic acid (foscarnet), is only available as an intravenous preparation and can cause nephrotoxicity and significant electrolyte disturbances [70]. PFA can be considered as a second-line therapy and its use is reserved to HSV, VZV, and HCMV patients that have failed ACV or GCV therapy due to viral resistance or that cannot be treated with GCV due to side effects of the drug [62]. PFA has been used as antiviral for the treatment of KSHV and, despite its lower activity against KSHV replication *in vitro* compared to GCV and CDV, this drug has shown efficacy in KS patients [35,36].

4.4. DNA Polymerase Inhibitors: Non-Nucleoside Analogs

The 4-oxo-dihydroquinolines derivatives have been reported to have activity against most herpesviruses, but not against other DNA or RNA viruses. They were found to inhibit the polymerases of HSV, VZV, CMV, EBV and KSHV *in vitro*, and were shown active against a variety of ACV-, GCV-, and PFA-resistant HSV and HCMV mutants [99]. By means of flow cytometry, the antiviral activity (EC50 values) of four 4-oxo-dihydroquinolines against KSHV were calculated and ranged from 1.9 μM to 11 μM (Table 2) [99].

4.5. Non-DNA Polymerase Inhibitors that Target Viral DNA Synthesis

KSHV encodes for its own DNA polymerase processivity factor, which is required for lytic viral replication and allows the viral DNA polymerase to synthesize extended stretches of DNA without dissociating from the template [117]. The highly specific interaction between the polymerase and the processivity factor may be effectively targeted by small molecules to inhibit (i) the enzymatic activity of the polymerase, (ii) the interaction between the two proteins or (iii) the function of the processivity factor itself. Dorjsuren and colleagues employed an *in vitro* assay to screen compounds inhibiting KSHV DNA synthesis through targeting the viral DNA polymerase/processivity factor complex [100]. Of 18 active compounds examined, NSC 373989 was shown to specifically block lytic KSHV DNA replication in phorbol-12-myristate-13-acetate (PMA)-stimulated KSHV-infected BCBL-1 cells (Table 2) [100]. The active compounds have structures similar to various classes of topoisomerase II inhibitors, and these results suggested that certain agents may serve as dual inhibitors of human DNA topoisomerase II as well as of KSHV DNA synthesis [100,117]. For example, (+)-Rutamarin, a topoisomerase II inhibitor isolated from plants, such as Ruta graveolens L, was also found to selectively inhibit KSHV replication [118].

4.6. Other Inhibitors of KSHV Replication

A number of compounds purified from plants are known to inhibit KSHV replication. For instance, angelicin, isolated from the seeds of *Psoralea corylifolia*, is able to inhibit lytic replication of γ-herpesviruses during the early stage of *de novo* infection and/or reactivation [119]. Other drugs inhibit the immediately-early Rta promoter of KSHV, or alter the interaction of cellular transcription factors with Rta. They include resveratrol (a non-flavonoid polyphenol present in *Polygonum cuspidatum*) and the major cannabinoid compound of marijuana, delta-9 tetrahydrocannabinol [120,121].

Zhang and colleagues have demonstrated that inhibition of KSHV replication could be achieved by the use of phosphorodiamidate morpholin oligomers (PMO) which are short single-stranded DNA oligomers with a modified backbone conferring resistance to nucleases [122]. In this study, Rta (KSHV replication and transcription activator) and LANA (latency-associated nuclear antigen) mRNAs were targeted by antisense peptide-conjugated PMO in PEL cells, resulting is a significant decrease in viral DNA levels as well as in the expression of several KSHV early and late genes.

Herpesviruses express a structurally and functionally conserved dimeric protease required for capsid assembly during lytic replication. Herpesvirus proteases do not resemble to any known protease fold pattern and are thus classified into a separate family of serine proteases [123]. Initial attempts to inhibit herpesvirus proteases targeted the active site of the enzyme, relying heavily on chemical structures for covalent inhibition and/or peptidomimetic scaffolds. Specifically targeting the active site of herpesvirus proteases have not yet result in pharmacologically viable lead compounds despite some *in vitro* success [124–126]. Craik's group reported on a small molecule, DD2 (a benzyl-substituted 4-(pyridine-2-amido) benzoic acid), able to disrupt dimerization of KSHV protease by trapping an inactive monomeric conformation [127,128]. Two DD2 analogues generated through carboxylate biosteric replacement were shown to inhibit proteases of all three herpesvirus subfamilies (*i.e.*, α, β, and γ herpesvirinae) [129].

4.7. Potential Novel Drug Targets in KSHV

Nucleoside analogs are the leading compounds for treating or suppressing herpesvirus infections for more than 50 years now. ACV, PCV and their orally bioavailable prodrugs may not be fully effective, but they have been remarkably free from any toxic side-effects. While it will be very difficult for new compounds to match these favorable properties, not only novel potential viral targets are being explored as therapy for herpesvirus infections, including immediately-early viral proteins, the viral helicase-primase and the viral terminase, but also cellular proteins that are important for viral replication. While several studies have provided proof-of-principle that helicase-primase and terminase inhibitors can be effective antiviral against α- and β-herpesviruses in cell culture and in humans [130–132], these new viral targets have not yet been explored for γ-herpesviruses, such as KSHV.

5. Cellular Targets

There is an abundance of evidence that host cell protein kinases, and the downstream pathways that they control, play a critical role in herpesvirus infection [133]. Inhibitors that target these host proteins might act as antiviral agents, yet, the risk for cytotoxicity and side effects increases by targeting host protein kinases [134]. Suppression of virus replication by a number of small-molecule inhibitors of cellular protein kinases has been demonstrated *in vitro* and several inhibitors have been incorporated into clinical trials examining their efficacies for the treatment of cancers [133]. Cellular serine/threonine protein kinases, which play an important role during the course of a KSHV infection, are mTOR, cyclic-dependent kinases, casein kinase 2, p90 ribosomal S6 kinases and PI3K, as well as tyrosine kinases, such as vascular-endothelial growth factor receptor, ephrin A2 and platelet-derived growth factor receptors [133]. Rapamycin, an mTOR inhibitor, was shown to prevent and induce regression of KS by inhibiting the expression of immediately-early proteins (Zta and Rta) of KSHV [135]. Dasatinib, an ATP-competitive tyrosine kinase inhibitor that inhibits multiple tyrosine kinases including EphA2, significantly reduces KSHV infection when cells are pretreated [133]. Because ephrin A2 functions as a cellular receptor for KSHV infection of endothelial cells, inhibitors of this tyrosine kinase may show promise as anti-KSHV agents.

6. Animal Models for Antiviral Efficacy Evaluation

MHV-68 infection in immunocompetent (BALB/c) mice has been well studied as an animal model for addressing fundamental aspects of KSHV pathogenesis and/or immunity [136–138]. This mouse model proved to be adequate for vaccination studies and for the investigation of strategies that modulate the tumorigenicity of virus-infected cells [139].

Infection of mice with murine γ-herpesvirus 68 (MHV-68) has been exploited as an experimental model to explore proof-of-principle vaccination strategies, such as MHV-68 subunit vaccines targeting lytic and latency-associated viral proteins, heat-inactivated MHV-68 virions and MHV-68 replication-deficient viruses. These vaccines were able to reduce the level of MHV-68 acute infection, but had little impact on long-term latency establishment [140–151]. Developing a therapeutic vaccine to increase the immune control of KSHV lytic replication and to decrease the KSHV viral load in people already infected may reduce the risk of KS and even virus shedding and transmission [152]. Since disease incidence in the majority of KSHV-infected people is low, scientific interests and efforts to develop a KSHV vaccine have been limited. Another major obstacle is the lack of an amenable animal model to evaluate the protective effects [153]. However, the access to a KSHV vaccine would have an impact on people that are at high risk of developing tumors, such as in HIV patients, immunosuppressed individuals, or for persons living in endemic African areas [153].

Additionally, the MHV-68 mouse model is particularly useful for the evaluation of the efficacy of antiviral agents that target the viral lytic cycle, since viral replication occurs in the lung of infected mice. Previously, distinct endpoints have been used to evaluate the efficacy of antiviral agents *in vivo*, such as mortality in MHV-68 immunocompromised mice [93] or inhibition of viral replication in lungs of immunocompetent mice [154].

Different ways have been used to set up KSHV infection in mice. In one model, purified virus is injected intravenously to NOD/SCID mice with severe combined immunodeficiency affecting T- and B-lymphocyte development as well as with Natural Killer (NK) cell, macrophage and granulocyte numbers and function reduced [155]. This model is suited to evaluate longitudinal patterns of viral gene expression, cell tropism and immune responses. Some NOD/SCID mice implanted with functional human hematopoietic tissue grafts (NOD/SCID-hu) were shown to produce human KSHV-specific antibodies [155]. Furthermore, GCV treatment of these chimeric mice at the time of inoculation led to prolonged but reversible suppression of viral DNA and RNA levels. A second model used NOD/SCID mice reconstituted with KSHV-infected $CD34^+$ hematopoietic progenitor cells (HPC) where it was shown that the virus establishes persistent infection in NOD/SCID mice and disseminated following differentiation of infected HPCs into the B-cell and monocytes linkages [156].

Growth of PEL derived cells lines as xenografts in immune deficient mice has been used to study the *in vivo* effects of therapeutic strategies for KSHV-associated malignancies [157–160]. A concern of this xenograft model is whether they entirely reflect clinical presentations of KSHV PEL. A recent study using a PEL xenograft model by intraperitoneal injection of KSHV PEL cells into the peritoneal cavity of NOD/SCID mice found that these animals not only developed massive ascites but also single or multiple solid tumors on various tissues in ~70%–80% of animals. Although this xenograft model can be used for the study of effusion and solid lymphoma observed in patients, tumor cells grown *in vivo* displayed unique features (including viral lytic gene expression profile, rate of solid tumor development and tumor microenvironment) to those passed *in vitro* [161].

7. Conclusions

Since the discovery of KSHV 20 years ago, little progress has been made towards therapies directed against this oncogenic virus. In the absence of molecules targeting viral latency and FDA-approved antiviral agents for the treatment of KSHV infections, few compounds have been evaluated, mostly *in vitro*, and all targeting the viral DNA polymerase. Previously, GCV was shown to be the most effective drug among the marketed antiviral agents for the treatment of KS patients. In contrast to GCV, other PK-dependent drugs, such as ACV, PCV, H2G and A-5021, showed weak or no anti-KSHV activity, while TK-dependent drugs, such as HDVD, may be suitable drug candidates. Most importantly, because of the unclear role of antiviral therapies targeting lytic phase in the prevention or treatment of KSHV-induced diseases, randomized controlled clinical trials are needed to determine their true efficacy in different clinical settings. Furthermore, it would be of interest to investigate the role and potential of novel ANPs in the treatment of KSHV infections, since CDV therapy has shown unclear outcomes in patients. Finally, more efforts should be invested to examine the potential of non-nucleoside inhibitors against KSHV replication, as well as drugs that target viral proteins other than the viral DNA polymerase, since this proof-of-principle has been shown beneficial for other herpesviruses, such as HSV and HCMV.

Acknowledgements

This work was supported by a grant of the "Geconcertreerde Onderzoeksacties" (Krediet No. 10/014).

References and Notes

1. Stanescu, L.; Foarfa, C.; Georgescu, A.C.; Georgescu, I. Kaposi's sarcoma associated with AIDS. *Rom. J. Morphol. Embryol.* **2007**, *4*, 181–187.
2. Chang, Y.; Cesarman, E.; Pessin, M.S.; Lee, F.; Culpepper, J.; Knowles, D.M.; Moore, P.S. Identification of herpesvirus-like DNA sequences in AIDS-associated Kaposi's sarcoma. *Science* **1994**, *266*, 1865–1869.
3. Ablashi, D.V.; Chatlynne, L.G.; Whitman, J.E., Jr.; Cesarman, E. Spectrum of Kaposi's sarcoma-associated herpesvirus, or human herpesvirus 8, diseases. *Clin. Microbiol. Rev.* **2002**, *15*, 439–464.
4. Dourmishev, L.A.; Dourmishev, A.L.; Palmeri, D.; Schwartz, R.A.; Lukac, D.M. Molecular genetics of Kaposi's sarcoma-associated herpesvirus (human herpesvirus-8) epidemiology and pathogenesis. *Microbiol. Mol. Biol. Rev.* **2003**, *67*, 175–212.
5. Lacoste, V.; Lavergne, A.; de Thoisy, B.; Pouliquen, J.F.; Gessain, A. Genetic diversity and molecular evolution of human and non-human primate Gammaherpesvirinae. *Infect. Genet. Evol.* **2010**, *10*, 1–13.
6. Laurent, C.; Meggetto, F.; Brousset, P. Human herpesvirus 8 infections in patients with immunodeficiencies. *Hum. Pathol.* **2008**, *39*, 983–993.
7. Sullivan, R.J.; Pantanowitz, L.; Casper, C.; Stebbing, J.; Dezube, B.J. HIV/AIDS: Epidemiology, pathophysiology, and treatment of Kaposi sarcoma-associated herpesvirus disease: Kaposi sarcoma, primary effusion lymphoma, and multicentric Castleman disease. *Clin. Infect. Dis.* **2008**, *47*, 1209–1215.
8. Fatahzadeh, M. Kaposi sarcoma: Review and medical management update. *Oral Surg. Oral Med. Oral Pathol. Oral Radiol.* **2012**, *113*, 2–16.
9. Schwartz, R.A. Kaposi's sarcoma: An update. *J. Surg. Oncol.* **2004**, *87*, 146–151.
10. Hengge, U.R.; Ruzicka, T.; Tyring, S.K.; Stuschke, M.; Roggendorf, M.; Schwartz, R.A.; Seeber, S. Update on Kaposi's sarcoma and other HHV8 associated diseases. Part 1:

Epidemiology, environmental predispositions, clinical manifestations, and therapy. *Lancet Infect. Dis.* **2002**, *2*, 281–292.

11. Szajerka, T.; Jablecki, J. Kaposi's sarcoma revisited. *AIDS Rev.* **2007**, *9*, 230–236.
12. Martorelli, D.; Muraro, E.; Merlo, A.; Turrini, R.; Fae, D.A.; Rosato, A.; Dolcetti, R. Exploiting the interplay between innate and adaptive immunity to improve immunotherapeutic strategies for Epstein-Barr-virus-driven disorders. *Clin. Dev. Immunol.* **2012**, *2012*, 931952.
13. Von Roenn, J.H. Clinical presentations and standard therapy of AIDS-associated Kaposi's sarcoma. *Hematol. Oncol. Clin. North Am.* **2003**, *17*, 747–762.
14. Vanni, T.; Sprinz, E.; Machado, M.W.; Santana, R.C.; Fonseca, B.A.; Schwartsmann, G. Systemic treatment of AIDS-related Kaposi sarcoma: Current status and perspectives. *Cancer Treat. Rev.* **2006**, *32*, 445–455.
15. Uldrick, T.S.; Whitby, D. Update on KSHV epidemiology, Kaposi Sarcoma pathogenesis, and treatment of Kaposi Sarcoma. *Cancer Lett.* **2011**, *305*, 150–162.
16. Casper, C.; Wald, A. The use of antiviral drugs in the prevention and treatment of Kaposi sarcoma, multicentric Castleman disease and primary effusion lymphoma. *Curr. Top. Microbiol. Immunol.* **2007**, *312*, 289–307.
17. Joffe, M.M.; Hoover, D.R.; Jacobson, L.P.; Kingsley, L.; Chmiel, J.S.; Visscher, B.R. Effect of treatment with zidovudine on subsequent incidence of Kaposi's sarcoma. *Clin. Infect. Dis.* **1997**, *25*, 1125–1133.
18. Sgadari, C.; Monini, P.; Barillari, G.; Ensoli, B. Use of HIV protease inhibitors to block Kaposi's sarcoma and tumour growth. *Lancet Oncol.* **2003**, *4*, 537–547.
19. Lock, M.J.; Thorley, N.; Teo, J.; Emery, V.C. Azidodeoxythymidine and didehydrodeoxythymidine as inhibitors and substrates of the human herpesvirus 8 thymidine kinase. *J. Antimicrob. Chemother.* **2002**, *49*, 359–366.
20. Gantt, S.; Casper, C.; Ambinder, R.F. Insights into the broad cellular effects of nelfinavir and the HIV protease inhibitors supporting their role in cancer treatment and prevention. *Curr. Opin. Oncol.* **2013**, *25*, 495–502.
21. Gantt, S.; Carlsson, J.; Ikoma, M.; Gachelet, E.; Gray, M.; Geballe, A.P.; Corey, L.; Casper, C.; Lagunoff, M.; Vieira, J. The HIV protease inhibitor nelfinavir inhibits Kaposi's sarcoma-associated herpesvirus replication *in vitro*. *Antimicrob. Agents Chemother.* **2011**, *55*, 2696–2703.
22. Ruocco, E.; Ruocco, V.; Tornesello, M.L.; Gambardella, A.; Wolf, R.; Buonaguro, F.M. Kaposi's sarcoma: Etiology and pathogenesis, inducing factors, causal associations, and treatments: Facts and controversies. *Clin. Dermatol.* **2013**, *31*, 413–422.
23. Simonelli, C.; Spina, M.; Cinelli, R.; Talamini, R.; Tedeschi, R.; Gloghini, A.; Vaccher, E.; Carbone, A.; Tirelli, U. Clinical features and outcome of primary effusion lymphoma in HIV-infected patients: A single-institution study. *J. Clin. Oncol.* **2003**, *21*, 3948–3954.
24. Dittmer, D.P.; Richards, K.L.; Damania, B. Treatment of Kaposi sarcoma-associated herpesvirus-associated cancers. *Front Microbiol.* **2012**, *3*, 141.
25. Lim, S.T.; Rubin, N.; Said, J.; Levine, A.M. Primary effusion lymphoma: Successful treatment with highly active antiretroviral therapy and rituximab. *Ann. Hematol.* **2005**, *84*, 551–552.

26. Sullivan, R.J.; Pantanowitz, L.; Dezube, B.J. Targeted therapy for Kaposi sarcoma. *BioDrugs* **2009**, *23*, 69–75.

27. Song, S.N.; Tomosugi, N.; Kawabata, H.; Ishikawa, T.; Nishikawa, T.; Yoshizaki, K. Down-regulation of hepcidin resulting from long-term treatment with an anti-IL-6 receptor antibody (tocilizumab) improves anemia of inflammation in multicentric Castleman disease. *Blood* **2010**, *116*, 3627–3634.

28. Williams, S.C. First IL-6-blocking drug nears approval for rare blood disorder. *Nat. Med.* **2013**, *19*, 1193.

29. Marcelin, A.G.; Aaron, L.; Mateus, C.; Gyan, E.; Gorin, I.; Viard, J.P.; Calvez, V.; Dupin, N. Rituximab therapy for HIV-associated Castleman disease. *Blood* **2003**, *102*, 2786–2788.

30. Ocio, E.M.; Sanchez-Guijo, F.M.; Diez-Campelo, M.; Castilla, C.; Blanco, O.J.; Caballero, D.; San Miguel, J.F. Efficacy of rituximab in an aggressive form of multicentric Castleman disease associated with immune phenomena. *Am. J. Hematol.* **2005**, *78*, 302–305.

31. Casper, C.; Nichols, W.G.; Huang, M.L.; Corey, L.; Wald, A. Remission of HHV-8 and HIV-associated multicentric Castleman disease with ganciclovir treatment. *Blood* **2004**, *103*, 1632–1634.

32. Gessain, A.; Duprez, R. Spindle cells and their role in Kaposi's sarcoma. *Int. J. Biochem. Cell Biol.* **2005**, *37*, 2457–2465.

33. Gantt, S.; Casper, C. Human herpesvirus 8-associated neoplasms: The roles of viral replication and antiviral treatment. *Curr. Opin. Infect. Dis.* **2011**, *24*, 295–301.

34. Asahi-Ozaki, Y.; Sato, Y.; Kanno, T.; Sata, T.; Katano, H. Quantitative analysis of Kaposi sarcoma-associated herpesvirus (KSHV) in KSHV-associated diseases. *J. Infect. Dis.* **2006**, *193*, 773–782.

35. Glesby, M.J.; Hoover, D.R.; Weng, S.; Graham, N.M.; Phair, J.P.; Detels, R.; Ho, M.; Saah, A.J. Use of antiherpes drugs and the risk of Kaposi's sarcoma: Data from the Multicenter AIDS Cohort Study. *J. Infect. Dis.* **1996**, *173*, 1477–1480.

36. Mocroft, A.; Youle, M.; Gazzard, B.; Morcinek, J.; Halai, R.; Phillips, A.N. Anti-herpesvirus treatment and risk of Kaposi's sarcoma in HIV infection. Royal Free/Chelsea and Westminster Hospitals Collaborative Group. *AIDS* **1996**, *10*, 1101–1105.

37. Spector, S.A.; McKinley, G.F.; Lalezari, J.P.; Samo, T.; Andruczk, R.; Follansbee, S.; Sparti, P.D.; Havlir, D.V.; Simpson, G.; Buhles, W.; Wong, R.; Stempien, M. Oral ganciclovir for the prevention of cytomegalovirus disease in persons with AIDS. Roche Cooperative Oral Ganciclovir Study Group. *N. Engl. J. Med.* **1996**, *334*, 1491–1497.

38. Casper, C.; Krantz, E.M.; Corey, L.; Kuntz, S.R.; Wang, J.; Selke, S.; Hamilton, S.; Huang, M.L.; Wald, A. Valganciclovir for suppression of human herpesvirus-8 replication: A randomized, double-blind, placebo-controlled, crossover trial. *J. Infect. Dis.* **2008**, *198*, 23–30.

39. Little, R.F.; Merced-Galindez, F.; Staskus, K.; Whitby, D.; Aoki, Y.; Humphrey, R.; Pluda, J.M.; Marshall, V.; Walters, M.; Welles, L.; Rodriguez-Chavez, I.R.; Pittaluga, S.; Tosato, G.; Yarchoan, R. A pilot study of cidofovir in patients with kaposi sarcoma. *J. Infect Dis.* **2003**, *187*, 149–153.

40. Mazzi, R.; Parisi, S.G.; Sarmati, L.; Uccella, I.; Nicastri, E.; Carolo, G.; Gatti, F.; Concia, E.; Andreoni, M. Efficacy of cidofovir on human herpesvirus 8 viraemia and Kaposi's sarcoma progression in two patients with AIDS. *AIDS* **2001**, *15*, 2061–2062.

41. Simonart, T.; Noel, J.C.; De Dobbeleer, G.; Parent, D.; Van Vooren, J.P.; De Clercq, E.; Snoeck, R. Treatment of classical Kaposi's sarcoma with intralesional injections of cidofovir: Report of a case. *J. Med. Virol.* **1998**, *55*, 215–218.

42. Lu, M.; Suen, J.; Frias, C.; Pfeiffer, R.; Tsai, M.H.; Chuang, E.; Zeichner, S.L. Dissection of the Kaposi's sarcoma-associated herpesvirus gene expression program by using the viral DNA replication inhibitor cidofovir. *J. Virol.* **2004**, *78*, 13637–13652.

43. Crum-Cianflone, N.F.; Wallace, M.R.; Looney, D. Successful secondary prophylaxis for primary effusion lymphoma with human herpesvirus 8 therapy. *AIDS* **2006**, *20*, 1567–1569.

44. Pastore, R.D.; Chadburn, A.; Kripas, C.; Schattner, E.J. Novel association of haemophagocytic syndrome with Kaposi's sarcoma-associated herpesvirus-related primary effusion lymphoma. *Br. J. Haematol.* **2000**, *111*, 1112–1115.

45. Hocqueloux, L.; Agbalika, F.; Oksenhendler, E.; Molina, J.M. Long-term remission of an AIDS-related primary effusion lymphoma with antiviral therapy. *AIDS* **2001**, *15*, 280–282.

46. Luppi, M.; Trovato, R.; Barozzi, P.; Vallisa, D.; Rossi, G.; Re, A.; Ravazzini, L.; Potenza, L.; Riva, G.; Morselli, M.; *et al.* Treatment of herpesvirus associated primary effusion lymphoma with intracavity cidofovir. *Leukemia* **2005**, *19*, 473–476.

47. Boulanger, E. Human herpesvirus 8 (HHV8). II. Pathogenic role and sensitivity to antiviral drugs. *Ann. Biol. Clin. (Paris)* **1999**, *57*, 19–28.

48. Moyo, T.K.; Richards, K.L.; Damania, B. Use of cidofovir for the treatment of HIV-negative human herpes virus-8-associated primary effusion lymphoma. *Clin. Adv. Hematol. Oncol.* **2010**, *8*, 372–374.

49. Valencia, M.E.; Moreno, V.; Martinez, P.; Casado, I. Favorable outcome of Castleman's disease treated with oral valganciclovir. *Med. Clin. (Barc.)* **2005**, *125*, 399.

50. Berezne, A.; Agbalika, F.; Oksenhendler, E. Failure of cidofovir in HIV-associated multicentric Castleman disease. *Blood* **2004**, *103*, 4368–4369.

51. Uldrick, T.S.; Polizzotto, M.N.; Aleman, K.; O'Mahony, D.; Wyvill, K.M.; Wang, V.; Marshall, V.; Pittaluga, S.; Steinberg, S.M.; Tosato, G.; *et al.* High-dose zidovudine plus valganciclovir for Kaposi sarcoma herpesvirus-associated multicentric Castleman disease: A pilot study of virus-activated cytotoxic therapy. *Blood* **2011**, *117*, 6977–6986.

52. Kenney, S.C.; Fingeroth, J.D. The Molecular Basis of Lytic Induction Therapy in Relation to Gamma herpesvirus (KSHV, EBV)-Associated, AIDS-Related Tumors. In *Molecular Basis for Therapy if AIDS-defining Cancers*; Dittmer, D., Krown, S.E., Eds.; Springer: New York, NY, USA, 2010; pp. 111–135.

53. Klass, C.M.; Offermann, M.K. Targeting human herpesvirus-8 for treatment of Kaposi's sarcoma and primary effusion lymphoma. *Curr. Opin. Oncol.* **2005**, *17*, 447–455.

54. Klass, C.M.; Krug, L.T.; Pozharskaya, V.P.; Offermann, M.K. The targeting of primary effusion lymphoma cells for apoptosis by inducing lytic replication of human herpesvirus 8 while blocking virus production. *Blood* **2005**, *105*, 4028–4034.
55. Lechowicz, M.; Dittmer, D.P.; Lee, J.Y.; Krown, S.E.; Wachsman, W.; Aboulafia, D.; Dezube, B.J.; Ratner, L.; Said, J.; Ambinder, R.F. Molecular and clinical assessment in the treatment of AIDS Kaposi sarcoma with valproic Acid. *Clin. Infec. Dis.* **2009**, *49*, 1946–1949.
56. Fu, D.X.; Tanhehco, Y.; Chen, J.; Foss, C.A.; Fox, J.J.; Chong, J.M.; Hobbs, R.F.; Fukayama, M.; Sgouros, G.; Kowalski, J.; Pomper, M.G.; Ambinder, R.F. Bortezomib-induced enzyme-targeted radiation therapy in herpesvirus-associated tumors. *Nature Med.* **2008**, *14*, 1118–1122.
57. Reid, E.G. Bortezomib-induced Epstein-Barr virus and Kaposi sarcoma herpesvirus lytic gene expression: Oncolytic strategies. *Curr. Opin. Oncol.* **2011**, *23*, 482–487.
58. Brown, H.J.; McBride, W.H.; Zack, J.A.; Sun, R. Prostratin and bortezomib are novel inducers of latent Kaposi's sarcoma-associated herpesvirus. *Antivir. Ther.* **2005**, *10*, 745–751.
59. Bhatt, S.; Ashlock, B.M.; Toomey, N.L.; Diaz, L.A.; Mesri, E.A.; Lossos, I.S.; Ramos, J.C. Efficacious proteasome/HDAC inhibitor combination therapy for primary effusion lymphoma. *J. Clin. Invest* **2013**, *123*, 2616–2628.
60. Jamieson, A.T.; Gentry, G.A.; Subak-Sharpe, J.H. Induction of both thymidine and deoxycytidine kinase activity by herpes viruses. *J. Gen. Virol.* **1974**, *24*, 465–480.
61. Miller, W.H.; Miller, R.L. Phosphorylation of acyclovir diphosphate by cellular enzymes. *Biochem. Pharmacol.* **1982**, *31*, 3879–3884.
62. Andrei, G.; De Clercq, E.; Snoeck, R. Viral DNA Polymerase Inhibitors. In *Viral Genome Replication*; Cameron, C.E., Gotte, M., Rancy, K., Eds.; Springer: New York, NY, USA, 2009; pp. 481–526.
63. Gustafson, E.A.; Schinazi, R.F.; Fingeroth, J.D. Human herpesvirus 8 open reading frame 21 is a thymidine and thymidylate kinase of narrow substrate specificity that efficiently phosphorylates zidovudine but not ganciclovir. *J. Virol* **2000**, *74*, 684–692.
64. Cannon, J.S.; Hamzeh, F.; Moore, S.; Nicholas, J.; Ambinder, R.F. Human herpesvirus 8-encoded thymidine kinase and phosphotransferase homologues confer sensitivity to ganciclovir. *J. Virol* **1999**, *73*, 4786–4793.
65. Lurain, N.S.; Chou, S. Antiviral drug resistance of human cytomegalovirus. *Clin. Microbiol. Rev.* **2010**, *23*, 689–712.
66. De Clercq, E. Discovery and development of BVDU (brivudin) as a therapeutic for the treatment of herpes zoster. *Biochem. Pharmacol.* **2004**, *68*, 2301–2315.
67. Gustafson, E.A.; Chillemi, A.C.; Sage, D.R.; Fingeroth, J.D. The Epstein-Barr virus thymidine kinase does not phosphorylate ganciclovir or acyclovir and demonstrates a narrow substrate specificity compared to the herpes simplex virus type 1 thymidine kinase. *Antimicrob. Agents Chemother.* **1998**, *42*, 2923–2931.
68. De Clercq, E. Antiviral drug discovery and development: Where chemistry meets with biomedicine. *Antivir. Res.* **2005**, *67*, 56–75.
69. Reardon, J.E.; Spector, T. Herpes simplex virus type 1 DNA polymerase. Mechanism of inhibition by acyclovir triphosphate. *J. Biol. Chem.* **1989**, *264*, 7405–7411.

70. Sellar, R.S.; Peggs, K.S. Management of multidrug-resistant viruses in the immunocompromised host. *Br. J. Hematol.* **2012**, *156*, 559–572.

71. De Clercq, E.; Naesens, L.; De Bolle, L.; Schols, D.; Zhang, Y.; Neyts, J. Antiviral agents active against human herpesviruses HHV-6, HHV-7 and HHV-8. *Rev. Med. Virol* **2001**, *11*, 381–395.

72. Neyts, J.; De Clercq, E. Antiviral drug susceptibility of human herpesvirus 8. *Antimicrob. Agents Chemother.* **1997**, *41*, 2754–2756.

73. Friedrichs, C.; Neyts, J.; Gaspar, G.; De Clercq, E.; Wutzler, P. Evaluation of antiviral activity against human herpesvirus 8 (HHV-8) and Epstein-Barr virus (EBV) by a quantitative real-time PCR assay. *Antivir. Res.* **2004**, *62*, 121–123.

74. Bounaadja, L.; Piret, J.; Goyette, N.; Boivin, G. Evaluation of Epstein-Barr virus, human herpesvirus 6 (HHV-6), and HHV-8 antiviral drug susceptibilities by use of real-time-PCR-based assays. *J. Clin. Microbiol.* **2013**, *51*, 1244–1246.

75. Sergerie, Y.; Boivin, G. Evaluation of susceptibility of human herpesvirus 8 to antiviral drugs by quantitative real-time PCR. *J. Clin. Microbiol.* **2003**, *41*, 3897–3900.

76. Medveczky, M.M.; Horvath, E.; Lund, T.; Medveczky, P.G. *In vitro* antiviral drug sensitivity of the Kaposi's sarcoma-associated herpesvirus. *AIDS* **1997**, *11*, 1327–1332.

77. Bacon, T.H.; Boyd, M.R. Activity of penciclovir against Epstein-Barr virus. *Antimicrob. Agents Chemother.* **1995**, *39*, 1599–1602.

78. Bacon, T.H.; Howard, B.A.; Spender, L.C.; Boyd, M.R. Activity of penciclovir in antiviral assays against herpes simplex virus. *J. Antimicrob. Chemother.* **1996**, *37*, 303–313.

79. Neyts, J.; Naesens, L.; Ying, C.; De, B.L.; De, C.E. Anti-herpesvirus activity of (1'S,2'R)-9-[[1',2'-bis(hydroxymethyl)-cycloprop-1'-yl]methyl] x guanine (A-5021) *in vitro* and *in vivo*. *Antivir. Res.* **2001**, *49*, 115–120.

80. Faulds, D.; Heel, R.C. Ganciclovir. A review of its antiviral activity, pharmacokinetic properties and therapeutic efficacy in cytomegalovirus infections. *Drugs* **1990**, *39*, 597–638.

81. De Clercq, E. Antiviral agents: Characteristic activity spectrum depending on the molecular target with which they interact. *Adv. Virus Res.* **1993**, *42*, 1–55.

82. Biron, K.K. Antiviral drugs for cytomegalovirus diseases. *Antiviral Res.* **2006**, *71*, 154–163.

83. Neyts, J.; Balzarini, J.; Andrei, G.; Chaoyong, Z.; Snoeck, R.; Zimmermann, A.; Mertens, T.; Karlsson, A.; De Clercq, E. Intracellular metabolism of the N7-substituted acyclic nucleoside analog 2-amino-7-(1,3-dihydroxy-2-propoxymethyl)purine, a potent inhibitor of herpesvirus replication. *Mol. Pharmacol.* **1998**, *53*, 157–165.

84. Zimmermann, A.; Michel, D.; Pavic, I.; Hampl, W.; Luske, A.; Neyts, J.; De Clercq, E.; Mertens, T. Phosphorylation of aciclovir, ganciclovir, penciclovir and S2242 by the cytomegalovirus UL97 protein: A quantitative analysis using recombinant vaccinia viruses. *Antiviral Res.* **1997**, *36*, 35–42.

85. Neyts, J.; Andrei, G.; Snoeck, R.; Jahne, G.; Winkler, I.; Helsberg, M.; Balzarini, J.; De Clercq, E. The N-7-substituted acyclic nucleoside analog 2-amino-7-[(1,3-dihydroxy-2-propoxy)methyl]purine is a potent and selective inhibitor of herpesvirus replication. *Antimicrob. Agents Chemother.* **1994**, *38*, 2710–2716.

86. Andrei, G.; De Clercq, E.; Snoeck, R. Drug targets in cytomegalovirus infection. *Infec. Dis. Drug Targets* **2009**, *9*, 201–222.

87. Gentry, B.G.; Kamil, J.P.; Coen, D.M.; Zemlicka, J.; Drach, J.C. Stereoselective phosphorylation of cyclopropavir by pUL97 and competitive inhibition by maribavir. *Antimicrob. Agents Chemother.* **2010**, *54*, 3093–3098.

88. Li, C.; Quenelle, D.C.; Prichard, M.N.; Drach, J.C.; Zemlicka, J. Synthesis and antiviral activity of 6-deoxycyclopropavir, a new prodrug of cyclopropavir. *Bioorg. Med. Chem.* **2012**, *20*, 2669–2674.

89. Prichard, M.N.; Williams, J.D.; Komazin-Meredith, G.; Khan, A.R.; Price, N.B.; Jefferson, G.M.; Harden, E.A.; Hartline, C.B.; Peet, N.P.; Bowlin, T.L. Synthesis and antiviral activities of methylenecyclopropane analogs with 6-alkoxy and 6-alkylthio substitutions that exhibit broad-spectrum antiviral activity against human herpesviruses. *Antimicrob. Agents Chemother.* **2013**, *57*, 3518–3527.

90. Coen, N.; Singh, U.; Vuyyuru, V.; van den Oord, J.J.; Balzarini, J.; Duraffour, S.; Snoeck, R.; Cheng, Y.C.; Chu, C.K.; Andrei, G. Activity and mechanism of action of HDVD, a novel pyrimidine nucleoside derivative with high levels of selectivity and potency against gammaherpesviruses. *J. Virol.* **2013**, *87*, 3839–3851.

91. Lin, J.C.; Machida, H. Comparison of two bromovinyl nucleoside analogs, 1-beta-D-arabinofuranosyl-E-5-(2-bromovinyl)uracil and E-5-(2-bromovinyl)-2'-deoxyuridine, with acyclovir in inhibition of Epstein-Barr virus replication. *Antimicrob. Agents Chemother.* **1988**, *32*, 1068–1072.

92. Choi, Y.; Li, L.; Grill, S.; Gullen, E.; Lee, C.S.; Gumina, G.; Tsujii, E.; Cheng, Y.C.; Chu, C.K. Structure-activity relationships of (E)-5-(2-bromovinyl)uracil and related pyrimidine nucleosides as antiviral agents for herpes viruses. *J. Med. Chem.* **2000**, *43*, 2538–2546.

93. Neyts, J.; De Clercq, E. *In vitro* and *in vivo* inhibition of murine gamma herpesvirus 68 replication by selected antiviral agents. *Antimicrob. Agents Chemother.* **1998**, *42*, 170–172.

94. De Clercq, E.; Zhang, Z.X.; Sim, I.S. Treatment of experimental herpes simplex virus encephalitis with (E)-5-(2-bromovinyl)-2'-deoxyuridine in mice. *Antimicrob. Agents Chemother.* **1982**, *22*, 421–425.

95. Weber, O. Novel mouse models for the investigation of experimental drugs with activity against human varicella-zoster virus. *Antiviral Chem. Chemother.* **2000**, *11*, 283–290.

96. Zhu, W.; Burnette, A.; Dorjsuren, D.; Roberts, P.E.; Huleihel, M.; Shoemaker, R.H.; Marquez, V.E.; Agbaria, R.; Sei, S. Potent antiviral activity of north-methanocarbathymidine against Kaposi's sarcoma-associated herpesvirus. *Antimicrob. Agents Chemother.* **2005**, *49*, 4965–4973.

97. Coen, N.; Duraffour, S.; Haraguchi, K.; Balzarini, J.; van den Oord, J.J.; Snoeck, R.; Andrei, G. Anti-herpesvirus Activity of Two Novel 4'-Thiothymidine Derivatives KAY-2-41 and KAH-39-149 is dependent on viral and cellular thymidine kinases. *Antimicrob. Agents Chemother.* **2014**, *58*, 4328–4340.

98. Coen, N.; Duraffour, S.; Naesens, L.; Krecmerova, M.; Van den Oord, J.; Snoeck, R.; Andrei, G. Evaluation of Novel Acyclic Nucleoside Phosphonates against Human and Animal Gammaherpesviruses Revealed an Altered Metabolism of Cyclic Prodrugs upon Epstein-Barr Virus Reactivation in P3HR-1 Cells. *J. Virol.* **2013**, *87*, 12422–12432.
99. Hartline, C.B.; Harden, E.A.; Williams-Aziz, S.L.; Kushner, N.L.; Brideau, R.J.; Kern, E.R. Inhibition of herpesvirus replication by a series of 4-oxo-dihydroquinolines with viral polymerase activity. *Antivir. Res.* **2005**, *65*, 97–105.
100. Dorjsuren, D.; Burnette, A.; Gray, G.N.; Chen, X.; Zhu, W.; Roberts, P.E.; Currens, M.J.; Shoemaker, R.H.; Ricciardi, R.P.; Sei, S. Chemical library screen for novel inhibitors of Kaposi's sarcoma-associated herpesvirus processive DNA synthesis. *Antivir. Res.* **2006**, *69*, 9–23.
101. De Clercq, E.; Holy, A.; Rosenberg, I.; Sakuma, T.; Balzarini, J.; Maudgal, P.C. A novel selective broad-spectrum anti-DNA virus agent. *Nature* **1986**, *323*, 464–467.
102. Lee, W.A.; Martin, J.C. Perspectives on the development of acyclic nucleotide analogs as antiviral drugs. *Antivir. Res.* **2006**, *71*, 254–259.
103. De Clercq, E.; Field, H.J. Antiviral prodrugs—The development of successful prodrug strategies for antiviral chemotherapy. *Br. J. Pharmacol* **2006**, *147*, 1–11.
104. De Clercq, E.; Holy, A. Acyclic nucleoside phosphonates: A key class of antiviral drugs. *Nat. Rev. Drug Discov.* **2005**, *4*, 928–940.
105. De Clercq, E. Antivirals and antiviral strategies. *Nat. Rev. Microbiol.* **2004**, *2*, 704–720.
106. Neyts, J.; Snoeck, R.; Balzarini, J.; De Clercq, E. Particular characteristics of the anti-human cytomegalovirus activity of (S)-1-(3-hydroxy-2-phosphonylmethoxypropyl)cytosine (HPMPC) *in vitro*. *Antivir. Res.* **1991**, *16*, 41–52.
107. Ho, H.T.; Woods, K.L.; Bronson, J.J.; De, B.H.; Martin, J.C.; Hitchcock, M.J. Intracellular metabolism of the antiherpes agent (S)-1-[3-hydroxy-2-(phosphonylmethoxy)propyl]cytosine. *Mol. Pharmacol.* **1992**, *41*, 197–202.
108. De Clercq, E. The clinical potential of the acyclic (and cyclic) nucleoside phosphonates. The magic of the phosphonate bond. *Biochem. Pharmacol.* **2011**, *2*, 99–109.
109. Krecmerova,M. Nucleoside and Nucleotide Analogues for the Treatment of Herpesvirus Infections: Current Stage and New Prospects in the Field of Acyclic Nucleoside Phosphonates. In *Herpesviridae—A Look Into This Unique Family of Viruses*; Magel, G., Tyring, S., Eds.; InTech: Rijeka, Croatia, 2012; pp. 245–270.
110. Holy, A. Phosphonomethoxyalkyl analogs of nucleotides. *Cur. Pharmac. Des.* **2003**, *9*, 2567–2592.
111. Krecmerova, M.; Holy, A.; Piskala, A.; Masojidkova, M.; Andrei, G.; Naesens, L.; Neyts, J.; Balzarini, J.; De, C.E.; Snoeck, R. Antiviral activity of triazine analogues of 1-(S)-[3-hydroxy-2-(phosphonomethoxy)propyl]cytosine (cidofovir) and related compounds. *J. Med. Chem.* **2007**, *50*, 1069–1077.
112. Tichy, T.; Andrei, G.; Dracinsky, M.; Holy, A.; Balzarini, J.; Snoeck, R.; Krecmerova, M. New prodrugs of Adefovir and Cidofovir. *Bioorg. Med. Chem.* **2011**, *19*, 3527–3539.
113. Rice, A.D.; Adams, M.M.; Wallace, G.; Burrage, A.M.; Lindsey, S.F.; Smith, A.J.; Swetnam, D.; Manning, B.R.; Gray, S.A.; Lampert, B.; *et al.* Efficacy of CMX001 as a post exposure antiviral in New Zealand White rabbits infected with rabbitpox virus, a model for orthopoxvirus infections of humans. *Viruses* **2011**, *3*, 47–62.

114. Painter, W.; Robertson, A.; Trost, L.C.; Godkin, S.; Lampert, B.; Painter, G. First pharmacokinetic and safety study in humans of the novel lipid antiviral conjugate CMX001, a broad-spectrum oral drug active against double-stranded DNA viruses. *Antimicrob. Agents Chemother.* **2012**, *56*, 2726–2734.
115. Marty, F.M.; Winston, D.J.; Rowley, S.D.; Vance, E.; Papanicolaou, G.A.; Mullane, K.M.; Brundage, T.M.; Robertson, A.T.; Godkin, S.; Mommeja-Marin, H.; *et al.* CMX001 to prevent cytomegalovirus disease in hematopoietic-cell transplantation. *N. Engl. J. Med.* **2013**, *369*, 1227–1236.
116. Crumpacker, C.S. Mechanism of action of foscarnet against viral polymerases. *Am. J. Med.* **1992**, *92*, 3S–7S.
117. Lin, K.; Ricciardi, R.P. A rapid plate assay for the screening of inhibitors against herpesvirus DNA polymerases and processivity factors. *J. Virol. Methods* **2000**, *88*, 219–225.
118. Xu, B.; Wang, L.; Gonzalez-Molleda, L.; Wang, Y.; Xu, J.; Yuan, Y. Antiviral activity of (+)-rutamarin against Kaposi's sarcoma-associated herpesvirus by inhibition of the catalytic activity of human topoisomerase II. *Antimicrob. Agents Chemother.* **2014**, *58*, 563–573.
119. Cho, H.J.; Jeong, S.G.; Park, J.E.; Han, J.A.; Kang, H.R.; Lee, D.; Song, M.J. Antiviral activity of angelicin against gammaherpesviruses. *Antivir. Res.* **2013**, *100*, 75–83.
120. Dyson, O.F.; Walker, L.R.; Whitehouse, A.; Cook, P.P.; Akula, S.M. Resveratrol inhibits KSHV reactivation by lowering the levels of cellular EGR-1. *PLoS One* **2012**, *7*, e33364.
121. Medveczky, M.M.; Sherwood, T.A.; Klein, T.W.; Friedman, H.; Medveczky, P.G. Delta-9 tetrahydrocannabinol (THC) inhibits lytic replication of gamma oncogenic herpesviruses *in vitro*. *BMC. Med.* **2004**, *2*, 34.
122. Zhang, Y.J.; Stein, D.A.; Fan, S.M.; Wang, K.Y.; Kroeker, A.D.; Meng, X.J.; Iversen, P.L.; Matson, D.O. Suppression of porcine reproductive and respiratory syndrome virus replication by morpholino antisense oligomers. *Vet. Microbiol.* **2006**, *117*, 117–129.
123. Skorenski, M.; Sienczyk, M. Anti-herpesvirus agents: A patent and literature review (2003 to present). *Expert Opin. Ther. Pat.* **2014**, *24*, 925–941.
124. Borthwick, A.D.; Davies, D.E.; Ertl, P.F.; Exall, A.M.; Haley, T.M.; Hart, G.J.; Jackson, D.L.; Parry, N.R.; Patikis, A.; Trivedi, N.; *et al.* Design and synthesis of pyrrolidine-5,5'-*trans*-lactams (5-*oxo*-hexahydropyrrolo[3,2-b]pyrroles) as novel mechanism-based inhibitors of human cytomegalovirus protease. 4. Antiviral activity and plasma stability. *J. Med. Chem.* **2003**, *46*, 4428–4449.
125. Gopalsamy, A.; Lim, K.; Ellingboe, J.W.; Mitsner, B.; Nikitenko, A.; Upeslacis, J.; Mansour, T.S.; Olson, M.W.; Bebernitz, G.A.; Grinberg, D.; *et al.* Design and syntheses of 1,6-naphthalene derivatives as selective HCMV protease inhibitors. *J. Med. Chem.* **2004**, *47*, 1893–1899.
126. Borthwick, A.D. Design of translactam HCMV protease inhibitors as potent antivirals. *Med. Res. Rev.* **2005**, *25*, 427–452.
127. Shahian, T.; Lee, G.M.; Lazic, A.; Arnold, L.A.; Velusamy, P.; Roels, C.M.; Guy, R.K.; Craik, C.S. Inhibition of a viral enzyme by a small-molecule dimer disruptor. *Nat. Chem. Biol.* **2009**, *5*, 640–646.
128. Lee, G.M.; Shahian, T.; Baharuddin, A.; Gable, J.E.; Craik, C.S. Enzyme inhibition by allosteric capture of an inactive conformation. *J Mol. Biol.* **2011**, *411*, 999–1016.

129. Gable, J.E.; Lee, G.M.; Jaishankar, P.; Hearn, B.R.; Waddling, C.A.; Renslo, A.R.; Craik, C.S. Broad-spectrum allosteric inhibition of herpesvirus proteases. *Biochemistry* **2014**, *53*, 4648–4660.
130. Field, H.J.; Mickleburgh, I. The helicase-primase complex as a target for effective herpesvirus antivirals. *Adv. Exp. Med. Biol.* **2013**, *767*, 145–159.
131. Underwood, M.R.; Ferris, R.G.; Selleseth, D.W.; Davis, M.G.; Drach, J.C.; Townsend, L.B.; Biron, K.K.; Boyd, F.L. Mechanism of action of the ribopyranoside benzimidazole GW275175X against human cytomegalovirus. *Antimicrobial. Agents Chemother.* **2004**, *48*, 1647–1651.
132. Reefschlaeger, J.; Bender, W.; Hallenberger, S.; Weber, O.; Eckenberg, P.; Goldmann, S.; Haerter, M.; Buerger, I.; Trappe, J.; Herrington, J.A.; *et al.* Novel non-nucleoside inhibitors of cytomegaloviruses (BAY 38–4766): *In vitro* and *in vivo* antiviral activity and mechanism of action. *J. Antimicrob. Chemother.* **2001**, *48*, 757–767.
133. Li, R.; Hayward, S.D. Potential of protein kinase inhibitors for treating herpesvirus-associated disease. *Trends Microbiol.* **2013**, *21*, 286–295.
134. Andrei, G.; Snoeck, R. Advances in the treatment of varicella-zoster virus infections. *Adv. Pharmacol.* **2013**, *67*, 107–168.
135. Nichols, L.A.; Adang, L.A.; Kedes, D.H. Rapamycin blocks production of KSHV/HHV8: insights into the anti-tumor activity of an immunosuppressant drug. *PLoS One* **2011**, *6*, e14535.
136. Nash, A.A.; Dutia, B.M.; Stewart, J.P.; Davison, A.J. Natural history of murine gamma-herpesvirus infection. *Philos. Trans. R. Soc. Lond. B. Biol. Sci.* **2001**, *356*, 569–579.
137. Simas, J.P.; Efstathiou, S. Murine gammaherpesvirus 68: A model for the study of gammaherpesvirus pathogenesis. *Trends Microbiol.* **1998**, *6*, 276–282.
138. Speck, S.H.; Virgin, H.W. Host and viral genetics of chronic infection: A mouse model of gamma-herpesvirus pathogenesis. *Curr. Opin. Microbiol.* **1999**, *2*, 403–409.
139. Cipkova-Jarcuskova, J.; Chalupkova, A.; Hrabovska, Z.; Wagnerova, M.; Mistrikova, J. Biological and pathogenetic characterization of different isolates of murine gammaherpesvirus 68 (MHV-68) in the context of study of human oncogenic gammaherpesviruses. *Acta Virol.* **2013**, *57*, 105–112.
140. Arico, E.; Robertson, K.A.; Belardelli, F.; Ferrantini, M.; Nash, A.A. Vaccination with inactivated murine gammaherpesvirus 68 strongly limits viral replication and latency and protects type I IFN receptor knockout mice from a lethal infection. *Vaccine* **2004**, *22*, 1433–1440.
141. Boname, J.M.; Coleman, H.M.; May, J.S.; Stevenson, P.G. Protection against wild-type murine gammaherpesvirus-68 latency by a latency-deficient mutant. *J. Gen. Virol.* **2004**, *85*, 131–135.
142. Fowler, P.; Efstathiou, S. Vaccine potential of a murine gammaherpesvirus-68 mutant deficient for ORF73. *J. Gen. Virol.* **2004**, *85*, 609–613.
143. Jia, Q.; Freeman, M.L.; Yager, E.J.; McHardy, I.; Tong, L.; Martinez-Guzman, D.; Rickabaugh, T.; Hwang, S.; Blackman, M.A.; Sun, R.; *et al.* Induction of protective immunity against murine gammaherpesvirus 68 infection in the absence of viral latency. *J. Virol.* **2010**, *84*, 2453–2465.
144. Kayhan, B.; Yager, E.J.; Lanzer, K.; Cookenham, T.; Jia, Q.; Wu, T.T.; Woodland, D.L.; Sun, R.; Blackman, M.A. A replication-deficient murine gamma-herpesvirus blocked in late viral gene expression can establish latency and elicit protective cellular immunity. *J. Immunol.* **2007**, *179*, 8392–8402.

145. Obar, J.J.; Donovan, D.C.; Crist, S.G.; Silvia, O.; Stewart, J.P.; Usherwood, E.J. T-cell responses to the M3 immune evasion protein of murid gammaherpesvirus 68 are partially protective and induced with lytic antigen kinetics. *J. Virol.* **2004**, *78*, 10829–10832.
146. Rickabaugh, T.M.; Brown, H.J.; Martinez-Guzman, D.; Wu, T.T.; Tong, L.; Yu, F.; Cole, S.; Sun, R. Generation of a latency-deficient gammaherpesvirus that is protective against secondary infection. *J. Virol.* **2004**, *78*, 9215–9223.
147. Stevenson, P.G.; Doherty, P.C. Kinetic analysis of the specific host response to a murine gammaherpesvirus. *J. Virol.* **1998**, *72*, 943–949.
148. Stewart, J.P.; Micali, N.; Usherwood, E.J.; Bonina, L.; Nash, A.A. Murine gamma-herpesvirus 68 glycoprotein 150 protects against virus-induced mononucleosis: A model system for gamma-herpesvirus vaccination. *Vaccine* **1999**, *17*, 152–157.
149. Tibbetts, S.A.; McClellan, J.S.; Gangappa, S.; Speck, S.H.; Virgin, H.W. Effective vaccination against long-term gammaherpesvirus latency. *J. Virol.* **2003**, *77*, 2522–2529.
150. Usherwood, E.J.; Ward, K.A.; Blackman, M.A.; Stewart, J.P.; Woodland, D.L. Latent antigen vaccination in a model gammaherpesvirus infection. *J. Virol.* **2001**, *75*, 8283–8288.
151. Woodland, D.L.; Usherwood, E.J.; Liu, L.; Flano, E.; Kim, I.J.; Blackman, M.A. Vaccination against murine gamma-herpesvirus infection. *Viral Immunol.* **2001**, *14*, 217–226.
152. Wu, T.T.; Qian, J.; Ang, J.; Sun, R. Vaccine prospect of Kaposi sarcoma-associated herpesvirus. *Cur. Opin. Virol.* **2012**, *2*, 482–488.
153. Wu, T.T.; Blackman, M.A.; Sun, R. Prospects of a novel vaccination strategy for human gamma-herpesviruses. *Immunol. Res.* **2010**, *48*, 122–146.
154. Barnes, A.; Dyson, H.; Sunil-Chandra, N.P.; Collins, P.; Nash, A.A. 2'-Deoxy-5-ethyl-beta-4'-thiouridine inhibits replication of murine gammaherpesvirus and delays the onset of virus latency. *Antivir. Chem. Chemother.* **1999**, *10*, 321–326.
155. Parsons, C.H.; Adang, L.A.; Overdevest, J.; O'Connor, C.M.; Taylor, J.R., Jr.; Camerini, D.; Kedes, D.H. KSHV targets multiple leukocyte lineages during long-term productive infection in NOD/SCID mice. *J. Clin. Invest.* **2006**, *116*, 1963–1973.
156. Wu, W.; Vieira, J.; Fiore, N.; Banerjee, P.; Sieburg, M.; Rochford, R.; Harrington, W., Jr.; Feuer, G. KSHV/HHV-8 infection of human hematopoietic progenitor (CD34+) cells: Persistence of infection during hematopoiesis *in vitro* and *in vivo*. *Blood* **2006**, *108*, 141–151.
157. Lan, K.; Murakami, M.; Bajaj, B.; Kaul, R.; He, Z.; Gan, R.; Feldman, M.; Robertson, E.S. Inhibition of KSHV-infected primary effusion lymphomas in NOD/SCID mice by gamma-secretase inhibitor. *Cancer Biol. Ther.* **2009**, *8*, 2136–2143.
158. Ishikawa, C.; Tanaka, J.; Katano, H.; Senba, M.; Mori, N. Hippuristanol reduces the viability of primary effusion lymphoma cells both *in vitro* and *in vivo*. *Mar. Drugs* **2013**, *11*, 3410–3424.
159. Kariya, R.; Taura, M.; Suzu, S.; Kai, H.; Katano, H.; Okada, S. HIV protease inhibitor Lopinavir induces apoptosis of primary effusion lymphoma cells via suppression of NF-kappaB pathway. *Cancer Lett.* **2014**, *342*, 52–59.
160. Wu, W.; Rochford, R.; Toomey, L.; Harrington, W.J.; Feuer, G. Inhibition of HHV-8/KSHV infected primary effusion lymphomas in NOD/SCID mice by azidothymidine and interferon-alpha. *Leukemia Res.* **2005**, *29*, 545–555.

161. Dai, L.; Trillo-Tinoco, J.; Bai, L.; Kang, B.; Xu, Z.; Wen, X.; Del Valle, L.; Qin, Z. Systematic analysis of a xenograft mice model for KSHV+ primary effusion lymphoma (PEL). *PLoS One* **2014**, *9*, e90349.

7

Glycosylation of KSHV Encoded vGPCR Functions in its Signaling and Tumorigenicity

Hui Wu [1,†], Liqun Liu [2,†], Jun Xiao [1], Mengdie Chi [1], Yixiao Qu [1] and Hao Feng [1,*]

[1] Key Laboratory of Protein Chemistry and Developmental Biology of Ministry of Education of China, College of Life Science, Hunan Normal University, Changsha 410081, China; E-Mails: wu2hui218@sina.com.cn (H.W.); zzz54120@hotmail.com (J.X.); chimengdie@163.com (M.C.); 15873156916@163.com (Y.Q.)

[2] Division of Pediatric Neurology, Children's Medical Center, The Second Xiangya Hospital, Central South University, Changsha 410011, China; E-Mail: llq91217@163.com

† These authors contributed equally to this work.

* Author to whom correspondence should be addressed; E-Mail: fenghao@hunnu.edu.cn;

Academic Editor: Zhi-Ming Zheng

Abstract: Kaposi's sarcoma-associated herpesvirus (KSHV) is a tumor virus and the etiologic agent of Kaposi's Sarcoma (KS). KSHV G protein-coupled receptor (vGPCR) is an oncogene that is implicated in malignancies associated with KHSV infection. In this study, we show that vGPCR undergoes extensive N-linked glycosylation within the extracellular domains, specifically asparagines 18, 22, 31 and 202. An immunofluorescence assay demonstrates that N-linked glycosylation are necessary for vGPCR trafficking to the cellular membrane. Employing vGPCR mutants whose glycosylation sites were ablated, we show that these vGPCR mutants failed to activate downstream signaling in cultured cells and were severely impaired to induce tumor formation in the xenograph nude mouse model. These findings support the conclusion that glycosylation is critical for vGPCR tumorigenesis and imply that chemokine regulation at the plasma membrane is crucial for vGPCR mediated signaling.

Keywords: KSHV; vGPCR; Glycosylation

---

1. Introduction

Kaposi's sarcoma (KS) was initially described in the late 19th Century as an indo-lent dermal vascular tumor predominantly affecting the lower extremities of elderly Mediterranean men. KS is a tumor which is characterized with abnormal vascular proliferation and the infection of the Kaposi's sarcoma-associated herpesvirus (KSHV, also known as human herpesvirus 8/HHV-8). KSHV is classified into γ-herpesvirus family and believed to be the etiologic agent of KS and two other B-cell related tumors known as primary effusion lymphoma (PEL) and multicentric Castleman's disease [1,2]. The KSHV genome encodes an array of proteins that are implicated in modulating host inflammatory responses, angiogenesis and tumor formation, most notably the G protein-coupled receptor (vGPCR or ORF74) [3,4].

KSHV encoded G protein-coupled receptor (vGPCR) is a seven trans-membrane receptor and one of the several known KSHV onco-proteins. vGPCR is constitutively activated and triggers downstream signaling components including the phospholipase C pathway, and PI3 kinase/AKT axis even without ligand binding. It has broad signaling effects *in vitro*, activating NF-κB, NF-AT and AP-1 [5–7]. However, chemokine association regulates vGPCR mediated signaling. For example, GRO-a up-regulates and IP-10 down-regulates vGPCR signaling. vGPCR expression *in vivo* caused the tumor formation in nude mice and vGPCR transgenic mice developed disease that resembled human KS lesions [8,9].

vGPCR is a constitutively activated receptor and its expression *ex vivo* triggers a broad spectrum of signaling pathways. Through downstream signaling, this viral oncoprotein promotes the expression of plentiful host and viral functional genes including cytokines, signaling molecules, and transcription factors that result in promoting cell proliferation and endothelial tube formation [10–12]. However, vGPCR triggered COS-1 cells to death and caused toxicity in PEL cells when this viral receptor was over-expressed in these cells, which imply that this viral oncogene needs tightly regulated expression and signaling *in vivo* for its function in KSHV pathogenesis. Extensive studies had been done on its signaling and tumorigenecity; however, there were few reports about its post-translational regulation.

vGPCR is predominantly translated from a bi-cistronic mRNA transcript downstream of K14 (vOX2), presumably reducing vGPCR protein expression. Our previous study discovered that KSHV encoded small membrane protein K7 can bind vGPCR through its TM domain and accelerate vGPCR's degradation by the proteasome through maintaining this onco-protein in the endoplasmic reticulum (ER). Based on this mechanism, K7 dampened vGPCR mediated signaling *ex vivo* and greatly suppressed vGPCR's tumorigenicity *in vivo* [13]. Our data demonstrated that KSHV has evolved mechanisms such as the posttranslational degradation to achieve a temporary expression of the constitutively active vGPCR during lytic infection.

Posttranslational modification (PTM) is the major regulation of G protein-coupled receptor, such as sulfation, hydroxylation, acylation, *etc.* Actually, the refinement modification of vGPCR is very important for this viral oncoprotein function *in vivo* and *in vitro*. We previously revealed that vGPCR incorporates sulfate groups within its extracellular N-terminal tyrosine residues (Y26 and Y28) and that the tyrosine sulfation is crucial for its tumorigenicity in nude mice although it does not impact vGPCR mediated signaling [14,15].

vGPCR contains four N-X-T/S motifs (asparagine residues 18, 22, 31 and 202) in its extracellular domains, which are predicted to be the potential sites of N-linked glycosylation. In this paper, we demonstrate that vGPCR is modified with N-linked glycosylation at the Asn residues 18, 22, 31 and 202, and all the N-linked glycan chains exist in the extra-cellular domains of this seven membrane-spaning viral protein. Immunofluorescence assay and luciferase report assay verified that glycosylation played important roles in vGPCR trafficking to the cellular membrane and vGPCR mediated signaling transduction. Tumor formation in the xenograph nude mouse model revealed that vGPCR mutants devoid of glycosylation had diminished tumor formation ability *in vivo*. All data supported that vGPCR possesses another posttranslational modification, N-linked glycosylation, which regulates vGPCR-mediated signaling and vGPCR-dependent tumorgenesis.

2. Materials and Methods

2.1. Cells and Plasmids

HEK293T (293T), HeLa and NIH3T3 cells were obtained from the American Type Culture Collection (ATCC). All the cells were grown in DMEM supplemented with 10% fetal bovine serum, 100 units/mL penicillin and 100 mg/mL streptomycin. pcDNA5/FRT-TO-HA-vGPCR and pCDH-HA-vGPCR-EF-Puro were kept in the lab [15]. vGPCR mutants with single site of N-linked glycosylation (N18, N22, N31, N202), with three sites of glycosylation (N18Q, N22Q, N31Q, N202Q) or without glycosylation site (N0) were site-mutated based on the HA-vGPCR and cloned into pcDNA5/FRT-TO or pCDH-EF-Puro empty vectors (System Bioscience, Dallas, TX, USA) separately. Lenti-virus containing vGPCR or its glycosylation mutants were made as before and NIH3T3 stably expressing vGPCR or its glycosylation mutants were established as previously described [13].

2.2. Endo H and PNGase F Digestion

293T cells were transfected with FRT-TO-HA-vGPCR or FRT-TO-HA-vGPCR-N0 separately and transfected cells were harvested at 48 h post-transfection and lysed in lysis buffer (50 mM Tris-Hcl/pH7.4, 150 mM NaCl, 1% NP-40, 5 mM EDTA, 0.05% Tween-20) containing protease inhibitor cocktail (Roche). Whole cell lysates (WCL) were divided into two aliquots, one aliquot digested with Endoglycosidase H (Endo H) or Peptide -*N*-Glycosidase F (PNGase F) according to the instruction of the manufacturer (NEB, Beijing, China); the other aliquot was treated with reaction buffer (without glycosidase) and used as control. Briefly, the sample mixed with the glycoprotein denaturing buffer and denatured at 100 °C for 10 min; then Endo H or PNGase F was added with the G5 or G7 reaction buffer and the samples were incubated at 37 °C for 1 h. The glycosidase digested sample and the control were isolated by 10% SDS-PAGE and examined by immunoblot assay.

2.3. Immunoblotting

Immunoblot (IB) assays were performed as previously described [14]. Briefly, the samples were isolated by 10% SDS-PAGE and transferred to PVDF membrane. The membrane was blocked with 3% BSA in TBST and probed with primary antibody. The membrane was incubated with goat anti-mouse IgG peroxidase conjugate or goat anti-rabbit IgG peroxidase conjugate (1:5000) and proteins were detected with ECL system (Bio-Rad, Beijing, China). Alternatively, the membrane was incubated with goat anti-mouse IgG alkaline phosphatase conjugate (1:5000) and proteins were visualized with

NBT/BCIP system (Pierce). For the commercial primary antibodies, mouse anti-HA antibody (Sigma, Shanghai, China) was used at the dilution ratio of 1:1000; both mouse anti-phospho-IκBα (Ser32/36, cell signaling) antibody and mouse anti-IκBα antibody (Santa Cruz, Shanghai, China) were diluted at 1:1000.

2.4. Luciferase Reporter Assay

The reporter cocktail consists of plasmids expressing fire fly luciferase (50 ng/mL) and β-galactosidase (400ng/mL). The expression of β-galactosidase is driven by a housekeeping glucophosphokinase promoter and the fire fly luciferase expression is under the control of response elements of NF-kB (Luci-NF-κB) or NF-AT (Luci-NFAT) transcription factor. 293T cells (in 6-well plate) were transiently transfected with 2.5 mL of luciferase reporter cocktail, 300ng of FRT-TO-HA-vGPCR or vGPCR glycosylation mutants. The cells were harvested and lysed on ice at 36 h post transfection. Centrifuged supernatant was used to measure luciferase and β-galactosidase activity according to manufacturer's protocol (Promega, Beijing, China).

2.5. Immunofluorescence Microscopy

HeLa cells were transfected with FRT-TO-HA-vGPCR or vGPCR glycosylation mutants. The transfected cells were fixed with paraformaldehyde at 24 h post transfection and permeabilized with Triton X-100 (0.2% in PBS) or not (for detecting plasma membrane protein). After stained with primary and secondary antibodies, cells were analyzed by immunofluorescence microscopy as previously described [13]. For commercial antibodies, mouse monoclonal anti-HA antibody (Sigma) was used at the dilution ratio of 1:400 and Alexa 488-conjugated secondary antibody (Molecular Probes, Shanghai, China) was diluted at 1:1000.

2.6. Apoptosis Assay

NIH3T3 cells stably expression vGPCR, its glycosylation mutants (N18, N22, N31, N202, N0) or the empty vector were seeded in the six-well plate (5×10^5/well) and the cells were washed 3 times with PBS 18 h post seeding. The washed cells were cultured in serum-free DMEM or Hank's Balanced Salt Solution (HBSS) for serum starvation-induced apoptosis. The cell viability was determined by typane blue staining at 24 h (HBSS) or 48 h (DMEM) post starvation treatment as previously described [13].

2.7. Tumor Formation Assays in Nude Mice

The animal experiments were undertaken according to the institutional ethical guidelines for animal experiments at the Guangzhou Institute of Medicine and Health, Chinese Academy of Sciences. NIH3T3 cells stably expressing the empty vector (pCDH-EF-Puro) and NIH3T3 cells stably expressing vGPCR or its glycosylation mutants (N18, N22, N31, N202, N0) were injected subcutaneously into the flanks of 3 to 5-week-old mice (athymic, nude/nude) separately. Nude mice were euthanized 3 weeks after inoculation, and the tumor weights were determined as described before [14].

3. Results

3.1. vGPCR Undergoes N-linked Glycosylation within Its Extracellular Sequences

Glycosylation is one of the major modifications of GPCRs and serves as a major post-translational regulatory mechanism for GPCRs. N-linked glycosylation is a common type of glycosidic bond in which N-linked glycans are almost always attached to the nitrogen atom of an asparagine (Asn) side chain that is present as a part of N-X-S/T consensus sequence.

vGPCR contains four Asn residues (18, 22, 31 and 202), which lie within the N-X-S/T consensus site. All the four Asn residues are located within the extracellular sequences of this viral protein: Asn 18, 22 and 31 are located in the N-terminal extracellular motif and Asn 202 is located in the 2nd extracellular loop (Figure 1A). The vGPCR mutant N0, in which all the four Asn (N) residues were mutated into glutamine (Q), was generated and used as the un-glycosylated control. As analyzed by immunoblot assay, the anti-HA antibody detected multiple bands from the whole-cell lysates of HEK293T cells transfected with a plasmid expressing vGPCR. These multiple bands showed that molecular weight of vGPCR varied from ~37 KDa to 50 KDa and the major band was approximately 45 KDa. The major band of vGPCR is larger in molecular weight than the predicted molecular weight (~37 KDa). To test whether these species are glycosylated, we used Endo H and PNGase F to digest and remove glycan chains. While Endo H removes high mannose sugar chain, PNGase F cleaves all glycans, including complex carbohydrates. After digestion with Endo H or PNGase F, there was only one band (~37 KDa) was detected, which indicated that vGPCR was modified with N-linked glycosylation. As expected, digestion of the two glycosidase did not impact the migration of the vGPCR-N0 mutant, which was supposed to have no N-linked glycosylation (Figure 1B). The glycosidase digestion data clearly demonstrates that vGPCR undergoes the modification of N-linked glycosylation.

3.2. vGPCR Possesses Four Sites of N-Linked Glycosylation

To exploit which Asn residue (18, 22, 31 and 202) was modified with N-linked glycosylation, vGPCR glycosylation mutants including N18, N22, N31 and N202 were generated, in which three of the four possible glycosylated Asn residues (except the indicated Asn residue) were mutated into glutamine. As shown by immunoblot analysis, all mutants had multiple bands and the major band was around 45 KDa, which migrated more slowly compared with the un-glycosylated N0 mutant (Figure 2A). The data demonstrated that all four Asn residues (18, 22, 31 and 202) are modified with N-linked glycosylation. Accordingly, mutants with three sites of glycosylation (N18Q, N22Q, N31Q and N202Q), in which the indicated Asn residues were mutated into glutamine, migrated more slowly than vGPCR-N0 and slightly faster than wild type vGPCR (Figure 2B). Thes results indicated that each of the four N-X-S/T sites was glycosylated and collectively contributed to the mass increase of vGPCR.

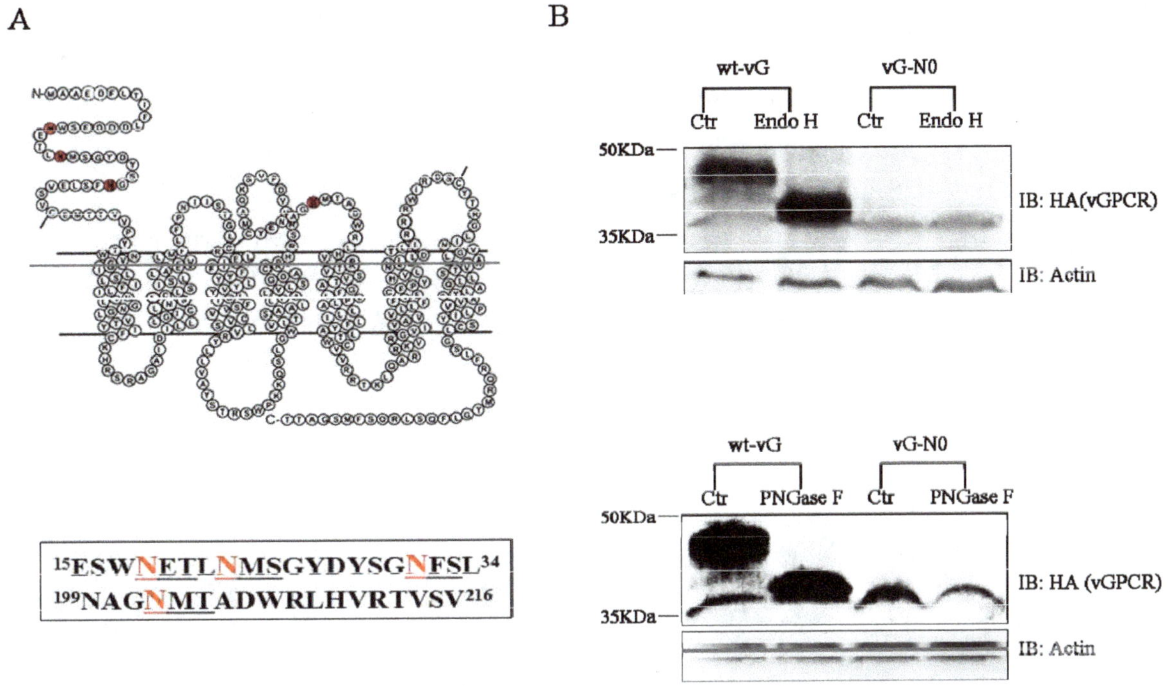

Figure 1. Kaposi's sarcoma-associated herpesvirus (KSHV) G protein-coupled receptor (vGPCR) undergoes post-translational modification of N-linked glycosylation within its extracellular sequences. (**A**) There are four potential N-linked glycosylation sites in vGPCR. Upper panel: sketch-map of vGPCR and four Asn residues in red (18, 22, 31 and 202); Lower panel: N-X-S/T consensus sequence (underlined) for N-linked glycosylation in vGPCR. (**B**) Glycosidase digestion of vGPCR and vGPCR-N0. HEK293T cells were transfected with FRT-TO-HA-vGPCR or un-glycosylated vGPCR mutant N0, cells were harvested at 48 h posttransfection and lysed, whole cell lysates were digested with Endo H (upper panel) or PGNas F separately. wt-vG: FRT-TO-HA-vGPCR; vG-N0: FRT-TO-HA-vGPCR-N0; Endo H:Endoglycosidase H; PNGase F:Peptide -*N*-Glycosidase F; IB: immunoblot; Ctr:control, the aliquot without glycosidase treatment.

3.3. Glycosylation of vGPCR Impacts Its Signal Transduction

Independent of ligand binding, vGPCR is sufficient to activate diverse signaling cascades that culminate in up-regulating cellular gene expression, including those driven by NF-κB and NFAT transcription factors. To probe the role of glycosylation in vGPCR mediated signal transduction, HEK293T cells were transfected with plasmids expressing vGPCR or its glycosylation mutants separately and the transfected cells were applied to luciferase reporter assay at 36 h post transfection. For NF-κB activation, the fold induction by un-glycosylated mutant N0 was less than 20% of that by wild type vGPCR. The NF-κB fold induction by vGPCR mutants with single site of glycosylation (N18, N22, N31 and N022) was decreased more or less (Figure 3A). For NFAT activation, the fold induction by N0 was only 25% of that by wild type vGPCR. The NFAT fold induction by vGPCR mutants with single site of glycosylation (N18, N22, N31 and N022) was decreased more or less (Figure 3B.). To further elucidate the functional mechanism of glycosylation of vGPCR in it mediated NF-κB activation, HEK293T cells were transfected with plasmids expressing vGPCR or its glycosylation mutants separately and the transfected cells were applied to immunoblot assay to examine the I-κB phosphorylation, which reflects NF-κB activation directly. As shown by immunoblot analysis, vGPCR over-expression triggered strong phosphorylation of I-κB alpha (IκB-α), which is an inhibitor of NF-κB (Figure 3C, indicated by the arrow). The phosphorylation of IκB-α triggered by N0 was dramatically attenuated and IκB-α phosphorylation mediated by vGPCR glycosylation mutants (N18, N22, N31 and N022) was decreased more or less (Figure 3C), which matched the NF-κB reporter assay (Figure 3A) very well. All the data demonstrates that N-linked glycosylation of vGPCR is necessary for the activation of NFAT and NF-κB downstream of vGPCR.

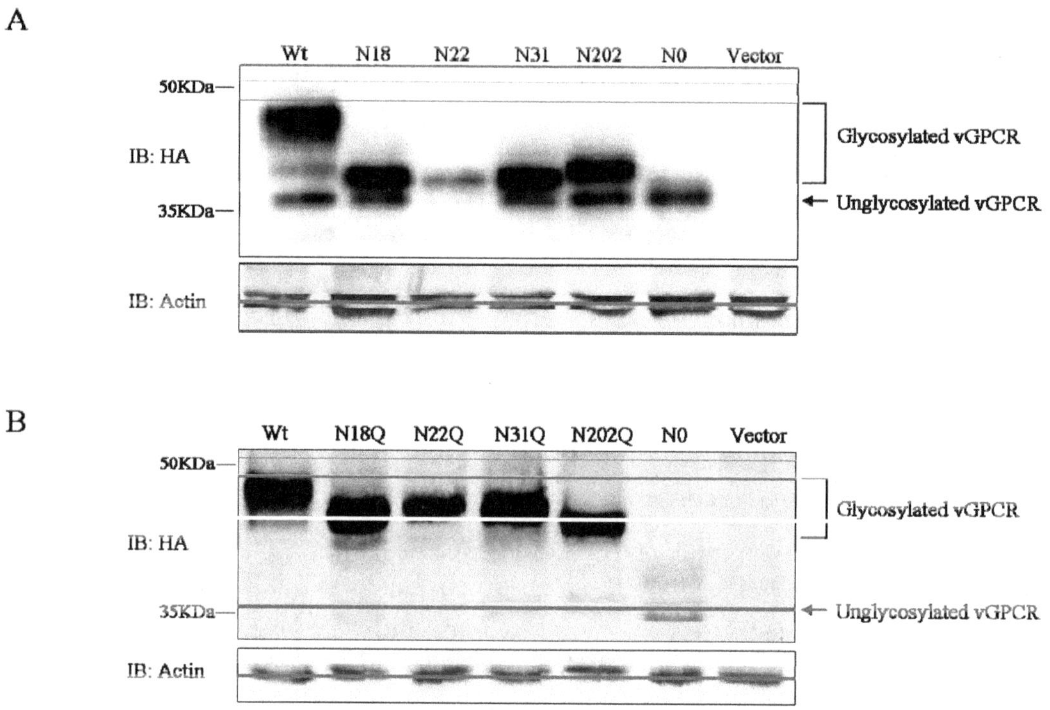

Figure 2. There exist four sites of N-linked glycosylation in vGPCR. (**A**) HEK293T cells were transfected with FRT-TO-HA-vGPCR, un-glycosylated vGPCR mutant N0 or vGPCR mutants with single site of glycosylation (N18, N22, N31 and N202) separately. Transfected cells were harvested at 48 h post-transfection and lysed, whole cell lysates were examined by immunoblot analysis. (**B**) HEK293T cells were transfected with FRT-TO-HA-vGPCR, un-glycosylated vGPCR N0 or vGPCR mutants with three sites of glycosylation (N18Q, N22Q, N31Q and N202Q) separately. Transfected cells were harvested at 48 h post-transfection and lysed, whole cell lysates were examined by immunoblot analysis. Wt: FRT-TO-HA-vGPCR; Vector: FRT-TO empty vector; N18, N22, N31, N202, N18Q, N22Q, N31Q, N22Q, N202Q, N0: FRT-TO plasmids expressing corresponding vGPCR mutants with HA tag.

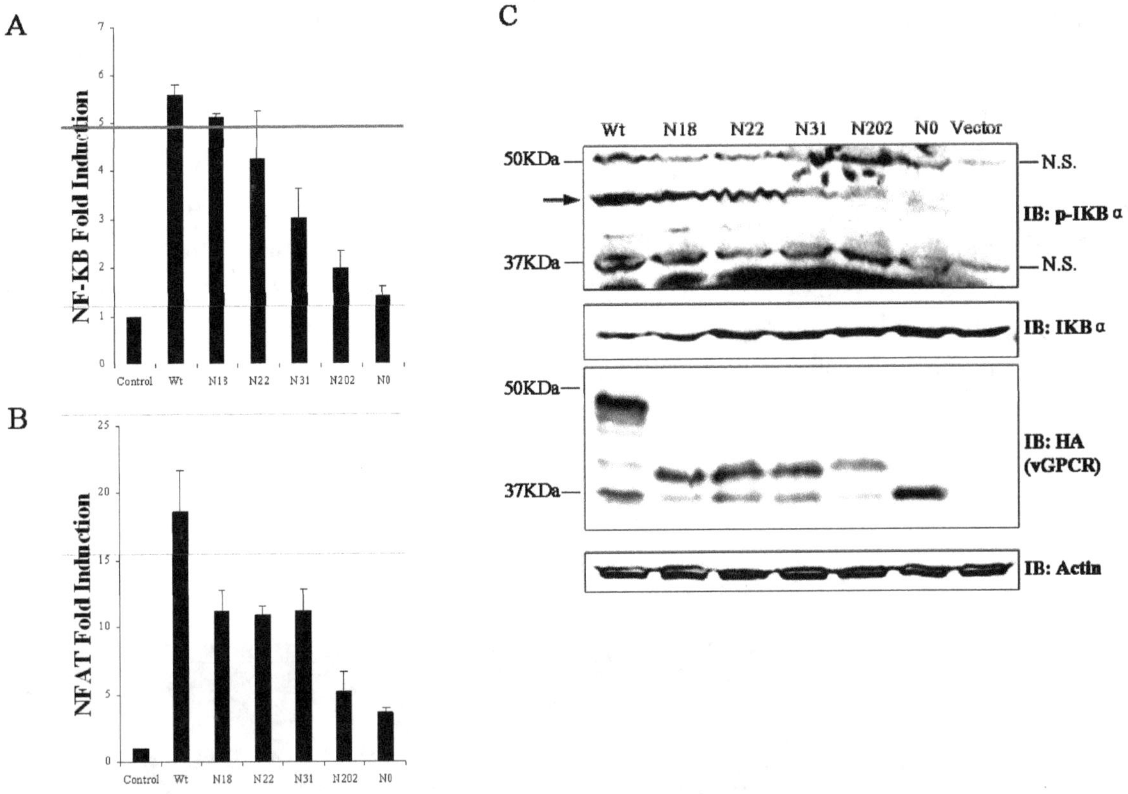

Figure 3. Glycosylation is necessary for vGPCR-mediated signaling. (**A**) and (**B**). Glycosylation is necessary for the induction of NFAT and NF-κB by vGPCR. HEK293T cells were co-transfected with Luci-NF-κB or Luci-NFAT, B-Gal reporter construct, and plasmids expressing vGPCR or vGPCR glycosylation mutants. Transfected cells were harvested and whole cell lysates were used for reporter assay according to the methods. (**A**) NF-κB fold induction by vGPCR and its glycosylation mutants. (**B**) NFAT fold induction by vGPCR and its glycosylation mutants. Data represent the average value of three independent experiments and error bars represent standard deviation. (**C**). Glycosylation is necessary for vGPCR triggered phosphorylation of I-κB. HEK293T cells were transfected with plasmids expressing vGPCR or vGPCR glycosylation mutants separately, Transfected cells were harvested at 24 h post transfection and whole cell lysates were examined by western blot to detect the IκBα phosphorylation. p-IκBα: phospho-IκBα; N.S.: non-specific band; Wt: FRT-TO-HA-vGPCR; Vector: FRT-TO empty vector; N18, N22, N31, N202, N0: FRT-TO plasmids expressing corresponding vGPCR mutants with HA tag; the arrow indicates the phosphorylated IκBα.

3.4 Glycosylation of vGPCR Plays an Important Role in Its Membrane Traffic

Glycosylation is a major regulation factor for the traffic of eukaryotic proteins, especially for GPCR's membrane traffic. vGPCR is a lytic gene and was detected on the cellular surface of the KSHV latent infected BCBL-1 cells after TPA induction. In addition to the surface presentation, vGPCR resides predominantly in the trans-Golgi network (TGN) when it was expressed in cells [13,14]. To determine whether glycosylation impacts vGPCR's surface expression, HeLa cells were transfected with plasmids expressing vGPCR or its glycosylation mutants (N18, N22, N31, N202 and N0) separately. The transfected HeLa cells were fixed and analyzed by immunofluorescence (IF) staining. The HeLa cells were permeabilized with Triton-X 100 to detect intracellular vGPCR or without detergent to detect cellular surface vGPCR, in which HeLa cells transfected with plasmid expressing wild type vGPCR as positive control. IF data showed that all the mutants could be detected in the permeabilized HeLa cells. However, vGPCR-N0 was not detected in un-permeabilized HeLa cells while all the mutants with single site of glycosylation (N18, N22, N31, N202) were detected in un-permeabilized HeLa cells. The data showed that vGPCR could not be delivered to cellular surface without the modification of N-linked glycosylation. The mutants with single site of glycosylation (N18, N22, N31 and N202) were delivered to plasma membrane, although their cellular surface expression level was much lower than that of wild type vGPCR. Taken together, the modification of N-linked glycosylation on each of the four Asn residues could help vGPCR traffic to the cellular surface (Figure 4).

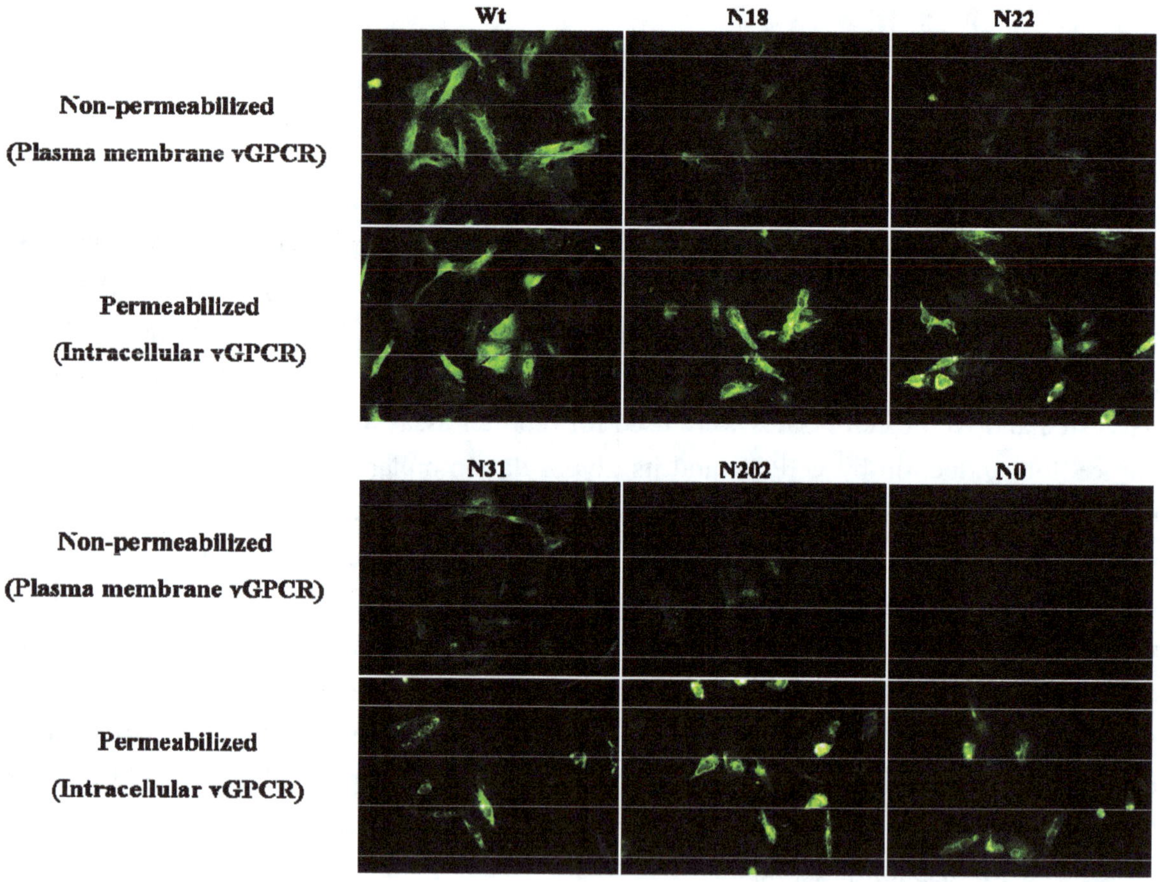

Figure 4. Glycosylation of vGPCR is crucial for its trafficking to plasma membrane. HeLa cells were transfected with plasmids expressing vGPCR or vGPCR glycosylation mutants (N18, N22, N31, N202 or N0) separately and the transfected cells were fixed at 24 h post transfection. The fixed cells were permeabilized (for detecting intracellular vGPCR) or not (for detecting vGPCR on the plasma membrane) and examined by immunoflurescence microscopy as described in methods. WT: FRT-TO-HA-vGPCR; N18, N22, N31, N202, N0: FRT-TO plasmids expressing corresponding vGPCR mutants with HA tag.

3.5. Glycosylation of vGPCR Impacts Its Tumorigenicity in Nude Mice

vGPCR possesses tyrosine sulfation in its N-terminal extracellular domain, which is crucial for vGPCR's tumorigenicity although it is not important for vGPCR mediated signaling. NIH3T3 cells stably expression vGPCR or its glycosylation mutants including N18, N22, N31, N202 and N0 were established and applied to tumor formation assay in nude mice in order to determine whether glycosylation of vGPCR functions in its tumorigenecity. Visual tumors were found in the nude mice inoculated with NIH3T3 cells expressing vGPCR at 10 days post inoculation and visual tumors were found in the nude mice inoculated with NIH3T3 cells expression vGPCR glycosylation mutants at 13~15 days post inoculation separately (data not shown). The nude mice were euthanized 21 days post inoculation and the tumor weight was measured. The average tumor weight of wild type vGPCR group was 2.122 gram; the average tumor weight of N18 group was 1.269 gram, N22 group 0.973 gram, N31 group 0.337 gram and N202 group 0.413 gram. The average tumor weight of N18, N22, N31 and N202 group was 60%, 46%, 16% and 19% of that of the wild type vGPCR group accordingly. Obviously, the tumor formation ability of vGPCR mutants with single site of glycosylation (N18, N22, N31, N202) in nude mice was much decreased compared with that of wild type vGPCR. It was interesting that nude mice of N0 group developed tumor and the average tumor weight was 0.594 gram, which was 28% of that of wild type vGPCR group (Figure 5A,B). To further elucidate the impact of glycosylation on vGPCR mediated tumorigenesis, NIH3T3 cells stably expression vGPCR or its glycosylation mutants (N18, N22, N31, N202 and N0) were applied to apoptosis assay, in which serum-free DMEM or HBSS were used to induce serum starvation-induced apoptosis. NIH3T3 cells expressing wild type vGPCR showed highest viability in both DMEM and HBSS environment after serum deprivation and NIH3T3 cells expressing vGPCR glycosylation mutants (N18, N22, N31, N202 and N0) demonstrated obviously attenuated viability in the condition of serum deprivation, which correlated with their tumorigenesis in nude mice. All the data demonstrates that glycosylation plays an important role in vGPCR mediated tumorigenesis.

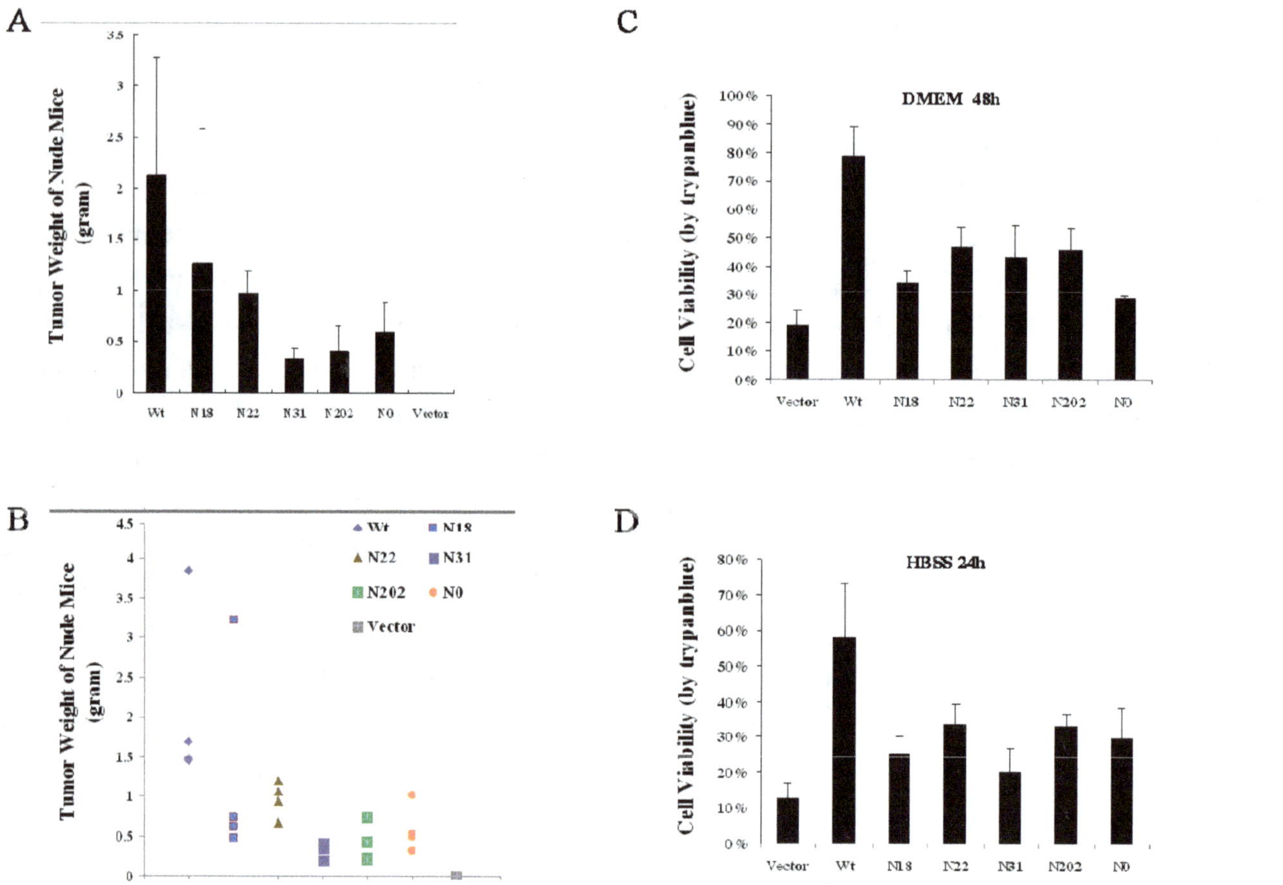

Figure 5. Glycosylation of vGPCR plays a key role in its tumorigenicity. (**A**) and (**B**): Glycosylation is crucial for vGPCR-dependent tumor formation in nude mice. NIH3T3 cells stably expressing vGPCR, its glycosylation mutants (N18, N22, N31, N202, N0) or empty vector (1×10^6 cells/mouse), together with NIH3T3 cells (2×10^6 cells/mouse) were inoculated into the flanks of nude mice separately and nude mice were euthanized at 3 weeks after inoculation. The average tumor weight of each group was shown in (**A**) and the individual tumor weight distribution of each group was shown in (**B**). Data represent 4 independent measurements for each group and error bars denote standard deviation. (**C**) and (**D**): Glycosylation is important for vGPCR-expressing cells' viability in serum starvation. NIH3T3 cells stably expressing vGPCR or its glycosylation mutants were used for serum induced apoptosis assay according to methods. Data represent three independent assays. DMEM 48 h: serum-free DMEM treatment for 48 h; HBSS 24 h: HBSS treatment for 24 h; Vector: NIH3T3 cells stably expressing the empty vector (pCDH-EF-Puro).

4. Discussion

KSHV GPCR (vGPCR) is a seven-transmembrane receptor and an acknowledged viral oncogene. vGPCR activates NF-κB and NFAT through downstream signaling and causes tumor formation in mice. Extensive studies have been done on its signaling and tumorigenicity; however, there are few reports about its post-translational modification, such as protein sulfation and protein glycosylation.

Glycosylation is a form of co-translational and post-translational modification for eukaryotic proteins and glycans serve a variety of structural and functional roles in membrane and secreted proteins.

vGPCR contains four Asn residues (residues 18, 22, 31 and 202) within its extracellular sequences, which lie within the N-X-S/T consensus sites. Data generated in this paper showed that vGPCR possesses N-linked glycosylation and all the four asparagines in vGPCR are modified with N-linked glycosylation. The glycan chains of vGPCR are all located in its extracellular domain and increase the molecular weight of vGPCR (from ~37 KDa to ~45 KDa). The immunoblot assay showed that N202 mutant migrated most slowly among the mutants with single site of glycosylation (N18, N22, N31 and N202) and N202Q migrated fastest among the mutants with three sites of glycosylation (N18Q, N22Q, N31Q and N202Q), which means the glycan chain on the Asn residue of 202 has the largest molecular weight among the four glycan chains of vGPCR (Figure 2).

Membrane glycoproteins are cotranslationally N-glycosylated in the endoplasmic reticulum (ER) and when properly folded, traffic via the secretory pathway to their final destination such as the plasma membrane. Thus, glycosylation is a key regulator of membrane traffic for many proteins, such as such as Kv1.4, PAR2, SLC4 and SLC26 family [16–18]. vGPCR is a lytic gene and was detected on the cellular surface of the KSHV latent infected BCBL-1 cells after TPA induction and vGPCR expressing NIH3T3 cells through immunofluoresent staining [13,14]. The un-glycosylated N0 mutrant was not detected on the cellular surface of transfected HeLa cells, which demonstrates glycosylation is necessary for vGPCR's traffic to plasma membrane. Any of the four Asn residues with glycosylation is sufficient for vGPCR's membrane traffic since mutants including N18, N22, N31 and N202 were all detected on the plasma membrane although cellular surface expression level was obviously lower than that of wild type vGPCR (especially N202). The cellular surface expression level of vGPCR mutants with three glycosylation sites (N18Q, N22Q, N31Q and N202Q) was much higher than that of single glycosylation site (N18, N22, N31 and N202), even nearly to that of wild type vGPCR (data not shown here), which implies that cellular surface expression level of vGPCR glycosylation mutants is directly proportional to the number of glycan chains.

vGPCR is a constitutively activated viral receptor and mediates a broad spectrum of signaling pathway; however, ligand association regulates vGPCR mediated signaling, which suggest that cellular surface expression and post-translational modification are crucial vGPCR's function [19,20]. Actually, our previous data showed vGPCR possesses tyrosine sulfation within its N-terminal extracellular domain. The refinement modification of tyrosine sulfation is specific for vGPCR's association with GRO-α and crucial for vGPCR-dependent tumorigenesis, although it does not impact vGPCR's membrane traffic and vGPCR mediated signaling [14]. The data that N0 was not detected on plasma membrane might explain why this un-glycosylated vGPCR mutant is almost deprived of the ability to activate NFAT and NF-κB in the reporter assay. The glycan chains on the site of 18, 22 and 31 are within the extracellular N-terminus of vGPCR, which is the crucial domain for chemokine binding. However, the glycan chain on the site of 202 is located in the 2rd extracellular loop of vGPCR, which might explain why N202 mutant showed the lowest fold induction of NF-KB and NFAT among the four mutants with single glycosyltiaon site (N18, N22, N31 and N202) in reporter assay (Figure 2). All the data emphasize that regulation of vGPCR on the plasma membrane is crucial for the function of this viral G protein-coupled receptor.

The immunoblot assay showed that expression level of N0 mutant in tissue culture was lower than that of wild type vGPCR and other glycosylation mutants when HEK293T cells were transfected with plasmids expressing vGPCR or its glycosylation mutants, which implied the un-glycosylated vGPCR mutant was not as stable as wild type vGPCR. Our data suggests that N-linked glycosylation plays an important role in protein folding and stability of vGPCR. It is imaginable that a N0 mutant, because of being deprived of modification of N-linked glycosylation, was not folded correctly and subjected to being retained in ER instead of being delivered to cellular membrane when this mutant was expressed in tissue culture. Misfolded N0 was targeted for degradation by the proteasome and might cause ER stress during rapid degradation when expressed in tissue culture [21]. That might be the reason why NIH3T3 cells expressing N0 mutant triggered tumor development in nude mice.

In conclusion, this paper demonstrates that vGPCR possesses modification of N-linked glycosylation at the Asn residues of 18, 22, 31 and 202, which is necessary for vGPCR's membrane traffic. *In vitro* and *in vivo* data elucidate that N-linked glycosylation of vGPCR plays crucial roles in vGPCR mediated signaling and in vGPCR-dependent tumorigenesis.

Acknowledgments

This work was supported by the National Natural Science Foundation of China (81171583, 81471963), Program for New Century Excellent Talents in University (NCET-11-0971), Hunan Provincial Natural Science Foundation (12JJ1005), Fund of Hunan Provincial Education Department (12A088) and the Cooperative Innovation Center of Engineering and New Products for Developmental Biology of Hunan Province (20134486).

Author Contributions

Hao Feng and Hui Wu conceived and designed the experiments; Hui Wu, Liqun Liu, Jun Xiao, Mengdie Chi, Yixiao Qu and Hao Feng performed the experiments; Hui Wu and Hao Feng analyzed the data; Hui Wu and Hao Feng wrote the paper.

References

1. Chang, Y.; Cesarman, E.; Pessin, M.S.; Lee, F.; Culpepper, J.; Knowles, D.M.; Moore, P.S. Identification of herpesvirus-like DNA sequences in AIDS-associated Kaposi's sarcoma. *Science* **1994**, *266*, 1865–1869.
2. Cesarman, E.; Chang, Y.; Moore, P.S.; Said, J.W.; Knowles, D.M. Kaposi's sarcoma-associated herpesvirus-like DNA sequences in AIDS-related body-cavity-based lymphomas. *N. Engl. J. Med.* **1995**, *332*, 1186–1191.
3. Bais, C.; Santomasso, B.; Coso, O.; Arvanitakis, L.; Raaka, E.G.; Gutkind, J.S.; Asch, A.S.; Cesarman, E.; Gershengorn, M.C.; Mesri, E.A. G-protein-coupled receptor of Kaposi's sarcoma-associated herpesvirus is a viral oncogene and angiogenesis activator. *Nature* **1998**, *391*, 86–89.

4. Jham, B.C.; Montaner, S. The Kaposi's sarcoma-associated herpesvirus G protein-coupled receptor: Lessons on dysregulated angiogenesis from a viral oncogene. *J. Cell. Biochem.* **2010**, *110*, 1–9.
5. Cannon, M.; Philpott, N.J.; Cesarman, E. The Kaposi's sarcoma-associated herpesvirus G protein-coupled receptor has broad signaling effects in primary effusion lymphoma cells. *J. Virol.* **2003**, *77*, 57–67.
6. Emuss, V.; Lagos, D.; Pizzey, A.; Gratrix, F.; Henderson, S.R.; Boshoff, C. KSHV manipulates notch signaling by DLL4 and JAG1 to alter cell cycle genes in lymphatic endothelia. *PLoS Pathog.* **2009**, *5*, e1000616.
7. Martin, D.; Galisteo, R.; Molinolo, A.A.; Wetzer, R.; Hirsch, E.; Gutkind, J.S. PI3Kγ mediates Kaposi's sarcoma-associated herpesvirus vGPCR-induced sarcomagenesis. *Cancer Cell* **2011**, *19*, 805–813.
8. Montaner, S.; Sodhi, A.; Molinolo, A.; Bugge, T.H.; Sawai, E.T.; He, Y.; Li, Y.; Ray, P.E.; Gutkind, J.S. Endothelial infection with KSHV genes *in vivo* reveals that vGPCR initiates Kaposi's sarcomagenesis and can promote the tumorigenic potential of viral latent genes. *Cancer Cell* **2003**, *3*, 23–36.
9. Sodhi, A.; Montaner, S.; Patel, V.; Gomez-Roman, J.J.; Li, Y.; Sausville, E.A.; Sawai, E.T.; Gutkind, J.S. Akt plays a central role in sarcomagenesis induced by Kaposi's sarcoma herpesvirusencoded G protein-coupled receptor. *Proc. Natl. Acad. Sci. USA* **2004**, *101*, 4821–4826.
10. Sodhi, A.; Chaisuparat, R.; Hu, J.; Ramsdell, A.K.; Manning, B.D.; Sausville, E.A.; Sawai, E.T.; Molinolo, A.; Gutkind, J.S.; Montaner, S. The TSC2/mTOR pathway drives endothelial cell transformation induced by the Kaposi's sarcoma-associated herpesvirus G protein-coupled receptor. *Cancer Cell* **2006**, *10*, 133–143.
11. Montaner, S.; Sodhi, A.; Servitja, J.M.; Ramsdell, A.K.; Barac, A.; Sawai, E.T.; Gutkind, J.S. The small GTPase Rac1 links the Kaposi sarcoma-associated herpesvirus vGPCR to cytokine secretion and paracrine neoplasia. *Blood* **2004**, *104*, 2903–2911.
12. Martin, D.; Galisteo, R.; Ji, Y.; Montaner, S.; Gutkind, J.S. An NF-κB gene expression signature contributes to Kaposi's sarcoma virus vGPCR-induced direct and paracrine neoplasia. *Oncogene* **2008**, *27*, 1844–1852.
13. Feng, H.; Dong, X.; Negaard, A.; Feng, P. Kaposi's sarcoma-associated herpesvirus K7 induces viral G protein-coupled receptor degradation and reduces its tumorigenicity. *PLoS Pathog.* **2008**, *4*, e1000157.
14. Feng, H.; Sun, Z.; Farzan, M.R.; Feng, P. Sulfotyrosines of the Kaposi's sarcoma-associated herpesvirus G protein-coupled receptor promote tumorigenesis through autocrine activation. *J. Virol.* **2010**, *84*, 3351–3361.
15. Wu, H.; Fu, Y.; Xiao, J.; Zhou, M.; Zhou, W.; Feng, H. The unsulfated extracellular N-terminus of vGPCR reduces the tumorigenicity of hGRO-α in nude mice. *Sci. China Life Sci.* **2013**, *56*, 26–31.
16. Watanabe, I.; Zhu, J.; Recio-Pinto, E.; Thornhill, W.B. Glycosylation affects the protein stability and cell surface expression of Kv1.4 but Not Kv1.1 potassium channels. A pore region determinant dictates the effect of glycosylation on trafficking. *J. Biol. Chem.* **2004**, *279*, 8879–8885.

17. Cunningham, M.R.; McIntosh, K.A.; Pediani, J.D.; Robben, J.; Cooke, A.E.; Nilsson, M.; Gould, G.W.; Mundell, S.; Milligan, G.; Plevin, R. Novel role for proteinase-activated receptor 2 (PAR2) in membrane trafficking of proteinase-activated receptor 4 (PAR4). *J. Biol. Chem.* **2012**, *287*, 16656–16669.

18. Cordat, E.; Reithmeier, R.A. Structure; function; and trafficking of SLC4 and SLC26 anion transporters. *Curr. Top. Membr.* **2014**, *73*, 1–67.

19. Bais, C.; van Geelen, A.; Eroles, P.; Mutlu, A.; Chiozzini, C.; Dias, S.; Silverstein, R.L.; Rafii, S.; Mersi, E.A. Kaposi's sarcoma associated herpesvirus G protein-coupled receptor immortalizes human endothelial cells by activation of the VEGF receptor-2/ KDR. *Cancer Cell* **2003**, *3*, 131–143.

20. Hu, J.; Jham, B.C.; Ma, T.; Friedman, E.R.; Ferreira, L.; Wright, J.M.; Accurso, B.; Allen, C.M.; Basile, J.R.; Montaner, S. Angiopoietin-like-4: A novel molecular hallmark in oral Kaposi's sarcoma. *Oral Oncol.* **2011**, *47*, 371–375.

21. Moenner, M.; Pluquet, O. ; Bouchecareilh, M. ; Chevet, E. Integrated endoplasmic reticulum stress responses in cancer. *Cancer Res.* **2007**, *67*, 10631–10634.

8

Molecular Biology of KSHV Lytic Reactivation

Pravinkumar Purushothaman †**, Timsy Uppal** † **and Subhash C. Verma** *

Department of Microbiology and Immunology, University of Nevada, Reno, School of Medicine, 1664 N Virginia Street, MS 320, Reno, NV 89557, USA;
E-Mails: pravinp@medicine.nevada.edu (P.P.); tuppal@medicine.nevada.edu (T.U.)

† These authors contributed equally to this work.

* Author to whom correspondence should be addressed; E-Mail: scverma@medicine.nevada.edu;

Academic Editor: Zhi-Ming Zheng

Abstract: Kaposi's sarcoma-associated herpesvirus (KSHV) primarily persists as a latent episome in infected cells. During latent infection, only a limited number of viral genes are expressed that help to maintain the viral episome and prevent lytic reactivation. The latent KSHV genome persists as a highly ordered chromatin structure with bivalent chromatin marks at the promoter-regulatory region of the major immediate-early gene promoter. Various stimuli can induce chromatin modifications to an active euchromatic epigenetic mark, leading to the expression of genes required for the transition from the latent to the lytic phase of KSHV life cycle. Enhanced replication and transcription activator (RTA) gene expression triggers a cascade of events, resulting in the modulation of various cellular pathways to support viral DNA synthesis. RTA also binds to the origin of lytic DNA replication to recruit viral, as well as cellular, proteins for the initiation of the lytic DNA replication of KSHV. In this review we will discuss some of the pivotal genetic and epigenetic factors that control KSHV reactivation from the transcriptionally restricted latent program.

Keywords: KSHV; reactivation; Lytic DNA replication; hypoxia; RTA; LANA

---

1. Introduction

Kaposi's sarcoma-associated herpesvirus (KSHV) or human herpesvirus 8 (HHV 8) is one of the seven human oncogenic viruses, and is the etiologic agent of Kaposi's sarcoma (KS, a multifocal, angiogenic and inflammatory malignancy of endothelial cell origin), as well as certain B-cell lymphomas, including primary effusion lymphoma (PEL) and multicentric Castleman's disease (MCD) [1–3]. KSHV has been consistently detected in all four clinical forms of KS, including: classical KS, endemic KS in Africa, epidemic AIDS-related KS, and iatrogenic/organ-transplant KS. Lately, a newly characterized KSHV-associated condition, abbreviated as KICS (KSHV Inflammatory Cytokine Syndrome) has been reported in patients with HIV and KSHV co-infection, displaying elevated levels of interleukin-6 (IL-6) production [4]. In healthy seropositive individuals, KSHV causes persistent infection by establishing latency in CD19+ peripheral B-lymphocytes [5].

Since its initial discovery, there has been intense research on understanding the molecular biology of KSHV-mediated tumorigenesis [6]. KSHV is a γ2-lymphotropic-oncogenic-herpesvirus and is genetically linked to the Epstein-Barr virus (EBV), murine γ-herpesvirus-68 (MHV-68), and herpesvirus saimiri (HVS) (reviewed in [7]). Similar to the other members of herpesvirus family, KSHV enters the host cell as a linear double-stranded DNA genome (160–175 kb), encapsidated in an icosahedral protein capsid that is surrounded by a tegument layer and an outer lipid bilayer envelope containing glycoproteins (reviewed in [8]). Upon infection, viral DNA is delivered to the nucleus, where it circularizes to a functional circular minichromosome and persists as a non-integrated episome for the lifetime of the host (reviewed in [9]).

Inside the host cell, KSHV exhibits a biphasic life cycle consisting of a life-long reversible latent phase and a transient lytic reactivation phase, which are distinguished by their virtually distinct gene expression profiles [10]. During latent infection, KSHV genome persists as a circular episome in the infected cell with a restricted latent gene expression without the production of progeny virions. The limited region within the KSHV genome, which is transcriptionally active during latency, encodes for four major open reading frames (ORFs), consisting of ORF73/Latency-associated nuclear antigen (LANA), ORF72/viral-cyclin (v-Cyc), ORF71/viral FLICE-inhibitory protein (v-FLIP), and ORFK12/Kaposins, along with 18 mature miRNAs (at last count) and viral interferon regulatory factor-3 (vIRF3) (reviewed in [11]). KSHV has a propensity to cause latent infection that is tightly regulated by the host immune responses and has been reported to play a significant role in the development of KSHV-associated malignancies. Since the KSHV genome does not encode any viral components required for latent DNA replication, Latency Associated Nuclear Antigen (LANA), the multifarious latent protein, is considered necessary (and sufficient) for latent viral episomal DNA replication and segregation, ensuring equal distribution of replicated episomes to each daughter cell during mitosis. To achieve this, LANA binds to the LANA-binding sequences within the terminal repeat (TR) region of the KSHV genome and tethers it to the host mitotic chromosomes via interaction, with several cellular chromatin-binding proteins, followed by replication of the viral genome using a *cis*-acting sequence in the TR region as a replication origin [12].

Global analysis of the viral gene expression of KS tumor cells indicated that KSHV predominantly expresses viral latent transcripts with only a few percent of cells being lytically active at any specific given time [13]. The latent phase of the viral life cycle is reversible and can be reactivated to enter into the second, well-ordered program of viral gene expression, *i.e.*, lytic reactivation. This phase predominantly consists of: (1) *KSHV Reactivation from latency*, followed by (2) *Lytic DNA Replication* and virion production. Upon reactivation from latency, a full repertoire of lytic viral genes are activated in a temporally regulated manner, leading to the transcriptional activation of three classes of lytic genes, namely, immediate early (IE), early (E), and late (L) genes [14–16]. The cellular machinery is switched on for an extensive viral DNA replication and gene expression, resulting in the assembly and release of infectious mature virion particles that egress out of the cell on disruption of the host-cell membrane. KSHV reactivation and lytic replication are not only important for viral propagation but also critical for KSHV-induced tumorigenesis.

Members of all three classes of lytic viral genes encode for proteins that assist in the formation of infectious virions [17]. The IE-lytic genes primarily govern the transition of KSHV genome from latent-to-lytic phase and consist of ORF50/RTA, ORF45, K8α, K8.2, K4.2, K4.1, K4, ORF48, ORF29b, K3, and ORF70. These genes are expressed within 10 h of induction and encode viral proteins that are directly involved in gene transcription and cellular modifications for viral replication. A series of studies have established that a single major IE-lytic protein-RTA acts as the quintessential latent-lytic switch that redirects KSHV to enter the productive transcriptional program required for viral spread and KS tumorigenesis. RTA protein (691 aa and 110 kDa) is the only viral lytic protein, both necessary and sufficient to disrupt latency and promote complete lytic cascade [18]. The RTA gene is reported to auto-activate its own promoter and transactivate the expression of multiple downstream lytic genes, including K8, K5, K2, K12, ORF6, ORF57, ORF74, K9, ORF59, K3, ORF37, K1, K8.1A, ORF21, vIL-6, PAN RNA, vIRF1, K1, and ORF65, either by itself (through RTA-responsive element, RRE) or in accord with other viral regulatory genes [19]. These E-lytic genes are expressed between 10–24 h post-induction and encode viral proteins primarily required for DNA replication and gene expression. The L-lytic genes that appear after 48 h post infection consist of viral structural proteins, including membrane glycoproteins (gB and K8.1), and a small viral capsid antigen required for assembly and maturation of the virions [20].

RTA plays an important role as both an initiator and a controller of KSHV lytic DNA replication [21]. Unlike latent DNA replication, lytic DNA replication: (1) depends on KSHV-encoded replication proteins; (2) initiates from a different origin (ori-Lyt); (3) replicates via a rolling-circle mechanism; and (4) leads to a multifold amplification of the viral DNA. The lytic origin of replication (ori-Lyt) consists of a specific origin binding protein (OBP) that plays a significant role in recruiting the core replication machinery to the site of replication. The two ori-Lyt domains, namely left ori-Lyt (ori-Lyt-L) and right ori-Lyt (ori-Lyt-R), are located between K4.2 and K5 and between ORF69 and ORF71, respectively, in the KSHV genome [22,23]. The ori-Lyts contain regions for various transcription factor-binding sites and RRE element that is essential for RTA-binding and ori-Lyt dependent DNA replication [22,23].

Despite the induction of lytic cycle following KSHV infection, there is a rapid inhibition of RTA promoter that further decelerates the full-blown KSHV reactivation [24]. The mechanisms that regulate the temporally ordered activation and genome-wide repression of lytic genes during primary infection are beginning to be resolved [25]. As both phases of KSHV life cycle are important for the development of KS and associated disorders, further understanding of the underlying mechanisms that coordinate regulation of gene expression may advance our knowledge of KSHV virology and assist in designing preventive therapeutic agents against KSHV lytic replication and associated tumorigenesis.

KSHV reactivation is an extremely complex process that involves a combination of both viral and cellular factors including but not limited to, temporary or permanent immune suppression, oxidative stress, inflammatory cytokines, hypoxia, viral co-infection and treatment with chromatin modifying agents. Thus far, a number of factors have been reported to stimulate or inhibit major viral proteins, however, the physiological relevance of these stimuli or repressors is far from being fully elucidated. In the following sections of this review, we will summarize recent studies that highlight the activation of KSHV lytic cycle and replication and will primarily focus on the relevant physiological, environmental, cellular, and viral regulatory factors involved in the regulation of KSHV's biphasic life cycle, gene expression, and viral infection.

2. LANA and KSHV Reactivation

The two major KSHV proteins-LANA and RTA are shown to interact with each other and control the switch between latency and lytic reactivation [26]. Studies from multiple research groups reported a tremendous increase in the expression of several IE-lytic genes including RTA, MTA, vIL-6, ORF59, and K8.1 in 293T cells following deletion of LANA, indicating LANA-associated repression of basal level of RTA promoter as well as other RTA-responsive promoters [27,28]. LANA is shown to interact with RTA promoter and inhibit RTA gene expression via functional interaction with a recombination signal binding protein for immunoglobulin κ J region (RBP-Jκ protein), which is a major transcriptional repressor of the Notch signaling pathway [29]. LANA-mediated repression of RTA promoter and RTA auto-activation depends on RBP-Jκ binding sites. LANA recruits RBP-Jκ protein to repress the expression of RTA gene and down-regulates RTA self-activation by competing with RTA in RBP-Jκ-binding. In addition, RTA protein itself contributes to the establishment of KSHV latency by activating LANA protein expression following *de novo* infection. Therefore, the molecular transition between latency and lytic reactivation is controlled by the interplay between LANA and RTA proteins in KSHV-infected cells.

DNA methylation or CpG dinucleotide methylation, associated with the transcriptional silencing, also plays a key role in the induction of KSHV lytic cycle as the treatment of PEL-derived cell lines with DNA methyltransferase inhibitor, 12-o-tetradecanoylphorbol-13-acetate (TPA) or 5-Azacytidine (5-AzaC) caused demethylation of lytic promoters and induced KSHV lytic phase *in vitro* [30]. Bisulfite sequencing of latently infected BCBL-1 cell lines revealed hypermethylation of functionally conserved RTA gene of KSHV genome by *de novo* methyltransferases DNMT3a/DNMT3b and establishment of methylation marks exclusively on RTA promoter, leading to gene silencing during latency [30–32]. Recent studies by Grundhoff's group reported a comprehensive tempo-spatial analysis of DNA methylation in several tumor-derived cell lines, as well as *de novo* infected endothelial cells using high resolution tiling microarrays together with immunoprecipitation of methylated DNA (MeDIP) [32]. These studies revealed that the KSHV genome is indeed subjected to hypermethylation at CpG dinucleotides, leading to the distinct, genome-wide DNA methylation patterns that include extensive methylation of lytic promoters followed by a poised state of repression during latency.

Interestingly, post-translational modifications of LANA, such as arginine methylation, phosphorylation and SUMOylation, have been shown to down-regulate the expression of lytic genes during the establishment of latency [28,33–35]. Treatment of BCBL-1 cells with histone deacetylase inhibitors, including sodium butyrate (NaB) and trichostatin A (TSA), caused a rapid dissociation of LANA from the RTA promoter and initiated transcription activation of RTA gene [28]. Furthermore, reports on phosphorylation of LANA by several kinases including glycogen synthase kinase (GSK-3β), DNA-PK/Ku and Pim-family kinase members, Pim-1 and Pim-3, have been reported to promote viral reactivation by negative modulation of LANA function [36–38]. LANA is also identified as a substrate for protein arginine methyltransferase 1 (PRMT1) and methylation at R20 site is found to influence strong binding of LANA to the KSHV genome and repression of lytic genes [39]. LANA is proposed to enhance histones (H2A and H2B) SUMOylation on the local chromatin by recruiting SUMO-Ubc9 complexes through SUMO-binding, resulting in a condensed chromatin and silencing of the KSHV genome (reviewed in [40]).

The early stage of KSHV infection is defined by the constitutive expression of latent genes, as well as temporally ordered expression of viral lytic genes. Recent genome-wide ChIP-seq studies described the epigenetic map of KSHV episomes during latency and indicated that chromatin of the KSHV genome is enriched with both active (H3ac or H3K4me3) and repressive histone marks (H3K9me3 and H3K27me3) [32,41]. Based on studies reported by several independent groups, KSHV-encoded latent genes are found to be associated with activating H3ac/H3K4me3-histone marks, whereas KSHV-encoded IE and E-lytic genes are found to possess either a H3ac/H3K4me3-rich euchromatin or a H3ac/H3K4me3 and H3K27me3-rich bivalent chromatin, and L-lytic genes are found to have increased levels of heterochromatin-associated repressive H3K9me3 and H3K27me3-histone marks. In addition, H3K9 histone demethylase JMJD2A, and H3K27 histone methyltransferase EZH2 of the Polycomb Repressive Complex 2 (PRC2), predominantly bind to the KSHV genome and their recruitment by LANA is shown to maintain H3K27me3-associated silencing marks on lytic genes and repress their expression during latency (reviewed in [9]). Decrease of H3K27me3 marks, by either transient expression of UTX/JMJD3, or by blocking with EZH2 of PRC2 complex, disrupts latency and induces lytic reactivation [32,41]. As LANA is continuously expressed following *de novo* infection/during latency and interacts with several transcriptional repressors (heterochromatin protein HP1α, methyl-CpG-binding protein MeCP2, histone deacetylase co-repressor mSin3 and DNA methyltransferases) and chromatin-remodeling proteins (H3K9me3 histone methyltransferase SUV39H1 and hSET1 complexes, H3K9 demethylase KDM3A, histone acetyltransferase CBP, histone deacetylase mSin3 and chromatin transcription complex FACT), it is evident that LANA helps to silence lytic gene expression and promotes KSHV latency through epigenetic control [42–47].

3. Stimulus Triggering KSHV Reactivation

Thus far, several PEL-infected cells, endothelial cells, CV-1, human fibroblasts and HEK cells are known to maintain KSHV in the latent form that can be induced to enter the complete productive cycle of KSHV, following treatment of cells with the broad-spectrum protein kinase C-activator (TPA) or histone deacetylase inhibitor (NaB) (reviewed in [48]). As a result, these cell lines serve as an authentic tumor model to study KSHV life cycle, providing several insights into the numerous cellular pathways that control viral reactivation. As these chemicals target numerous cellular and viral pathways, it appears that more than one mechanism is necessary to reactivate KSHV. More recently, KSHV was found to efficiently infect, immortalize, and transform, rat embryonic metanephric mesenchymal precursor (MM) cells [49]. KSHV-transformed MM cells (KMM) support the growth of KSHV-induced tumors, hence, providing a novel animal model to study the intrinsic oncogenic pathways underlying KSHV latency and reactivation.

3.1. Viral Co-Infection

While KSHV infection appears to be necessary for the development of KS, the immunodeficiency appears to be another significant factor, as the immunosuppressed patients are often susceptible to many other infectious agents [50]. Several viral proteins, including HIV-1 trans-activating protein (HIV-1 tat) [51], HIV-1 negative factor protein (HIV-1 Nef) [52], herpes simplex virus type 1 (HSV-1) [53], herpes simplex virus type 2 (HSV-2) [54], human cytomegalovirus (HCMV) [50], human herpesvirus-6 (HHV-6), herpes simplex virus type 2 (HSV-2) [54], human cytomegalovirus (HCMV) [50], human herpesvirus-6 (HHV-6) [55], human herpesvirus-7 (HHV-7) [56], and papillomavirus [57] are proven to be potent cofactors that can activate KSHV lytic replication and influence KSHV pathogenesis. In addition, it has been demonstrated that inflammatory cytokines, such as oncostatin M (OSM), hepatocyte growth factor (HGF), interferon-γ (IFN-γ) [58], and toll-like receptors 7 and 8 (TLR7/8), when stimulated by viral infections, can trigger KSHV reactivation (reviewed in [59]).

3.2. Hypoxia

As an important co-factor, hypoxia (low tissue oxygen concentration) is physiologically linked with the initiation and progression of KSHV-associated cancers and known to induce the accumulation of hypoxia-inducible factors (HIF-1α/2α) (reviewed in [59]). Hypoxic stress in PEL cells is shown to stimulate KSHV lytic reactivation through accumulation of HIF-1α within the hypoxia-responsive elements (HRE, 5'-RCGTCG-3') region of RTA promoter, and accumulation of HIF-1α/2α within the HRE2 regions of ORF34-37 promoters. Hypoxia also triggers the activation of plasma cell-differentiation factor X-box binding protein 1 (XBP-1) that trans-activates the KSHV RTA promoter with HIF-1α, leading to the expression of RTA protein and reactivation from latency [60]. Splicing of XBP-1 mRNA, an event that occurs during B-cell differentiation, is also critical for disrupting latency and promoting KSHV reactivation, with the possibility of integration of latter into the host cell differentiation program. In addition, under hypoxic conditions, LANA is reported to interact with HIF-1α bound to HREs within the RTA promoter to upregulate its gene expression [61]. Recent studies by Cai *et al.* have demonstrated that LANA interacts with a new host nuclear protein and hypoxia-sensitive chromatin remodeler, KAP1 (KRAB-associated protein 1), through its SUMO-2 interacting motif (LANA^{SIM}), and recruits it to the lytic promoter region of the KSHV genome for transcriptional repression [62]. Inhibition of KAP1 in KSHV-infected PEL cells enhanced the hypoxia-induced lytic reactivation through association of RBP-Jκ with HIF-1α within the RTA promoter region [63]. In KSHV-harboring cells, shRNA knockdown of KAP-1 resulted in the induction of lytic genes and a five-fold increase of RTA-mediated lytic reactivation.

3.3. Oxidative Stress and Reactive Oxygen Species (ROS)

As all forms of KS are characterized by increased levels of inflammation and oxidative stress, it is postulated that reactive oxygen species (ROS), such as hydrogen peroxide (H_2O_2), mediate KSHV reactivation from latency (reviewed in [59]). A recent report showed that hypoxia and pro-inflammatory cytokines-mediated spontaneous KSHV reactivation and lytic replication are supported by hydrogen peroxide (H_2O_2) through both autocrine and paracrine signaling [64]. H_2O_2 is sufficient for inducing and mediating KSHV lytic replication in KS tumors by activating ERK1/2, JNK, and p38 mitogen-activated protein kinase (p38 MAPK) pathways [65]. Significantly, treatment with antioxidant/H_2O_2 scavengers; *N*-acetyl-L-cysteine (NAC), catalase and glutathione peroxidase inhibits KSHV lytic replication and tumor progression *in vivo* and slows down the development of KSHV-induced lymphoma in a mouse xenograft model [64]. Another study reported that, in infected PEL cell lines BC-3 and BCBL-1, increased levels of reactive oxygen species (ROS) may induce oxidative stress that can trigger transcriptional activation of KSHV lytic cycle and promote cell death [66]. Additionally, ROS levels can be upregulated by NF-kB inhibition and treatment of infected cells with an increased amount of NF-kB inhibitor than used for inducing KSHV reactivation,,can further elevate ROS levels and induce apoptosis [66]. In addition, p38 signaling and anti-cancer drugs (cisplatin and arsenic trioxide) are found to induce KSHV lytic cycle and host cell death in an ROS-dependent manner [66]. These results directly relate KSHV reactivation to oxidative stress and inflammation, suggesting that the antioxidants and anti-inflammation drugs could be potential drugs for effectively targeting KSHV lytic replication and KSHV-associated tumorigenesis.

3.4. Histone Deacetylases and Histone Deacetylase Inhibitors (HDACs and HDACi)

Several research groups have reported that HDAC Class I, II, and III can regulate KSHV reactivation, and activation of lytic gene expression can be triggered by treatment of KSHV latent cells with HDAC inhibitors [9,67,68]. HDACs are a group of enzymes that remove acetyl groups from ε-*N*-acetyl lysine amino acids in histones/proteins and play an important role in the regulation of gene expression. As mentioned earlier, during latency, IE and E-lytic genes possess bivalent chromatin associated with both repressive (H3K9me3 and H3K27me3) and activating (H3K4me3, H3ac and H3K9/K14-ac)-histone marks. In addition, previous studies showed that demethylation of H3K27me3 using UTX or dissociation of the histone methyltransferase EZH2, counteracts PRC2 repression of the RTA promoter [32,41]. In order to determine which HDAC classes (Class I and II) regulate KSHV latency and reactivation, five latently infected Vero- and PEL-cell lines were treated with a series of HDACi, including Valproic acid (VPA), trichostatin A (TSA), nicotinamide, sirtinol, tubacin, and NaB (Figure 1) [67]. The results indicated that HDAC class I inhibitors of were sufficient enough to induce KSHV virus and lytic gene expression with varied reactivation potential. Out of all the HDACi tested, VPA was found to be the most effective inducer of lytic cycle gene expression, followed by TSA. The data suggested that inhibition of HDAC class I molecules, alone, is sufficient to reactivate KSHV but the inhibition of class I and IIa molecules, together, is optimal for reactivation [67].

Additionally, Gao's research group recently determined the role of Class III HDACs inhibitors or sirtuins (SIRTs) on the KSHV life cycle and reactivation by treatment of KSHV-positive PEL cell lines (BCP-1, BC-3 and BCBL-1) with three distinct HDACi, namely-nicotinamide (NAM), sirtinol and NaB [68]. The studies revealed that both NAM and sirtinol could efficiently reactivate KSHV from latency. In addition, it was shown that SIRT1 is involved in the control of latency and can prevent the expression of several downstream genes due to its interaction with the RTA promoter [68].

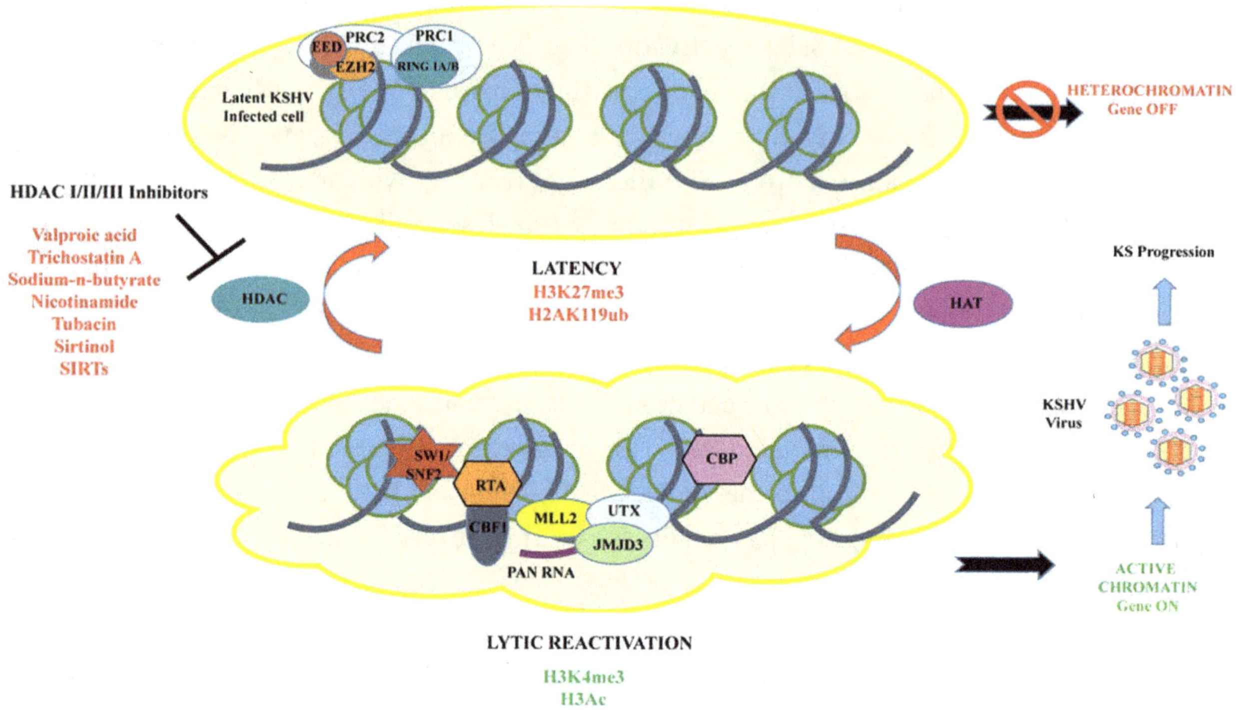

Figure 1. A model for the chromatin landscape of RTA promoter during KSHV latency and lytic reactivation. During latency, the chromatin of RTA promoter is enriched in both activating (H3ac/H3K4me3) and repressive (H3K27me3)-histone marks, as well as the transcription repressors (Polycomb Repressive Complex 2 and HDACs), hence, the RTA promoter is transcriptionally silent. Following reactivation, the bivalent chromatin of RTA promoter is remodeled into transcriptionally active euchromatin by histone modifying enzymes, such as histone acetylases (HAT/CBP), H3K27me3 demethylase (UTX/JMJD3), H3K4 methyltransferase (MLL complex), and inhibitors of HDACs (Valproic acid, trichostatin A, NaB, nicotinamide, sirtinol, tubacin, and SIRTs, leading to the production of infectious KSHV virious and progression of KSHV-induced malignancies.

3.5. Dietary Supplements

A recent report determined that Resveratrol (Rev), an important dietary supplement, inhibits KSHV reactivation by altering the interactions between early growth response-1 (Egr-1) and the RTA promoter [69]. Electrophoretic mobility shift assays (EMSA) and chromatin immunoprecipitation (ChIP) experiments revealed that Egr-1, a cellular transcription factor known to play a critical role in the replication of several viruses, may potentially bind to the KSHV RTA promoter via at least two different GC-rich binding regions and follow a similar expression profile during *de novo* KSHV infection. Elevated cellular Egr-1 expression is reported to enhance viral RTA expression in a Raf > MEK > ERK-dependent manner [69]. Further, Rev is found to lower ERK1/2 activity and expression of Egr-1 in KSHV-infected cells, resulting in the suppression of virus reactivation from latency, though the precise mechanism by which Rev regulates KSHV reactivation is still unclear [69].

4. Role of Viral and Cellular Proteins Important for Lytic DNA Replication

KSHV lifecycle undergoes a transition between a dormant, latent phase and an active lytic replication phase [15,59]. KSHV lytic DNA replication requires the expression of at least eight viral genes including: ORF9 (DNA polymerase), ORF6 (single-stranded DNA binding protein), ORF40/41 (primase-associated factor), ORF44 (helicase), ORF56 (primase), ORF59 (processivity factor), ORF50 (replication and transcription activator or RTA), and ORF K8 (K-bZIP) [70,71]. RTA, an immediate early protein, is the most important protein required for the activation of lytic replication, transcription initiation, as well as recruitment of additional factors (reviewed in [48,59,72]). This section will describe several viral, as well as cellular, proteins that are important for lytic reactivation.

4.1. Viral Factors

4.1.1. K-RTA (KSHV Replication and Transcription Activator)

KSHV encoded ORF50/ RTA (replication and transcription activator), is a key regulator for the lytic reactivation from viral latency [15,73]. Expression of RTA is both essential and sufficient for KSHV reactivation [71,73,74]. Genetic mutation of RTA results in impaired reactivation and lytic DNA replication [75]. RTA has been reported to be phosphorylated [76,77], Poly (ADP-ribosyl)ated [77] and ubiquitinated [78]. RTA also autoactivates its own promoter [19] and transactivates other important lytic genes, including vIL-6 [79,80] polyadenylated nuclear RNA (PAN) [81] ORF57 (MTA) [82], ORF59 (PF8) [83,84], K-bZIP [82], vIRF1 (ORF-K9) [85], ORF-K1 [86], small viral capsid protein (ORF65) [87], ORF56 [88], SOX (ORF37) [89], vOX [90], and ORF52 [79]. RTA binds and transactivate many promoters containing K-RTA response element (RRE) [91]. KSHV LANA is also known to repress lytic reactivation, as well as RTA-mediated autoactivation [92]. LANA-mediated suppression of RTA autoactivation is dependent on RBP-Jκ, which competes with RTA for binding to RBP-Jκ [29]. Lytic reactivation results in the acetylation of LANA, leading to the dissociation of LANA from the ORF50 promoter bound to Sp1 [28]. Genome-wide screening revealed a consensus RTA interaction motif, TTCCAGGAT(N)(0–16)TTCCTGGGA [93,94]. In addition, specific amplification of bound sequences *in vitro* showed a number of RTA direct binding targets [93], such as ORF8, ORFK4.1, ORFK5, PAN, ORF16, ORF29, ORF45, RTA, K-bZIP, ORFK10.1, ORF59, ORFK12, ORF71/72, vOX/vGPCR (ORF74), ORF-K15, the two oriLyts, and the miR cluster [94]. These variations indicate that RTA cooperatively binds to its targets by associating with other regulatory proteins [79,93]. Additionally, RTA activates its own promoter by binding to the Oct-1 transcription factor and RBP-Jκ [19,95]. Furthermore, RTA mediated transactivation of viral lytic promoters, such as MTA and thymidine kinase (TK, ORF21), depends on Sp1, octamer-binding protein-1 (Oct-1), and XBP-1 [96–98]. However, direct binding of RTA to its promoter is not critical for its autoactivation [95,99]. Several other recent studies have also shown that RTA is recruited to RREs through interaction with RBP-Jκ [100,101]. Similarly, in a recent study using recombinant viruses with deleted RBP-Jκ sites within RTA promoter showed an increased viral latency and a reduced efficiency for lytic replication [102]. In addition, RTA stimulates the Notch signaling pathway, RTA mediated intracellular-activation of Notch1 is sufficient to reactivate KSHV from latency to the lytic replication cycle [103,104].

RTA transactivation of viral promoters also depends on its interactions with other cellular proteins. RTA recruits CREB binding protein (CBP), the SWI/SNF chromatin-remodeling complex, and the TRAP/mediator coactivator into viral promoters [105]. RTA binding positively regulates Histone acetyl transferase (HAT) activity of CREB [106]. A recent report showed that RTA transactivates cellular Bcl-2 through targeting of CCN_9GG-like RTA responsive elements (RREs) for lytic reactivation and enhanced virion production [107]. Furthermore, it has been shown that K-RTA associates with a homologue of the Kruppel-associated box-zinc finger proteins (KRAB-ZFPs), for its transactivation function [108–110]. The co-repressor of K-RBP, Kruppel-associated box domain-associated protein-1 (KAP-1), is a cellular transcriptional repressor that regulates chromosomal remodeling, participates in the maintenance of latency by repressing lytic promoters [111]. During latency KAP-1 binds to viral lytic promoters to repress gene expression and depletion of KAP-1 is sufficient to induce KSHV reactivation [62]. Studies show that sumoylation and phosphorylation are required to regulate KAP-1 association with heterochromatin [62,111]. KAP-1 is phosphorylated at Ser 824, during lytic reactivation, resulting in decreased sumoylation and association to the condensed chromatin on viral promoters [111]. A recent study confirmed that KAP1 is targeted by KSHV-encoded latency-associated nuclear antigen (LANA) to repress the transactivation of K-RTA [112]. Additionally, knockdown of KAP1 in KSHV-infected primary effusion lymphoma (PEL) cells reduced viral episome stability and enhanced the efficiency of KSHV lytic reactivation by hypoxia, suggesting that both KAP1 and the cooperative interaction of RBS HRE within the RTA promoter are crucial for KSHV latency and hypoxia-induced lytic reactivation [63]. K-RTA interacts with K-bZIP, and increasing evidence indicates that repression of K-RTA transactivation by K-bZIP, a basic leucine zipper (bZIP) transcription factor encoded by KSHV, is essential for the modulation of lytic DNA replication by a feedback circuit [70,113,114]. RTA also interacts with C/EBPα, and the cooperative interaction of K-bZIP and RTA with C/EBPα is essential for the activation of K-bZIP promoter by binding to a proximal C/EBPα binding site [115]. The promoters of RTA, PAN, and MTA are activated through direct interaction of the C/EBPα and RTA complex [116]. K-RTA is also shown to be associated with viral ORF59, a processivity factor for viral DNA polymerase, and ORF45, a multifunctional tegument protein required for lytic replication [117,118].

Recent studies showed that K-RTA activity is regulated by its association with cellular peptidyl-prolyl cis/trans isomerases (PPIase), Pin1. Pin1 binds specifically to phosphorserine or phosphorthreonine-proline (pS/T-P) motifs in the K-RTA and enhances K-RTA transactivation [119]. Additionally, it has been shown that K-RTA is regulated by a 48aa small peptide, vSP-1, encoded by a polyadenylated RNA of 3.0 kb (T3.0), transcribed from the opposite strand of the KSHV RTA (ORF50) DNA template. vSP-1 associates with RTA at the protein abundance regulatory signal (PARS) motifs, and this interaction prevents RTA from degradation by ubiquitin-proteasome pathways, thus, facilitating KSHV lytic replication [120]. Apart from direct DNA interaction, RTA also cooperates with various host transcriptional factors to transactivate several downstream viral genes. Additionally, K-RTA exhibits an ubiquitin E3 ligase activity, RTA is auto-ubiquitinated and directs several cellular and viral proteins for proteasome-mediated degradation [121]. One of the cellular proteins targeted by RTA is Hey1, which interacts with repressor mSin3A. This, in turns, downregulates the expression of RTA by direct interaction with the RTA promoter [122]. RTA upregulates its own expression through ubiquitin-mediated targeting of Hey1 for degradation. Another cellular protein targeted by RTA mediated degradation is IRF-7, a critical modulator of type I IFN induction [78]. IFN signaling plays a crucial role in repressing KSHV lytic replication, therefore, this finding indicates that RTA might circumvent these cellular innate immune defenses during lytic reactivation. Direct association of RTA to the origin of lytic DNA replication (oriLyt) has been demonstrated [21]; there are two distinct oriLyt regions in the KSHV genome [23]. The left oriLyt (oriLyt-L) lies between ORFK4.2 and K5 and is comprised of a region encoding numerous transcription factor binding sites, an A+T-rich region, and a G+C repeat. Similarly, the right oriLyt (oriLyt-R) is situated between ORF69 and vFLIP and is an inverted duplication of oriLyt-L. Importantly, both oriLyts contain RREs, and [123] a direct interaction of RTA to RREs is critical for oriLyt-dependent DNA replication [23,70,123]. The presence of RREs and a downstream TATA box indicate that this region may serve as an RTA-dependent promoter, and a transcription event may be required for oriLyt-dependent DNA replication [123]. Additionally, recent studies have identified that K-RTA is able to function as a STUbL, which is capable of ubiquitylation of SUMO and SUMO conjugates *in vitro* and *in vivo*. Thus, K-RTA is an ubiquitin ligase, preferentially targeting SUMO-containing proteins for ubiquitylation; including sumoylated K-bZIP and promyelocytic leukemia (PML) nuclear bodies [124]. Together, these results suggest that RTA is a master regulator of viral lytic DNA replication.

4.1.2. ORF57-mRNA Transcript Accumulation (MTA)

ORF57 is a viral early protein, which favors viral intron-less transcript accumulation, transports, and enhances splicing of intron-containing viral RNA transcripts [125]. MTA is essential for KSHV lytic replication, moreover, genetic knockout of MTA disrupts KSHV productive lytic replication [125,126]. MTA protein carries domains with putative transcriptional and post-transcriptional functions [127]. MTA directly associate with RTA and both proteins are detected in the RTA promoter during lytic replication. KSHV MTA associates with DNA, which was identified by gel shift and chromatin immunoprecipitation assays [127,128]. In addition, it has been shown that MTA directly associates with K-bZIP protein and binds to promoter as well as transcribed regions of PAN RNA, K4, and K-bZIP [129]. These reports suggest that MTA stimulates RNA export through its association with Aly/REF, a cellular RNA-binding protein acting as an adaptor for the nuclear RNA export receptor NXF1/TAP [130]. Additionally, recent studies suggest that Aly/REF-ORF57 association does not necessarily play any significant role in the ORF57-mediated enhancement of ORF59 expression, as Aly/RE knockdown in host cells did not affect the function of ORF57 [131,132]. MTA enhances the expression of RTA or other lytic genes, most probably by binding to transcription regulatory proteins. Further, MTA cooperates with RTA to modulate the viral gene expression in a cell-line-specific manner [127,128]. It is suggested that a putative A/T hook domain within MTA arbitrates DNA binding and transcriptional initiation [127].

MTA modulates a cascade of viral gene expression and accumulation of specific viral and cellular mRNAs during lytic replication [132]. Physical association of MTA and RTA is essential for the synergistic regulatory effect of MTA. When RTA's transactivation function is removed, MTA no longer affects the expression of viral genes, indicating that their cooperative effect depends on RTA's transactivation function [128]. It has been shown that MTA regulates mRNA accumulation. Further, a recent study employing a genome-wide CLIP (cross-linking and immunoprecipitation) approach detected KSHV PAN, a long non-coding polyadenylated nuclear RNA, as an important target of ORF57 [133]. Genetic disruption of ORF57 affects PAN RNA expression. In co-transfection experiments, expression of exogenous ORF57 alone increased PAN RNA expression by 20–30-fold, which was due to the MRE (MTA responsive element) at the 5' PAN RNA, however, not as much on an ENE (expression and nuclear retention element) at the 3' end of PAN RNA. Further studies showed that the major function of the 5' PAN MRE is to increase the half-life of PAN in the presence of ORF57 [133]. Systematic mutational analyses identified a core motif, consisting of nine nucleotides, in MRE-II, which is essential for ORF57 interaction and function. The 9-nt core in MRE-II also interacts with cellular poly (A)-binding protein C1 (PABPC1) [134], but not E1B-AP5, which binds to another region of MRE-II [133]. In the presence of ORF57, PAN RNA is partially exportable, suggesting that ORF57 functions to accumulate a non-coding viral RNA during the course of lytic infection [133,134]. Additionally, MTA is also shown to stabilize RNAs and activates translation of mRNAs that carry internal ribosome entry sites [135]. It has been also shown that KSHV ORF57 specifically binds to ORF59 RNA and associates with cellular RNA export cofactors RBM15 and OTT3 to enhance the expression of ORF59 [136].

4.1.3. KSHV K8-K-bZIP—Lytic Replication-Associated Protein (RAP)

K-bZIP is a basic leucine zipper-containing protein encoded by KSHV K8 [137]. The K-bZIP gene locus consists of two promoters: one early promoter controlling K-bZIP and the second late promoter controlling K8.1 [91,138]. K-bZIP directly binds to K-RTA through K-bZIP's basic domain and a specific RTA region [139,140]. Further, association of K-bZIP suppresses K-RTA transactivation of the MTA promoter in a dose-dependent manner [109,110]. Recent studies suggest that K-bZIP is not required for lytic reactivation in KSHV BACmid systems [114,141,142], however, it was reported to be crucial for virus production in infected PEL cells [143]. It has been shown that K-bZIP interacts with oriLyt [22,144–146] and is critical for oriLyt-dependent DNA replication in a plasmid-based transient expression system [70], but its absence can be complemented by an over-expression of RTA [145]. Similarly, association of cellular transcription factor CCAAT/enhancer-binding protein α (C/EBPα) to K-bZIP has also been shown to increase the expression and stabilization of C/EBPα and p21CIP1 proteins, followed by G0/G1 cell cycle arrest [18,115]. Similarly, KSHV bZIP can also bind to the positive regulatory domain I/III region of the IFNb promoter to block IRF3-mediated IFNb transcription [147,148]. In addition, K-bZIP represses the RTA autoactivation [139] and colocalizes with HDAC1/2 through the leucine zipper domain without the requirement of sumoylation of K-bZIP [149]. K-bZIP is phosphorylated on residues Thr111 and Ser167 by a serine/threonine protein kinase (vPK) encoded by ORF36 [146,150]. However, phosphorylation at T111 has a negative effect on both the extent of sumoylation and the repressive activity of K-bZIP [150]. K-bZIP is sumoylated at residue lysine 158, and this sumoylation is essential for K-bZIP mediated transcription repression [146]. As a SUMO adaptor, KbZIP represses transcription by recruiting Ubc9 to specific viral promoters [146]. In addition, it has been shown that K-bZIP functions as the viral SIM-containing poly-SUMO-specific E3 ligase, with specificity for SUMO-2/3 [35]. Further, K-bZIP catalyzes its auto-sumoylation and the sumoylation of other K-bZIP-interacting proteins, such as p53 and pRB [148].

A genome-wide analysis of K-bZIP's transcriptional regulation on KSHV gene promoters showed that RTA activated 34 viral promoters whereas K-bZIP alone activated 21 promoters [140]. Nonetheless, when RTA and K-bZIP were combined together, K-bZIP was found to repress three RTA-responsive promoters, suggesting that K-bZIP might also transactivate some viral lytic genes during KSHV reactivation [140]. These data strongly suggest that K-bZIP plays a crucial during lytic gene expression and DNA replication in PEL cells [140,151]. Further, K-bZIP also directly binds to oriLyt, indicating that K-bZIP might be playing a crucial in lytic DNA replication [70]. Further, the interaction of K-bZIP with oriLyt is also modulated by LANA expression [145]. Taken together, these studies show that K-bZIP has dual independent functions in modulating the KSHV life cycle by facilitating lytic DNA replication or repressing the lytic gene expression as a feedback modulator [145]. Together with these results, knockdown of K-bZIP in latently infected BCBL-1 and BC-3 cells showed a significant reduction in the expression of RTA, MTA, and ORF26 transcripts, as well as decreased RTA and ORF-K8.1 protein levels, as well as defective viral DNA replication and virion production [143]. Collectively, these results suggest that K-bZIP regulates its own expression and possibly other RTA-transactivated lytic genes by a feedback loop.

4.1.4. ORF59- Viral Processivity Factor

Kaposi's sarcoma-associated herpesvirus (KSHV) ORF59 plays a critical role in viral lytic DNA replication as a DNA processivity factor to the viral DNA polymerase (ORF9) [70,83]. ORF59 is highly upregulated during lytic reactivation and *de novo* primary infection. ORF59 forms a homodimer in the cytoplasm and associates with ORF9 to translocate it to the nucleus during lytic DNA replication [152]. ORF59 associates with C/EBPα binding motifs within oriLyt and this binding is K-RTA dependent, where K-RTA acts as an initiator of lytic replication. Additionally, disruption of the K-RTA–ORF59 interaction by a dominant negative approach impairs oriLyt-dependent DNA replication [84]. This strongly suggests that the K-Rta-ORF59 interaction is crucial for lytic DNA replication. ORF59 is a phosphoprotein and is phosphorylated by KSHV viral Ser/Thr kinase, ORF36 primarily at Ser378, which is essential for ORF59's ability to bind to RTA and the oriLyt [83,84]. In a recent study, it has been shown that lytic infection of KSHV induces severe DNA double-strand breaks (DSBs) and impede non-homologous end joining (NHEJ) in host cells. Further, ORF59 was found to be associated with Ku70 and Ku86 and this association was dependent on DSBs, suggesting that KSHV lytic replication may induce tumorigenesis by causing DNA DSBs and interrupting the DSB repair of mechanism [153].

4.1.5. ORF6-Single Strand Binding Protein

KSHV ORF6, a delayed-early gene encodes for a 126 kDa ssDNA binding protein that has been shown to participate in origin-dependent DNA replication [74,154,155]. The expression of ORF6 is regulated by RTA, which could bind to RBP-Jκ recognition site on the ORF6 promoter via interaction with the RBP-Jκ protein [95,155]. Genetic disruption analysis of the ORF6 gene, using the bacterial artificial chromosome (BAC) system, identified the functional role of ORF6 in lytic DNA replication. The mutant virus showed impaired DNA synthesis and failed to make progeny virions. Additionally, transient expression of ORF6 has rescued both defects, suggesting that ORF6 is critical for KSHV lytic replication [155].

4.2. Cellular Factors

Several cellular signaling pathways are identified to be involved in the reactivation of KSHV from latency, such as PKCd [156], b-Raf/MEK/ERK [157], PKA [104], Notch and RBP-Jκ [95,158], p38 and JNK [159], Pim-1 and Pim-3 [160], PI3K and Akt [161], and TLR7/8 signaling [162]. Apart from these signaling pathways, a number of additional cellular factors also mediate KSHV reactivation [163–167] (Figure 2). It has been shown that intracellular calcium transport activates Ca++ dependent viral reactivation, and inhibition of calcineurin signaling, in turn, blocks KSHV reactivation [168]. Similarly, Protein kinase C delta (PKCdelta) plays a role in KSHV lytic replication [156]. Activation of the MEK/ERK, JNK, and p38 mitogen-activated protein kinase (MAPK) pathways play a central role during KSHV infection. Activation of the MAPK pathway, immediately after infection, enables the establishment of a successful KSHV infection [169,170]. Furthermore, MAPK pathways are induced during lytic reactivation [157,159,163]. Similarly, cellular MAP4K4 is also known to play a crucial role in inflammation, insulin resistance, and the invasiveness of several human malignancies [171,172]. Recently, it has been suggested that MAP4K4 act as a novel mediator of KSHV lytic reactivation from latency [172]. Similarly, yet another essential pathway mediating KSHV reactivation is the Raf/MEK/ERK/Ets-1 pathway [163]. Likewise, promoters of K-RTA, MTA, K-bZIP, and origins of lytic replication (oriLyt) have been shown to carry a functional DNA-binding site for AP-1 and are responsive to AP-1 activation [70,159,173]. During *de novo* infection, KSHV has been shown to induce MEK/ERK, JNK, and p38 MAPK pathways in human umbilical vascular endothelial cells (HUVEC). This, in turn, regulates AP-1 to facilitate its entry into the target cells and initiate a productive lytic replication at the early acute stage of infection [169,174]. Additionally, in latent KSHV-infected, cells these MAPK pathways modulate both spontaneous and TPA-induced KSHV reactivations and activate the expression of several transcription factors, such as AP-1 and Ets-1 [157,159,175].

KSHV lifecycle is also controlled by the viral protein, K-RTA by altering the Notch signaling pathway through binding with RBP-Jκ [95,101]. Additionally, Notch signaling and the expression of two Notch ligands (JAG1 and DLL4) are upregulated through KSHV genes, expressed during KSHV latent and lytic infection [176,177]. Similarly, Hypoxia-inducible factor (HIF) has been shown to induce numerous genes associated with angiogenesis and tumor growth, and the KSHV infected cells express elevated levels of HIF1α and HIF2α [177–179]. Furthermore, both LANA and vIRF3 have been shown to play roles in the stabilization of HIF1α via protein–protein interactions [61,178,180]. Secondary infections by other pathogens, such as HIV and bacteria, have been shown to trigger KSHV reactivation [162,181]. Similarly, short-chain fatty acids (SCFA) from periodontal pathogens suppress histone deacetylases HDAC1, EZH2, and SUV39H1 and downregulates the expression of silent information regulator-1 (SIRT1) to promote KSHV replication [182]. Cytokine-mediated JAK–STAT signaling also regulate various important biological processes, such as immune response, cell growth, and differentiation. KSHV infection has been shown to upregulate gp130 receptor expression, which leads to a constitutive phosphorylation of JAK2/STAT3 activation [183,184]. Further studies have revealed that both LANA and vGPCR play roles in the modulation of JAK2/STAT3 signaling to create angiogenic factors [185,186]. This is further confirmed by the LANA-mediated STAT6 phosphorylation through the inhibition of IL-4 for the maintenance of latency [187].

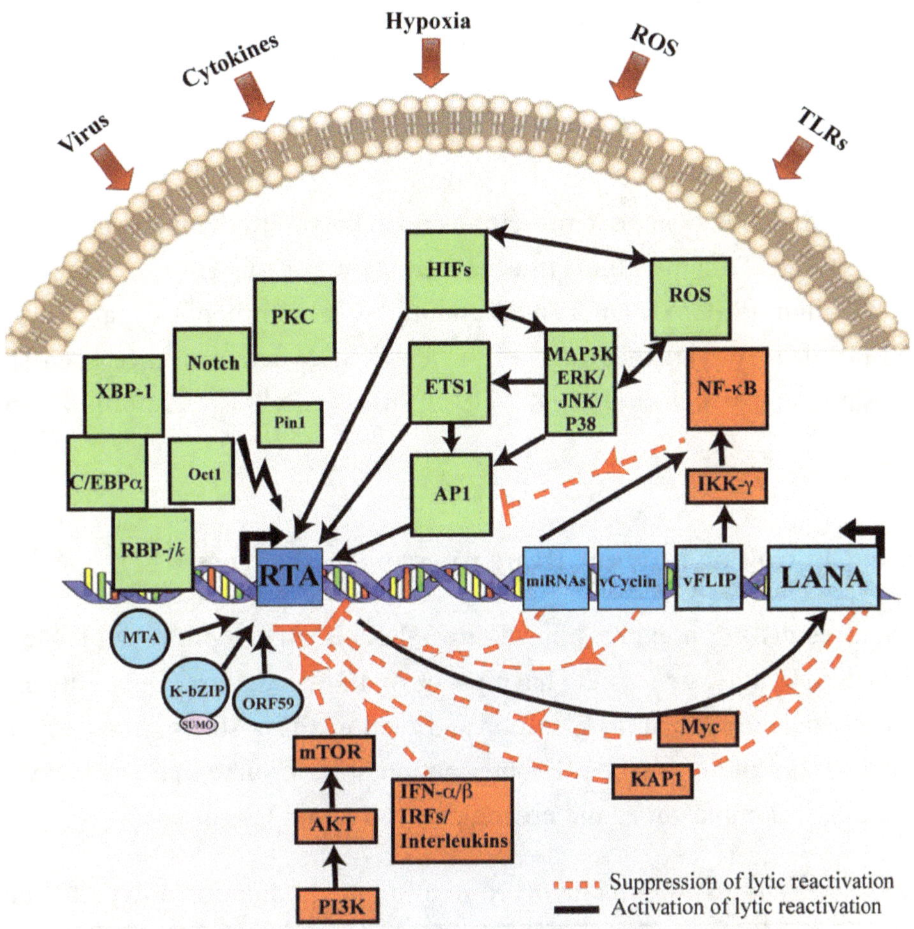

Figure 2. Schematic representation of cellular signaling pathways involved in KSHV latency and reactivation. During latency, KSHV latent genes, including LANA, vFLIP, miRNA, and vCyclin activate and maintain various cytokine-mediated cell proliferation and angiogenesis pathways, such as JAK/STAT, PI3K/AKT/mTOR, cMyc, and NF-κB, to suppress KSHV lytic reactivation. The red line represents the inhibitory pathways involved in the maintenance of KSHV latency. Disruption of these signaling pathways by various stimuli, such as secondary infection by bacteria, viruses, hypoxia, inflammatory cytokines, and oxidative stress upregulate RTA expression resulting in KSHV reactivation. The solid black arrows represent signaling pathways that are activated during KSHV lytic reactivation. Moreover, RTA, as well as RTA-induced KSHV genes MTA and K-bZIP, have been shown to interact with XBP-1 and C/EBPα to modulate various cellular signaling pathways. Deregulation of these cellular signaling pathways, such as MAPK, PKCd, b-Raf/MEK/ERK, PKA, Notch, RBP-Jκ, JNK, Pim-1/Pim-3, and TLR7/8 signaling by RTA lead to the reactivation of latently infected KSHV cells to lytic replication. This figure is adopted and modified from a previous review [59].

KSHV has evolved multiple mechanisms to manipulate cellular anti-apoptotic and survival pathways and disruption of these pathways reactivates KSHV [188–190]. Apart from AP-1, NF-κB also antagonizes RBP-Jκ to impair the expression and transactivation function of RTA [190]. Furthermore, inhibition of NF-κB pathway in latently infected cells disrupts viral latency and activates viral lytic replication [191].

However, the available data suggest that the role of the NF-κB pathway in the KSHV life cycle is context dependent [192]. It is very likely that the balance of AP-1 and NF-κB pathways decide the fate of virus replication status in a particular cell type [174,193,194]. Consistent with these findings, a recent study showed that inhibition of the pro-survival PI3K-Akt pathway favors KSHV reactivation from latency [195]. Furthermore, inhibition of the Akt pathway reactivates KSHV from latency by increasing the RTA expression [195]. KSHV encoded proteins are also known to modulate the cellular phosphatidyl inositol-3-kinase (PI3K)/AKT/mammalian target of the rapamycin (mTOR) signaling pathway to control cell proliferation. Cellular PI3K/AKT/mTOR signaling is a common to many growth factors and cytokine receptors [196]. However, thus far, only a few KSHV proteins have been shown to regulate PI3K/AKT/mTOR signaling, which include K1 [197,198], (vGPCR) [199,200], vIL-6 [183,201], and ORF45 [170,202].

5. Lytic Proteins in Controlling Immune Regulation and Pathogenesis

Lytic reactivation results in an expression of several KSHV lytic proteins (Table 1). Many of the proteins encoded by KSHV lytic genes also have pro-growth or transforming abilities. Major functions of KSHV lytic proteins include cellular proliferation and evading the host's immune response. The immune functions targeted by viral proteins include IFN production, interferon regulatory factor (IRF) activation, complement activation, inflammasome, and chemokine activation (Figure 3).

Table 1. KSHV lytic proteins involved in immune modulation and pathogenesis.

KSHV genes	KSHV proteins	Function	References
K1	Variable ITAM-Containing Protein (VIP)	Type I transmembrane signaling protein containing a functional immunoreceptor tyrosine-based activation motif. Regulate membrane transport in B cells.	[203]
K2	Viral Interleukin-6 (vIL-6)	Homologues of cellular IL-6. Activate JAK/STAT, MAPK, and PI3K/Akt signaling pathways to regulate B-cell proliferation.	[51,204]
K3/K5	Modulator of immune recognition (MIR1/MIR2)	Viral E3 ligases capable of ubiquitinating MHC-I, ICAM-1, B7-2, Tetherin (CD317/BST2), DC-SIGN, and DC-SIGNR.	[205,206]
K4/K4.1/K6	Viral CC-Chemokine Ligands (vCCLs)	Homologues of cellular chemokines: viral CC-chemokine ligand 1 vCCL1 (vMIP1), vCCL2 (vMIP2), and vCCL3 (vMIP3), respectively. Blocks signaling through chemokine receptors.	[207,208]
K7	Viral Inhibitor of Apoptosis (vIAP)	Interact with cellular proteins PLIC1, caspase 3/Bcl-2, CAML, Vps34, and promote cell survival during lytic replication.	[209,210]
K9/K10/K11	KSHV interferon regulatory factors (vIRF-1, vIRF-2, vIRF-3 and vIRF-4)	Homologues of cellular interferon: Inhibitor of IFN1, p53, NFκB RelA, and p300.	[211,212]
K14	vOX2 or vCD200	Homologues of cellular OX2. A negative regulator of inflammatory signaling and surface glycoproteins.	[213,214]

Table 1. *Cont.*

KSHV genes	KSHV proteins	Function	References
K15	Viral membrane protein	Regulation of cellular signaling to induce various pro-survival and paracrine-mediated pro- angiogenic cellular cytokines and chemokines, including IL6, IL8, IL-1a/b, CXCL3, and Cox2.	[215,216]
ORF4	KSHV complement Control protein (KCP)	Homologue to cellular RCA. Regulate complement activation by increasing the decay of the classical C3 convertase.	[217–219]
ORF45	ORF45	Inhibit type1 IFN induction by sequestering the cellular interferon regulatory factor-7 to cytoplasm.	[220,221]
ORF63	ORF63	Homologue to cellular inflammasome complex NLRP1.	[222]
ORF64	Viral deubiquitinase	A non specific deubiquitinase, shown to deubiquitinate RIG-I to suppress RIG-I-mediated activation of the IFNb.	[223]
ORF74	Viral G-protein-coupled receptor (vGPCR)	Homologue of cellular IL-8 receptor. vGPCR induce secretion of proinflammatory cytokines and angiogenic growth factors.	[200,224]
ORF75	ORF75	A viral effector for the degradation of ND10 proteins.	[225,226]
PAN RNA	Polyadenylated Nuclear RNA	Modulator of viral gene expression.	[227–230]

KSHV employs diverse mechanisms for controlling both IFN production and signaling as IFN is a potent antiviral defense that is critical for KSHV persistence [231]. The genomic region encompassing ORFs K9 to K11 encodes KSHV vIRFs 1-4 [232]. vIRF1 can bind to and disrupt the transcriptional activities of IRF1, IRF3, and IRF7 [211,212]. Additionally, vIRFs 1, 3, and 4 have been shown to inhibit p53 activity via, either direct binding to the tumor suppressor (vIRF-1 and vIRF-3), or through association with ATM kinase or via stabilization of MDM2, which induces ubiquitination and proteasomal degradation of p53 [233,234]. Similarly, KSHV viral interleukin-6 (vIL6), encoded by ORF K2, shares many functional characteristics with human IL6 and, as a result, the viral cytokine can activate gp130 and downstream signaling pathways, including the JAK/STAT, MAPK, and PI3K/Akt pathways [204,235]. These pathways regulate a variety of transcription factors and response elements (RE), such as the STAT1/3 and STAT5 IL6 RE, C/EBP, and c-jun promoter IL6 RE (JRE-IL-6) [236]. A viral homologue of the cellular angiogenic IL-8 receptor [224], vGPCR has been shown to activate a number of crucial signaling pathways, including PLC, PKC, MAPK, PI3K/Akt/mTOR, and NFκB [237]. Downstream signaling from these pathways activates the AP1, NFAT, NF-kB, HIF-1a, and CREB transcription factors, which, in turn, contribute to vGPCR-mediated production of pro-inflammatory cytokines and chemokines [237].

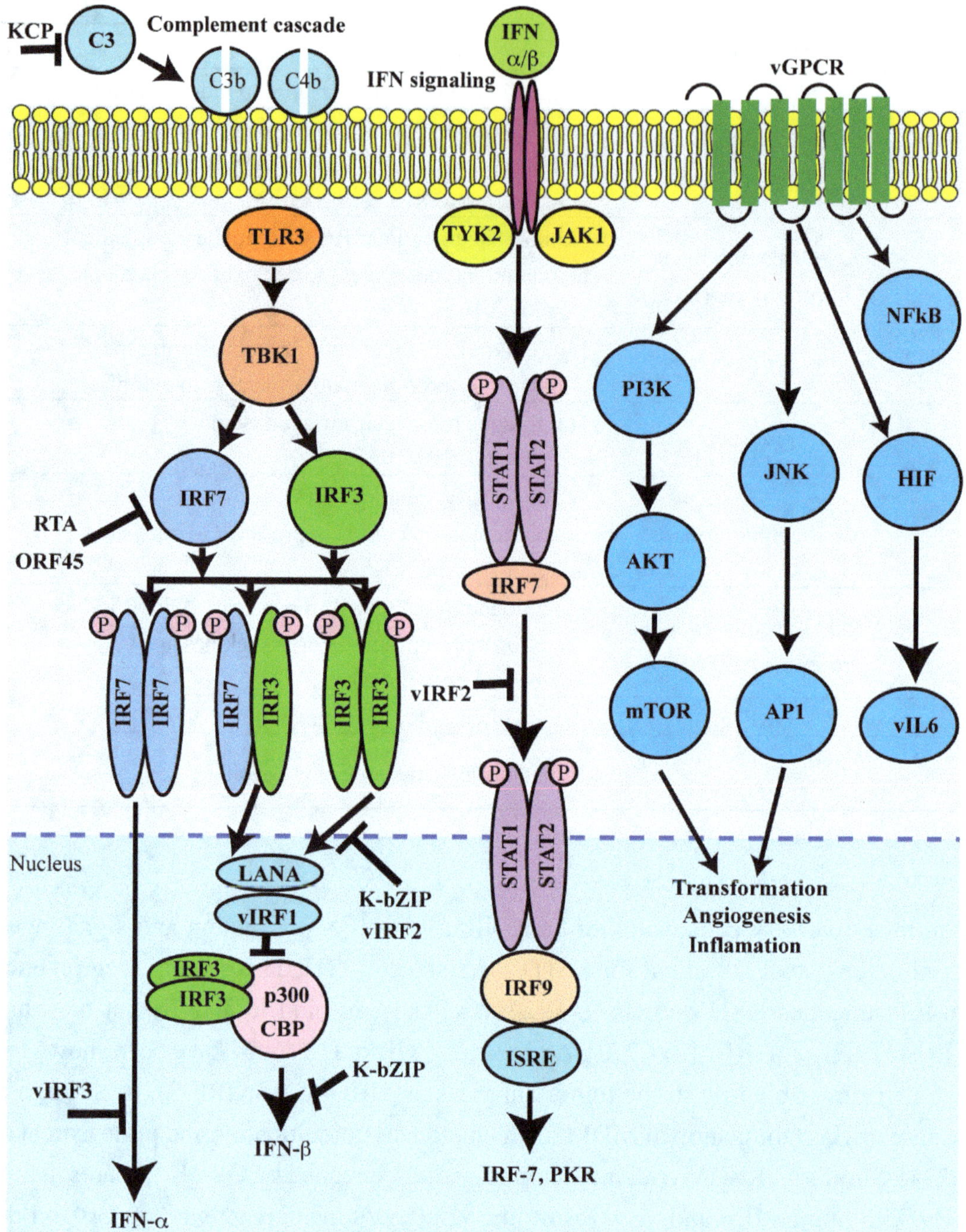

Figure 3. Schematic representation of lytic proteins in immune regulation and pathogenesis: The major immune functions targeted by viral lytic proteins include IFN production, interferon regulatory factor (IRF) activation, complement activation, inflammasome and chemokine activation. Regulating both IFN production and signaling is a potent antiviral defense, vIRF can bind and disrupt the transcriptional activities of IRF1, IRF3, and IRF7. Additionally, vGPCR is a constitutively active homologue of the IL8 receptor. vGPCR activates various cell signaling pathways and transcription factors to enhance the production of pro-inflammatory chemokines and cytokines, such as vIL-6. Furthermore, KSHV-encoded KCP regulates complement by increasing the decay of the classical C3 convertase.

Many of the KSHV K-gene encoded lytic proteins have also been shown to modulate KSHV infection and pathogenesis. Multifunctional transmembrane glycoprotein K1 encoded by the first ORF of KSHV can constitutively activate multiple pro-growth signaling pathways in KSHV-infected cells. [238]. Oligomerization of K1 trigger auto-phosphorylation of ITAM and activate various Src homology 2 (SH2) containing signaling proteins, including PI3K (p85)/Akt, PLCg, Vav, Syk, Lyn, RasGAP, and Grb2 [239,240]. Similarly, it has been shown that K15-activated cellular signaling pathways induce the transcription of a number of cellular cytokines and chemokines, including IL6, IL8, CCL20, CCL2, CXCL3, IL-1a/b, and Cox2 [215,216]. Additionally, KSHV K7 or viral inhibitor of apoptosis (vIAP), is a homologue of cellular Bcl-2 proteins and contains a putative mitochondrial-targeting signal and localizes to mitochondria and ER [210,241]. It has been reported that K7/vIAP inhibits caspase 3 activity by interacting with cellular Bcl-2 via its BIR (baculovirus IAP repeat) [210]. Furthermore, KSHV K3 and K5 (also called modulator of the immune recognition (MIR) 1 and 2, respectively) are viral E3 ligases capable of ubiquitinating the MHC-I cytoplasmic tail to trigger internalization and proteasomal degradation of the MHC-I complex [205,242,243]. K3 and K5 proteins also have been shown to downregulate both C-type lectins, DC-SIGN, and DC-SIGNR by ubiquitin mediated degradation [206]. Similarly, KSHV K6, K4, and K4.1 encode three homologues of cellular chemokines: viral CC-chemokine ligand 1 vCCL1(vMIP1), vCCL2 (vMIP2), and vCCL3 (vMIP3), respectively [154,244,245]. Apart from immune evasion properties, v-chemokines also have been shown to promote angiogenesis through the induction of VEGF [246,247]. KSHV-encoded early lytic protein K14 is another negative regulator of inflammatory signaling and surface glycoprotein (vOX2). K14 shows significant homology with OX2 or CD200, a member of the immunoglobulin superfamily that is broadly distributed on the cell surface [90]. vCD200 promotes the secretion of proinflammatory cytokines on stimulation of monocytes, macrophages, and DCs through a direct interaction with cellular CD200R, inhibiting myeloid cell activation and reducing Th1-cell-associated cytokine production [214,248].

Furthermore, KSHV-ORF4-encoded inhibitor of the complement system, designated as KSHV complement Control Protein (KCP) [217,249,250], regulates complement by increasing the decay of the classical C3 convertase and acting as cofactors for the inactivation of C3b and C4b, components of the C3 and C5 convertases [251,252]. Similarly, KSHV encoded ORF45, an immediate early gene product, plays a crucial role in lytic replication [253]. ORF45 has been shown to inhibit type1 IFN induction upon infection by sequestering the cellular interferon regulatory factor-7 (IRF-7) to the cytoplasm [220,221]. It has been shown that ORF45 can also regulate eIF4B phosphorylation in an mTOR and MAPK independent manner. Additionally, the ORF45 protein is also involved in the transport of the viral capsid-tegument complexes along the microtubule filaments [254].

It has been also been shown that KSHV encoded tegument protein, ORF75 is an essential protein as a new viral effector [255] for the degradation of ND10 proteins, thereby regulating lytic replication and KSHV infection [225]. In addition, the ORF75 also has been shown to induce the degradation of ATRX and relocalization of Daxx, as well as be involved in NF-kB coactivation with KSHV K13/vFLIP [225,256]. Similarly, KSHV encoded ORF64 is a deubiquitinase that non-specifically targets K48 or K63 ubiquitination. It has been shown that KSHV ORF64 is capable of deubiquitinating RIG-I to suppress RIG-I-mediated activation of the IFNb promoter [223]. Studies showed that KSHV ORF63 has homology to parts of cellular inflammasome complex NLRP1 [222,257]. This ORF63 function seems to be critical for supporting viral gene expression and genome replication, as well as suppressing IL-1b production [222,258]. Additionally, KSHV encoded structural PAN RNA has been also shown as a multifunctional transcript that can globally control viral and cellular gene expression during lytic reactivation [259] through direct interaction with chromatin modifying complexes, such as components of PRC2 [228,229]. PAN RNA interacts with demethylases, UTX, and JMJD3 to modify the suppressive H3K27me3 mark within the KSHV genome [260]. Moreover, PAN RNA expression decreased the expression of interferon γ, interleukin 18, interferon α16, and RNase L [229].

6. Conclusions

Kaposi's sarcoma associated herpesvirus (KSHV) modulates various cellular pathways by which it is able to establish and maintain persistent infection in the host to initiate tumorigenesis. Several of these latent viral and lytic proteins are known to transform host cells, linking KSHV with the development of severe human malignancies. These virus-induced cancers pose a large threat to global public health, specifically in areas that are still struggling with malignancies associated with HIV-AIDS with limited treatment options. Over the years, tremendous progress has been made in elucidating the molecular mechanisms of KSHV latency and lytic replication. Nonetheless, there are still vast aspects of viral infection and transformation that are not well explored. With the help of rapid advancements in modern technology, it is presumed that a thorough knowledge of the KSHV life cycle will be achieved over the next few years. Further understanding of the unique mechanisms that KSHV adopts for the establishment of successful lifetime persistence in the infected host will eventually pave the way for novel therapeutic approaches for the treatment of KSHV diseases.

Acknowledgments

We thank the Verma Lab members for all the constructive comments and helpful discussions. This work was supported by public health grants from the National Institute of Health (grant numbers CA174459 and AI105000) and the Research Scholar Grant (124389-RSG-13-230-01-MPC) from the American Cancer Society.

References

1. Sturzl, M.; Zietz, C.; Monini, P.; Ensoli, B. Human herpesvirus-8 and Kaposi's sarcoma: Relationship with the multistep concept of tumorigenesis. *Adv. Cancer Res.* **2001**, *81*, 125–159.
2. Cesarman, E.; Chang, Y.; Moore, P.S.; Said, J.W.; Knowles, D.M. Kaposi's sarcoma-associated herpesvirus-like DNA sequences in AIDS-related body-cavity-based lymphomas. *N. Engl. J. Med.* **1995**, *332*, 1186–1191.
3. Soulier, J.; Grollet, L.; Oksenhendler, E.; Cacoub, P.; Cazals-Hatem, D.; Babinet, P.; d'Agay, M.F.; Clauvel, J.P.; Raphael, M.; Degos, L.; *et al.* Kaposi's sarcoma-associated herpesvirus-like DNA sequences in multicentric Castleman's disease. *Blood* **1995**, *86*, 1276–1280.

4. Uldrick, T.S.; Wang, V.; O'Mahony, D.; Aleman, K.; Wyvill, K.M.; Marshall, V.; Steinberg, S.M.; Pittaluga, S.; Maric, I.; Whitby, D.; *et al.* An interleukin-6-related systemic inflammatory syndrome in patients co-infected with Kaposi sarcoma-associated herpesvirus and HIV but without Multicentric Castleman disease. *Clin. Infect. Dis.* **2010**, *51*, 350–358.
5. Dittmer, D.; Stoddart, C.; Renne, R.; Linquist-Stepps, V.; Moreno, M.E.; Bare, C.; McCune, J.M.; Ganem, D. Experimental transmission of Kaposi's sarcoma-associated herpesvirus (KSHV/HHV-8) to SCID-hu Thy/Liv mice. *J. Exp. Med.* **1999**, *190*, 1857–1868.
6. Chang, Y.; Cesarman, E.; Pessin, M.S.; Lee, F.; Culpepper, J.; Knowles, D.M.; Moore, P.S. Identification of herpesvirus-like DNA sequences in AIDS-associated Kaposi's sarcoma. *Science* **1994**, *266*, 1865–1869.
7. Verma, S.C.; Robertson, E.S. Molecular biology and pathogenesis of Kaposi sarcoma-associated herpesvirus. *FEMS Microbiol. Lett.* **2003**, *222*, 155–163.
8. Veettil, M.V.; Bandyopadhyay, C.; Dutta, D.; Chandran, B. Interaction of KSHV with host cell surface receptors and cell entry. *Viruses* **2014**, *6*, 4024–4046.
9. Knipe, D.M.; Lieberman, P.M.; Jung, J.U.; McBride, A.A.; Morris, K.V.; Ott, M.; Margolis, D.; Nieto, A.; Nevels, M.; Parks, R.J.; *et al.* Snapshots: Chromatin control of viral infection. *Virology* **2013**, *435*, 141–156.
10. Dourmishev, L.A.; Dourmishev, A.L.; Palmeri, D.; Schwartz, R.A.; Lukac, D.M. Molecular genetics of Kaposi's sarcoma-associated herpesvirus (human herpesvirus-8) epidemiology and pathogenesis. *Microbiol. Mol. Biol. Rev.* **2003**, *67*, 175–212.
11. Mesri, E.A.; Cesarman, E.; Boshoff, C. Kaposi's sarcoma and its associated herpesvirus. *Nat. Rev. Cancer* **2010**, *10*, 707–719.
12. Ballestas, M.E.; Kaye, K.M. The latency-associated nuclear antigen, a multifunctional protein central to Kaposi's sarcoma-associated herpesvirus latency. *Future Microbiol.* **2011**, *6*, 1399–1413.
13. Chen, L.; Lagunoff, M. Establishment and maintenance of Kaposi's sarcoma-associated herpesvirus latency in B cells. *J. Virol.* **2005**, *79*, 14383–14391.
14. Renne, R.; Zhong, W.; Herndier, B.; McGrath, M.; Abbey, N.; Kedes, D.; Ganem, D. Lytic growth of Kaposi's sarcoma-associated herpesvirus (human herpesvirus 8) in culture. *Nat. Med.* **1996**, *2*, 342–346.
15. Sun, R.; Lin, S.F.; Staskus, K.; Gradoville, L.; Grogan, E.; Haase, A.; Miller, G. Kinetics of Kaposi's sarcoma-associated herpesvirus gene expression. *J. Virol.* **1999**, *73*, 2232–2242.
16. Jenner, R.G.; Alba, M.M.; Boshoff, C.; Kellam, P. Kaposi's sarcoma-associated herpesvirus latent and lytic gene expression as revealed by DNA arrays. *J. Virol.* **2001**, *75*, 891–902.
17. Cai, Q.; Verma, S.C.; Lu, J.; Robertson, E.S. Molecular biology of Kaposi's sarcoma-associated herpesvirus and related oncogenesis. *Adv. Virus Res.* **2010**, *78*, 87–142.
18. Wang, S.E.; Wu, F.Y.; Yu, Y.; Hayward, G.S. CCAAT/enhancer-binding protein-α is induced during the early stages of Kaposi's sarcoma-associated herpesvirus (KSHV) lytic cycle reactivation and together with the KSHV replication and transcription activator (RTA) cooperatively stimulates the viral RTA, MTA, and PAN promoters. *J. Virol.* **2003**, *77*, 9590–9612.
19. Deng, H.; Young, A.; Sun, R. Auto-activation of the rta gene of human herpesvirus-8/Kaposi's sarcoma-associated herpesvirus. *J. Gen. Virol.* **2000**, *81*, 3043–3048.

20. Song, M.J.; Deng, H.; Sun, R. Comparative study of regulation of RTA-responsive genes in Kaposi's sarcoma-associated herpesvirus/human herpesvirus 8. *J. Virol.* **2003**, *77*, 9451–9462.
21. Wang, Y.; Tang, Q.; Maul, G.G.; Yuan, Y. Kaposi's sarcoma-associated herpesvirus ori-Lyt-dependent DNA replication: Dual role of replication and transcription activator. *J. Virol.* **2006**, *80*, 12171–12186.
22. Lin, C.L.; Li, H.; Wang, Y.; Zhu, F.X.; Kudchodkar, S.; Yuan, Y. Kaposi's sarcoma-associated herpesvirus lytic origin (ori-Lyt)-dependent DNA replication: Identification of the ori-Lyt and association of K8 bZip protein with the origin. *J. Virol.* **2003**, *77*, 5578–5588.
23. AuCoin, D.P.; Colletti, K.S.; Xu, Y.; Cei, S.A.; Pari, G.S. Kaposi's sarcoma-associated herpesvirus (human herpesvirus 8) contains two functional lytic origins of DNA replication. *J. Virol.* **2002**, *76*, 7890–7896.
24. Krishnan, H.H.; Naranatt, P.P.; Smith, M.S.; Zeng, L.; Bloomer, C.; Chandran, B. Concurrent expression of latent and a limited number of lytic genes with immune modulation and antiapoptotic function by Kaposi's sarcoma-associated herpesvirus early during infection of primary endothelial and fibroblast cells and subsequent decline of lytic gene expression. *J. Virol.* **2004**, *78*, 3601–3620.
25. Toth, Z.; Brulois, K.; Lee, H.R.; Izumiya, Y.; Tepper, C.; Kung, H.J.; Jung, J.U. Biphasic euchromatin-to-heterochromatin transition on the KSHV genome following *de novo* infection. *PLoS Pathog.* **2013**, *9*, e1003813.
26. Lan, K.; Kuppers, D.A.; Verma, S.C.; Sharma, N.; Murakami, M.; Robertson, E.S. Induction of Kaposi's sarcoma-associated herpesvirus latency-associated nuclear antigen by the lytic transactivator RTA: A novel mechanism for establishment of latency. *J. Virol.* **2005**, *79*, 7453–7465.
27. Li, Q.; Zhou, F.; Ye, F.; Gao, S.J. Genetic disruption of KSHV major latent nuclear antigen LANA enhances viral lytic transcriptional program. *Virology* **2008**, *379*, 234–244.
28. Lu, F.; Day, L.; Gao, S.J.; Lieberman, P.M. Acetylation of the latency-associated nuclear antigen regulates repression of Kaposi's sarcoma-associated herpesvirus lytic transcription. *J. Virol.* **2006**, *80*, 5273–5282.
29. Lan, K.; Kuppers, D.A.; Robertson, E.S. Kaposi's sarcoma-associated herpesvirus reactivation is regulated by interaction of latency-associated nuclear antigen with recombination signal sequence-binding protein Jκ, the major downstream effector of the Notch signaling pathway. *J. Virol.* **2005**, *79*, 3468–3478.
30. Chen, J.; Ueda, K.; Sakakibara, S.; Okuno, T.; Parravicini, C.; Corbellino, M.; Yamanishi, K. Activation of latent Kaposi's sarcoma-associated herpesvirus by demethylation of the promoter of the lytic transactivator. *Proc. Natl. Acad. Sci. USA* **2001**, *98*, 4119–4124.
31. Shamay, M.; Krithivas, A.; Zhang, J.; Hayward, S.D. Recruitment of the *de novo* DNA methyltransferase Dnmt3a by Kaposi's sarcoma-associated herpesvirus LANA. *Proc. Natl. Acad. Sci. USA* **2006**, *103*, 14554–14559.
32. Gunther, T.; Grundhoff, A. The epigenetic landscape of latent Kaposi sarcoma-associated herpesvirus genomes. *PLoS Pathog.* **2010**, *6*, e1000935.
33. Woodard, C.; Shamay, M.; Liao, G.; Zhu, J.; Ng, A.N.; Li, R.; Newman, R.; Rho, H.S.; Hu, J.; Wan, J.; *et al.* Phosphorylation of the chromatin binding domain of KSHV LANA. *PLoS Pathog.* **2012**, *8*, e1002972.

34. Chang, P.C.; Cheng, C.Y.; Campbell, M.; Yang, Y.C.; Hsu, H.W.; Chang, T.Y.; Chu, C.H.; Lee, Y.W.; Hung, C.L.; Lai, S.M.; *et al.* The chromatin modification by SUMO-2/3 but not SUMO-1 prevents the epigenetic activation of key immune-related genes during Kaposi's sarcoma associated herpesvirus reactivation. *BMC Genomics* **2013**, *14*, 824.

35. Chang, P.C.; Izumiya, Y.; Wu, C.Y.; Fitzgerald, L.D.; Campbell, M.; Ellison, T.J.; Lam, K.S.; Luciw, P.A.; Kung, H.J. Kaposi's sarcoma-associated herpesvirus (KSHV) encodes a SUMO E3 ligase that is SIM-dependent and SUMO-2/3-specific. *J. Biol. Chem.* **2010**, *285*, 5266–5273.

36. Bajaj, B.G.; Verma, S.C.; Lan, K.; Cotter, M.A.; Woodman, Z.L.; Robertson, E.S. KSHV encoded LANA upregulates Pim-1 and is a substrate for its kinase activity. *Virology* **2006**, *351*, 18–28.

37. Cha, S.; Lim, C.; Lee, J.Y.; Song, Y.J.; Park, J.; Choe, J.; Seo, T. DNA-PK/Ku complex binds to latency-associated nuclear antigen and negatively regulates Kaposi's sarcoma-associated herpesvirus latent replication. *Biochem. Biophys. Res. Commun.* **2010**, *394*, 934–939.

38. Liu, J.; Martin, H.; Shamay, M.; Woodard, C.; Tang, Q.Q.; Hayward, S.D. Kaposi's sarcoma-associated herpesvirus LANA protein downregulates nuclear glycogen synthase kinase 3 activity and consequently blocks differentiation. *J. Virol.* **2007**, *81*, 4722–4731.

39. Campbell, M.; Chang, P.C.; Huerta, S.; Izumiya, C.; Davis, R.; Tepper, C.G.; Kim, K.Y.; Shevchenko, B.; Wang, D.H.; Jung, J.U.; *et al.* Protein arginine methyltransferase 1-directed methylation of Kaposi sarcoma-associated herpesvirus latency-associated nuclear antigen. *J. Biol. Chem.* **2012**, *287*, 5806–5818.

40. Campbell, M.; Izumiya, Y. Post-Translational Modifications of Kaposi's Sarcoma-Associated Herpesvirus Regulatory Proteins—SUMO and KSHV. *Front. Microbiol.* **2012**, *3*, 31.

41. Toth, Z.; Maglinte, D.T.; Lee, S.H.; Lee, H.R.; Wong, L.Y.; Brulois, K.F.; Lee, S.; Buckley, J.D.; Laird, P.W.; Marquez, V.E.; *et al.* Epigenetic analysis of KSHV latent and lytic genomes. *PLoS Pathog.* **2010**, *6*, e1001013.

42. Lim, C.; Lee, D.; Seo, T.; Choi, C.; Choe, J. Latency-associated nuclear antigen of Kaposi's sarcoma-associated herpesvirus functionally interacts with heterochromatin protein 1. *J. Biol. Chem.* **2003**, *278*, 7397–7405.

43. Matsumura, S.; Persson, L.M.; Wong, L.; Wilson, A.C. The latency-associated nuclear antigen interacts with MeCP2 and nucleosomes through separate domains. *J. Virol.* **2010**, *84*, 2318–2330.

44. Krithivas, A.; Young, D.B.; Liao, G.; Greene, D.; Hayward, S.D. Human herpesvirus 8 LANA interacts with proteins of the mSin3 corepressor complex and negatively regulates Epstein-Barr virus gene expression in dually infected PEL cells. *J. Virol.* **2000**, *74*, 9637–9645.

45. Stuber, G.; Mattsson, K.; Flaberg, E.; Kati, E.; Markasz, L.; Sheldon, J.A.; Klein, G.; Schulz, T.F.; Szekely, L. HHV-8 encoded LANA-1 alters the higher organization of the cell nucleus. *Mol. Cancer* **2007**, *6*, 28.

46. Kim, K.Y.; Huerta, S.B.; Izumiya, C.; Wang, D.H.; Martinez, A.; Shevchenko, B.; Kung, H.J.; Campbell, M.; Izumiya, Y. Kaposi's sarcoma-associated herpesvirus (KSHV) latency-associated nuclear antigen regulates the KSHV epigenome by association with the histone demethylase KDM3A. *J. Virol.* **2013**, *87*, 6782–6793.

47. Hu, J.; Yang, Y.; Turner, P.C.; Jain, V.; McIntyre, L.M.; Renne, R. LANA binds to multiple active viral and cellular promoters and associates with the H3K4methyltransferase hSET1 complex. *PLoS Pathog.* **2014**, *10*, e1004240.

48. Guito, J.; Lukac, D.M. KSHV Rta Promoter Specification and Viral Reactivation. *Front. Microbiol.* **2012**, *3*, 30.
49. Jones, T.; Ye, F.; Bedolla, R.; Huang, Y.; Meng, J.; Qian, L.; Pan, H.; Zhou, F.; Moody, R.; Wagner, B.; *et al.* Direct and efficient cellular transformation of primary rat mesenchymal precursor cells by KSHV. *J. Clin. Investig.* **2012**, *122*, 1076–1081.
50. Vieira, J.; O'Hearn, P.; Kimball, L.; Chandran, B.; Corey, L. Activation of Kaposi's sarcoma-associated herpesvirus (human herpesvirus 8) lytic replication by human cytomegalovirus. *J. Virol.* **2001**, *75*, 1378–1386.
51. Zhou, F.; Xue, M.; Qin, D.; Zhu, X.; Wang, C.; Zhu, J.; Hao, T.; Cheng, L.; Chen, X.; Bai, Z.; *et al.* HIV-1 Tat promotes Kaposi's sarcoma-associated herpesvirus (KSHV) vIL-6-induced angiogenesis and tumorigenesis by regulating PI3K/PTEN/AKT/GSK-3beta signaling pathway. *PLoS One* **2013**, *8*, e53145.
52. Zhu, X.; Guo, Y.; Yao, S.; Yan, Q.; Xue, M.; Hao, T.; Zhou, F.; Zhu, J.; Qin, D.; Lu, C. Synergy between Kaposi's sarcoma-associated herpesvirus (KSHV) vIL-6 and HIV-1 Nef protein in promotion of angiogenesis and oncogenesis: Role of the AKT signaling pathway. *Oncogene* **2014**, *33*, 1986–1996.
53. Roupelieva, M.; Griffiths, S.J.; Kremmer, E.; Meisterernst, M.; Viejo-Borbolla, A.; Schulz, T.; Haas, J. Kaposi's sarcoma-associated herpesvirus Lana-1 is a major activator of the serum response element and mitogen-activated protein kinase pathways via interactions with the Mediator complex. *J. Gen. Virol.* **2010**, *91*, 1138–1149.
54. Tang, Q.; Qin, D.; Lv, Z.; Zhu, X.; Ma, X.; Yan, Q.; Zeng, Y.; Guo, Y.; Feng, N.; Lu, C. Herpes simplex virus type 2 triggers reactivation of Kaposi's sarcoma-associated herpesvirus from latency and collaborates with HIV-1 Tat. *PLoS One* **2012**, *7*, e31652.
55. Lu, C.; Zeng, Y.; Huang, Z.; Huang, L.; Qian, C.; Tang, G.; Qin, D. Human herpesvirus 6 activates lytic cycle replication of Kaposi's sarcoma-associated herpesvirus. *Am. J. Pathol.* **2005**, *166*, 173–183.
56. Blauvelt, A. Skin diseases associated with human herpesvirus 6, 7, and 8 infection. *J. Investig. Dermatol. Symp. Proc.* **2001**, *6*, 197–202.
57. Rady, P.L.; Yen, A.; Rollefson, J.L.; Orengo, I.; Bruce, S.; Hughes, T.K.; Tyring, S.K. Herpesvirus-like DNA sequences in non-Kaposi's sarcoma skin lesions of transplant patients. *Lancet* **1995**, *345*, 1339–1340.
58. Mercader, M.; Taddeo, B.; Panella, J.R.; Chandran, B.; Nickoloff, B.J.; Foreman, K.E. Induction of HHV-8 lytic cycle replication by inflammatory cytokines produced by HIV-1-infected T cells. *Am. J. Pathol.* **2000**, *156*, 1961–1971.
59. Ye, F.; Lei, X.; Gao, S.J. Mechanisms of Kaposi's Sarcoma-Associated Herpesvirus Latency and Reactivation. *Adv. Virol.* **2011**, *2011*, doi:10.1155/2011/193860.
60. Romero-Ramirez, L.; Cao, H.; Nelson, D.; Hammond, E.; Lee, A.H.; Yoshida, H.; Mori, K.; Glimcher, L.H.; Denko, N.C.; Giaccia, A.J.; *et al.* XBP1 is essential for survival under hypoxic conditions and is required for tumor growth. *Cancer Res.* **2004**, *64*, 5943–5947.
61. Cai, Q.; Lan, K.; Verma, S.C.; Si, H.; Lin, D.; Robertson, E.S. Kaposi's sarcoma-associated herpesvirus latent protein LANA interacts with HIF-1α to upregulate RTA expression during hypoxia: Latency control under low oxygen conditions. *J. Virol.* **2006**, *80*, 7965–7975.

62. Cai, Q.; Cai, S.; Zhu, C.; Verma, S.C.; Choi, J.Y.; Robertson, E.S. A unique SUMO-2-interacting motif within LANA is essential for KSHV latency. *PLoS Pathog.* **2013**, *9*, e1003750.
63. Zhang, L.; Zhu, C.; Guo, Y.; Wei, F.; Lu, J.; Qin, J.; Banerjee, S.; Wang, J.; Shang, H.; Verma, S.C.; *et al.* Inhibition of KAP1 enhances hypoxia-induced Kaposi's sarcoma-associated herpesvirus reactivation through RBP-Jκ. *J. Virol.* **2014**, *88*, 6873–6884.
64. Ye, F.; Zhou, F.; Bedolla, R.G.; Jones, T.; Lei, X.; Kang, T.; Guadalupe, M.; Gao, S.J. Reactive oxygen species hydrogen peroxide mediates Kaposi's sarcoma-associated herpesvirus reactivation from latency. *PLoS Pathog.* **2011**, *7*, e1002054.
65. Ruffels, J.; Griffin, M.; Dickenson, J.M. Activation of ERK1/2, JNK and PKB by hydrogen peroxide in human SH-SY5Y neuroblastoma cells: Role of ERK1/2 in H_2O_2-induced cell death. *Eur. J. Pharmacol.* **2004**, *483*, 163–173.
66. Li, X.; Feng, J.; Sun, R. Oxidative stress induces reactivation of Kaposi's sarcoma-associated herpesvirus and death of primary effusion lymphoma cells. *J. Virol.* **2011**, *85*, 715–724.
67. Shin, H.J.; DeCotiis, J.; Giron, M.; Palmeri, D.; Lukac, D.M. Histone deacetylase classes I and II regulate Kaposi's sarcoma-associated herpesvirus reactivation. *J. Virol.* **2014**, *88*, 1281–1292.
68. Li, Q.; He, M.; Zhou, F.; Ye, F.; Gao, S.J. Activation of Kaposi's sarcoma-associated herpesvirus (KSHV) by inhibitors of class III histone deacetylases: Identification of sirtuin 1 as a regulator of the KSHV life cycle. *J. Virol.* **2014**, *88*, 6355–6367.
69. Dyson, O.F.; Walker, L.R.; Whitehouse, A.; Cook, P.P.; Akula, S.M. Resveratrol inhibits KSHV reactivation by lowering the levels of cellular EGR-1. *PLoS One* **2012**, *7*, e33364.
70. AuCoin, D.P.; Colletti, K.S.; Cei, S.A.; Papouskova, I.; Tarrant, M.; Pari, G.S. Amplification of the Kaposi's sarcoma-associated herpesvirus/human herpesvirus 8 lytic origin of DNA replication is dependent upon a cis-acting AT-rich region and an ORF50 response element and the trans-acting factors ORF50 (K-Rta) and K8 (K-bZIP). *Virology* **2004**, *318*, 542–555.
71. Gradoville, L.; Gerlach, J.; Grogan, E.; Shedd, D.; Nikiforow, S.; Metroka, C.; Miller, G. Kaposi's sarcoma-associated herpesvirus open reading frame 50/Rta protein activates the entire viral lytic cycle in the HH-B2 primary effusion lymphoma cell line. *J. Virol.* **2000**, *74*, 6207–6212.
72. Cousins, E.; Nicholas, J. Molecular biology of human herpesvirus 8: Novel functions and virus-host interactions implicated in viral pathogenesis and replication. *Recent Results Cancer Res.* **2014**, *193*, 227–268.
73. Sun, R.; Lin, S.F.; Gradoville, L.; Yuan, Y.; Zhu, F.; Miller, G. A viral gene that activates lytic cycle expression of Kaposi's sarcoma-associated herpesvirus. *Proc. Natl. Acad. Sci. USA* **1998**, *95*, 10866–10871.
74. Lukac, D.M.; Renne, R.; Kirshner, J.R.; Ganem, D. Reactivation of Kaposi's sarcoma-associated herpesvirus infection from latency by expression of the ORF 50 transactivator, a homolog of the EBV R protein. *Virology* **1998**, *252*, 304–312.
75. Xu, Y.; AuCoin, D.P.; Huete, A.R.; Cei, S.A.; Hanson, L.J.; Pari, G.S. A Kaposi's sarcoma-associated herpesvirus/human herpesvirus 8 ORF50 deletion mutant is defective for reactivation of latent virus and DNA replication. *J. Virol.* **2005**, *79*, 3479–3487.
76. Lukac, D.M.; Kirshner, J.R.; Ganem, D. Transcriptional activation by the product of open reading frame 50 of Kaposi's sarcoma-associated herpesvirus is required for lytic viral reactivation in B cells. *J. Virol.* **1999**, *73*, 9348–9361.

77. Gwack, Y.; Nakamura, H.; Lee, S.H.; Souvlis, J.; Yustein, J.T.; Gygi, S.; Kung, H.J.; Jung, J.U. Poly(ADP-ribose) polymerase 1 and Ste20-like kinase hKFC act as transcriptional repressors for γ2-herpesvirus lytic replication. *Mol. Cell Biol.* **2003**, *23*, 8282–8294.
78. Yu, Y.; Wang, S.E.; Hayward, G.S. The KSHV immediate-early transcription factor RTA encodes ubiquitin E3 ligase activity that targets IRF7 for proteosome-mediated degradation. *Immunity* **2005**, *22*, 59–70.
79. Bu, W.; Palmeri, D.; Krishnan, R.; Marin, R.; Aris, V.M.; Soteropoulos, P.; Lukac, D.M. Identification of direct transcriptional targets of the Kaposi's sarcoma-associated herpesvirus Rta lytic switch protein by conditional nuclear localization. *J. Virol.* **2008**, *82*, 10709–10723.
80. Deng, H.; Song, M.J.; Chu, J.T.; Sun, R. Transcriptional regulation of the interleukin-6 gene of human herpesvirus 8 (Kaposi's sarcoma-associated herpesvirus). *J. Virol.* **2002**, *76*, 8252–8264.
81. Song, M.J.; Brown, H.J.; Wu, T.T.; Sun, R. Transcription activation of polyadenylated nuclear rna by rta in human herpesvirus 8/Kaposi's sarcoma-associated herpesvirus. *J. Virol.* **2001**, *75*, 3129–3140.
82. Byun, H.; Gwack, Y.; Hwang, S.; Choe, J. Kaposi's sarcoma-associated herpesvirus open reading frame (ORF) 50 transactivates K8 and ORF57 promoters via heterogeneous response elements. *Mol. Cells* **2002**, *14*, 185–191.
83. McDowell, M.E.; Purushothaman, P.; Rossetto, C.C.; Pari, G.S.; Verma, S.C. Phosphorylation of Kaposi's sarcoma-associated herpesvirus processivity factor ORF59 by a viral kinase modulates its ability to associate with RTA and oriLyt. *J. Virol.* **2013**, *87*, 8038–8052.
84. Rossetto, C.C.; Susilarini, N.K.; Pari, G.S. Interaction of Kaposi's sarcoma-associated herpesvirus ORF59 with oriLyt is dependent on binding with K-Rta. *J. Virol.* **2011**, *85*, 3833–3841.
85. Ueda, K.; Ishikawa, K.; Nishimura, K.; Sakakibara, S.; Do, E.; Yamanishi, K. Kaposi's sarcoma-associated herpesvirus (human herpesvirus 8) replication and transcription factor activates the K9 (vIRF) gene through two distinct cis elements by a non-DNA-binding mechanism. *J. Virol.* **2002**, *76*, 12044–12054.
86. Bowser, B.S.; Morris, S.; Song, M.J.; Sun, R.; Damania, B. Characterization of Kaposi's sarcoma-associated herpesvirus (KSHV) K1 promoter activation by Rta. *Virology* **2006**, *348*, 309–327.
87. Sathish, N.; Yuan, Y. Functional characterization of Kaposi's sarcoma-associated herpesvirus small capsid protein by bacterial artificial chromosome-based mutagenesis. *Virology* **2010**, *407*, 306–318.
88. Majerciak, V.; Yamanegi, K.; Zheng, Z.M. Gene structure and expression of Kaposi's sarcoma-associated herpesvirus ORF56, ORF57, ORF58, and ORF59. *J. Virol.* **2006**, *80*, 11968–11981.
89. Kronstad, L.M.; Brulois, K.F.; Jung, J.U.; Glaunsinger, B.A. Reinitiation after translation of two upstream open reading frames (ORF) governs expression of the ORF35–37 Kaposi's sarcoma-associated herpesvirus polycistronic mRNA. *J. Virol.* **2014**, *88*, 6512–6518.
90. Chung, Y.H.; Means, R.E.; Choi, J.K.; Lee, B.S.; Jung, J.U. Kaposi's sarcoma-associated herpesvirus OX_2 glycoprotein activates myeloid-lineage cells to induce inflammatory cytokine production. *J. Virol.* **2002**, *76*, 4688–4698.
91. Seaman, W.T.; Quinlivan, E.B. Lytic switch protein (ORF50) response element in the Kaposi's sarcoma-associated herpesvirus K8 promoter is located within but does not require a palindromic structure. *Virology* **2003**, *310*, 72–84.

92. Lan, K.; Kuppers, D.A.; Verma, S.C.; Robertson, E.S. Kaposi's sarcoma-associated herpesvirus-encoded latency-associated nuclear antigen inhibits lytic replication by targeting Rta: A potential mechanism for virus-mediated control of latency. *J. Virol.* **2004**, *78*, 6585–6594.
93. Ziegelbauer, J.; Grundhoff, A.; Ganem, D. Exploring the DNA binding interactions of the Kaposi's sarcoma-associated herpesvirus lytic switch protein by selective amplification of bound sequences *in vitro*. *J. Virol.* **2006**, *80*, 2958–2967.
94. Chen, J.; Ye, F.; Xie, J.; Kuhne, K.; Gao, S.J. Genome-wide identification of binding sites for Kaposi's sarcoma-associated herpesvirus lytic switch protein, RTA. *Virology* **2009**, *386*, 290–302.
95. Liang, Y.; Chang, J.; Lynch, S.J.; Lukac, D.M.; Ganem, D. The lytic switch protein of KSHV activates gene expression via functional interaction with RBP-Jκ (CSL), the target of the Notch signaling pathway. *Genes Dev.* **2002**, *16*, 1977–1989.
96. Wilson, S.J.; Tsao, E.H.; Webb, B.L.; Ye, H.; Dalton-Griffin, L.; Tsantoulas, C.; Gale, C.V.; Du, M.Q.; Whitehouse, A.; Kellam, P. X box binding protein XBP-1s transactivates the Kaposi's sarcoma-associated herpesvirus (KSHV) ORF50 promoter, linking plasma cell differentiation to KSHV reactivation from latency. *J. Virol.* **2007**, *81*, 13578–13586.
97. Ye, J.; Shedd, D.; Miller, G. An Sp1 response element in the Kaposi's sarcoma-associated herpesvirus open reading frame 50 promoter mediates lytic cycle induction by butyrate. *J. Virol.* **2005**, *79*, 1397–1408.
98. Carroll, K.D.; Khadim, F.; Spadavecchia, S.; Palmeri, D.; Lukac, D.M. Direct interactions of Kaposi's sarcoma-associated herpesvirus/human herpesvirus 8 ORF50/Rta protein with the cellular protein octamer-1 and DNA are critical for specifying transactivation of a delayed-early promoter and stimulating viral reactivation. *J. Virol.* **2007**, *81*, 8451–8467.
99. Chang, P.J.; Shedd, D.; Miller, G. Two subclasses of Kaposi's sarcoma-associated herpesvirus lytic cycle promoters distinguished by open reading frame 50 mutant proteins that are deficient in binding to DNA. *J. Virol.* **2005**, *79*, 8750–8763.
100. Chang, P.J.; Boonsiri, J.; Wang, S.S.; Chen, L.Y.; Miller, G. Binding of RBP-Jκ (CSL) protein to the promoter of the Kaposi's sarcoma-associated herpesvirus ORF47 (gL) gene is a critical but not sufficient determinant of transactivation by ORF50 protein. *Virology* **2010**, *398*, 38–48.
101. Persson, L.M.; Wilson, A.C. Wide-scale use of Notch signaling factor CSL/RBP-Jκ in RTA-mediated activation of Kaposi's sarcoma-associated herpesvirus lytic genes. *J. Virol.* **2010**, *84*, 1334–1347.
102. Lu, J.; Verma, S.C.; Cai, Q.; Saha, A.; Dzeng, R.K.; Robertson, E.S. The RBP-Jκ binding sites within the RTA promoter regulate KSHV latent infection and cell proliferation. *PLoS Pathog.* **2012**, *8*, e1002479.
103. Lan, K.; Murakami, M.; Choudhuri, T.; Kuppers, D.A.; Robertson, E.S. Intracellular-activated Notch1 can reactivate Kaposi's sarcoma-associated herpesvirus from latency. *Virology* **2006**, *351*, 393–403.
104. Chang, H.; Dittmer, D.P.; Shin, Y.C.; Hong, Y.; Jung, J.U. Role of Notch signal transduction in Kaposi's sarcoma-associated herpesvirus gene expression. *J. Virol.* **2005**, *79*, 14371–14382.
105. Gwack, Y.; Baek, H.J.; Nakamura, H.; Lee, S.H.; Meisterernst, M.; Roeder, R.G.; Jung, J.U. Principal role of TRAP/mediator and SWI/SNF complexes in Kaposi's sarcoma-associated herpesvirus RTA-mediated lytic reactivation. *Mol. Cell Biol.* **2003**, *23*, 2055–2067.

106. Gwack, Y.; Byun, H.; Hwang, S.; Lim, C.; Choe, J. CREB-binding protein and histone deacetylase regulate the transcriptional activity of Kaposi's sarcoma-associated herpesvirus open reading frame 50. *J. Virol.* **2001**, *75*, 1909–1917.
107. Gao, J.; Cai, Q.; Lu, J.; Jha, H.C.; Robertson, E.S. Upregulation of cellular Bcl-2 by the KSHV encoded RTA promotes virion production. *PLoS One* **2011**, *6*, e23892.
108. Wang, S.; Liu, S.; Wu, M.H.; Geng, Y.; Wood, C. Identification of a cellular protein that interacts and synergizes with the RTA (ORF50) protein of Kaposi's sarcoma-associated herpesvirus in transcriptional activation. *J. Virol.* **2001**, *75*, 11961–11973.
109. Yang, Z.; Wood, C. The transcriptional repressor K-RBP modulates RTA-mediated transactivation and lytic replication of Kaposi's sarcoma-associated herpesvirus. *J. Virol.* **2007**, *81*, 6294–6306.
110. Yang, Z.; Wen, H.J.; Minhas, V.; Wood, C. The zinc finger DNA-binding domain of K-RBP plays an important role in regulating Kaposi's sarcoma-associated herpesvirus RTA-mediated gene expression. *Virology* **2009**, *391*, 221–231.
111. Chang, P.C.; Fitzgerald, L.D.; Van Geelen, A.; Izumiya, Y.; Ellison, T.J.; Wang, D.H.; Ann, D.K.; Luciw, P.A.; Kung, H.J. Kruppel-associated box domain-associated protein-1 as a latency regulator for Kaposi's sarcoma-associated herpesvirus and its modulation by the viral protein kinase. *Cancer Res.* **2009**, *69*, 5681–5689.
112. Sun, R.; Liang, D.; Gao, Y.; Lan, K. Kaposi's sarcoma-associated herpesvirus-encoded LANA interacts with host KAP1 to facilitate establishment of viral latency. *J. Virol.* **2014**, *88*, 7331–7344.
113. Liao, W.; Tang, Y.; Lin, S.F.; Kung, H.J.; Giam, C.Z. K-bZIP of Kaposi's sarcoma-associated herpesvirus/human herpesvirus 8 (KSHV/HHV-8) binds KSHV/HHV-8 Rta and represses Rta-mediated transactivation. *J. Virol.* **2003**, *77*, 3809–3815.
114. Rossetto, C.; Gao, Y.; Yamboliev, I.; Papouskova, I.; Pari, G. Transcriptional repression of K-Rta by Kaposi's sarcoma-associated herpesvirus K-bZIP is not required for oriLyt-dependent DNA replication. *Virology* **2007**, *369*, 340–350.
115. Wang, S.E.; Wu, F.Y.; Fujimuro, M.; Zong, J.; Hayward, S.D.; Hayward, G.S. Role of CCAAT/enhancer-binding protein α (C/EBPα) in activation of the Kaposi's sarcoma-associated herpesvirus (KSHV) lytic-cycle replication-associated protein (RAP) promoter in cooperation with the KSHV replication and transcription activator (RTA) and RAP. *J. Virol.* **2003**, *77*, 600–623.
116. Wang, Y.; Yuan, Y. Essential role of RBP-Jκ in activation of the K8 delayed-early promoter of Kaposi's sarcoma-associated herpesvirus by ORF50/RTA. *Virology* **2007**, *359*, 19–27.
117. Liu, Y.; Cao, Y.; Liang, D.; Gao, Y.; Xia, T.; Robertson, E.S.; Lan, K. Kaposi's sarcoma-associated herpesvirus RTA activates the processivity factor ORF59 through interaction with RBP-Jκ and a cis-acting RTA responsive element. *Virology* **2008**, *380*, 264–275.
118. Chang, P.J.; Wang, S.S.; Chen, L.Y.; Hung, C.H.; Huang, H.Y.; Shih, Y.J.; Yen, J.B.; Liou, J.Y.; Chen, L.W. ORF50-dependent and ORF50-independent activation of the ORF45 gene of Kaposi's sarcoma-associated herpesvirus. *Virology* **2013**, *442*, 38–50.
119. Guito, J.; Gavina, A.; Palmeri, D.; Lukac, D.M. The cellular peptidyl-prolyl cis/trans isomerase Pin1 regulates reactivation of Kaposi's sarcoma-associated herpesvirus from latency. *J. Virol.* **2014**, *88*, 547–558.
120. Jaber, T.; Yuan, Y. A virally encoded small peptide regulates RTA stability and facilitates Kaposi's sarcoma-associated herpesvirus lytic replication. *J. Virol.* **2013**, *87*, 3461–3470.

121. Yang, Z.; Yan, Z.; Wood, C. Kaposi's sarcoma-associated herpesvirus transactivator RTA promotes degradation of the repressors to regulate viral lytic replication. *J. Virol.* **2008**, *82*, 3590–3603.

122. Gould, F.; Harrison, S.M.; Hewitt, E.W.; Whitehouse, A. Kaposi's sarcoma-associated herpesvirus RTA promotes degradation of the Hey1 repressor protein through the ubiquitin proteasome pathway. *J. Virol.* **2009**, *83*, 6727–6738.

123. Wang, Y.; Li, H.; Chan, M.Y.; Zhu, F.X.; Lukac, D.M.; Yuan, Y. Kaposi's sarcoma-associated herpesvirus ori-Lyt-dependent DNA replication: Cis-acting requirements for replication and ori-Lyt-associated RNA transcription. *J. Virol.* **2004**, *78*, 8615–8629.

124. Izumiya, Y.; Kobayashi, K.; Kim, K.Y.; Pochampalli, M.; Izumiya, C.; Shevchenko, B.; Wang, D.H.; Huerta, S.B.; Martinez, A.; Campbell, M.; *et al.* Kaposi's sarcoma-associated herpesvirus K-Rta exhibits SUMO-targeting ubiquitin ligase (STUbL) like activity and is essential for viral reactivation. *PLoS Pathog.* **2013**, *9*, e1003506.

125. Han, Z.; Swaminathan, S. Kaposi's sarcoma-associated herpesvirus lytic gene ORF57 is essential for infectious virion production. *J. Virol.* **2006**, *80*, 5251–5260.

126. Majerciak, V.; Pripuzova, N.; McCoy, J.P.; Gao, S.J.; Zheng, Z.M. Targeted disruption of Kaposi's sarcoma-associated herpesvirus ORF57 in the viral genome is detrimental for the expression of ORF59, K8α, and K8.1 and the production of infectious virus. *J. Virol.* **2007**, *81*, 1062–1071.

127. Palmeri, D.; Spadavecchia, S.; Carroll, K.D.; Lukac, D.M. Promoter- and cell-specific transcriptional transactivation by the Kaposi's sarcoma-associated herpesvirus ORF57/Mta protein. *J. Virol.* **2007**, *81*, 13299–13314.

128. Malik, P.; Blackbourn, D.J.; Cheng, M.F.; Hayward, G.S.; Clements, J.B. Functional co-operation between the Kaposi's sarcoma-associated herpesvirus ORF57 and ORF50 regulatory proteins. *J. Gen. Virol.* **2004**, *85*, 2155–2166.

129. Hunter, O.V.; Sei, E.; Richardson, R.B.; Conrad, N.K. Chromatin immunoprecipitation and microarray analysis suggest functional cooperation between Kaposi's Sarcoma-associated herpesvirus ORF57 and K-bZIP. *J. Virol.* **2013**, *87*, 4005–4016.

130. Malik, P.; Blackbourn, D.J.; Clements, J.B. The evolutionarily conserved Kaposi's sarcoma-associated herpesvirus ORF57 protein interacts with REF protein and acts as an RNA export factor. *J. Biol. Chem.* **2004**, *279*, 33001–33011.

131. Majerciak, V.; Zheng, Z.M. Kaposi's sarcoma-associated herpesvirus ORF57 in viral RNA processing. *Front. Biosci. (Landmark Ed)* **2009**, *14*, 1516–1528.

132. Nekorchuk, M.; Han, Z.; Hsieh, T.T.; Swaminathan, S. Kaposi's sarcoma-associated herpesvirus ORF57 protein enhances mRNA accumulation independently of effects on nuclear RNA export. *J. Virol.* **2007**, *81*, 9990–9998.

133. Massimelli, M.J.; Kang, J.G.; Majerciak, V.; Le, S.Y.; Liewehr, D.J.; Steinberg, S.M.; Zheng, Z.M. Stability of a long noncoding viral RNA depends on a 9-nt core element at the RNA 5' end to interact with viral ORF57 and cellular PABPC1. *Int. J. Biol. Sci.* **2011**, *7*, 1145–1160.

134. Massimelli, M.J.; Majerciak, V.; Kruhlak, M.; Zheng, Z.M. Interplay between polyadenylate-binding protein 1 and Kaposi's sarcoma-associated herpesvirus ORF57 in accumulation of polyadenylated nuclear RNA, a viral long noncoding RNA. *J. Virol.* **2013**, *87*, 243–256.

135. Sahin, B.B.; Patel, D.; Conrad, N.K. Kaposi's sarcoma-associated herpesvirus ORF57 protein binds and protects a nuclear noncoding RNA from cellular RNA decay pathways. *PLoS Pathog.* **2010**, *6*, e1000799.

136. Majerciak, V.; Uranishi, H.; Kruhlak, M.; Pilkington, G.R.; Massimelli, M.J.; Bear, J.; Pavlakis, G.N.; Felber, B.K.; Zheng, Z.M. Kaposi's sarcoma-associated herpesvirus ORF57 interacts with cellular RNA export cofactors RBM15 and OTT3 to promote expression of viral ORF59. *J. Virol.* **2011**, *85*, 1528–1540.

137. Lin, S.F.; Robinson, D.R.; Miller, G.; Kung, H.J. Kaposi's sarcoma-associated herpesvirus encodes a bZIP protein with homology to BZLF1 of Epstein-Barr virus. *J. Virol.* **1999**, *73*, 1909–1917.

138. Tang, S.; Zheng, Z.M. Kaposi's sarcoma-associated herpesvirus K8 exon 3 contains three 5'-splice sites and harbors a K8.1 transcription start site. *J. Biol. Chem.* **2002**, *277*, 14547–14556.

139. Izumiya, Y.; Lin, S.F.; Ellison, T.; Chen, L.Y.; Izumiya, C.; Luciw, P.; Kung, H.J. Kaposi's sarcoma-associated herpesvirus K-bZIP is a coregulator of K-Rta: Physical association and promoter-dependent transcriptional repression. *J. Virol.* **2003**, *77*, 1441–1451.

140. Ellison, T.J.; Izumiya, Y.; Izumiya, C.; Luciw, P.A.; Kung, H.J. A comprehensive analysis of recruitment and transactivation potential of K-Rta and K-bZIP during reactivation of Kaposi's sarcoma-associated herpesvirus. *Virology* **2009**, *387*, 76–88.

141. Kato-Noah, T.; Xu, Y.; Rossetto, C.C.; Colletti, K.; Papouskova, I.; Pari, G.S. Overexpression of the kaposi's sarcoma-associated herpesvirus transactivator K-Rta can complement a K-bZIP deletion BACmid and yields an enhanced growth phenotype. *J. Virol.* **2007**, *81*, 13519–13532.

142. Wang, Y.; Sathish, N.; Hollow, C.; Yuan, Y. Functional characterization of Kaposi's sarcoma-associated herpesvirus open reading frame K8 by bacterial artificial chromosome-based mutagenesis. *J. Virol.* **2011**, *85*, 1943–1957.

143. Lefort, S.; Flamand, L. Kaposi's sarcoma-associated herpesvirus K-bZIP protein is necessary for lytic viral gene expression, DNA replication, and virion production in primary effusion lymphoma cell lines. *J. Virol.* **2009**, *83*, 5869–5880.

144. Wu, F.Y.; Wang, S.E.; Tang, Q.Q.; Fujimuro, M.; Chiou, C.J.; Zheng, Q.; Chen, H.; Hayward, S.D.; Lane, M.D.; Hayward, G.S. Cell cycle arrest by Kaposi's sarcoma-associated herpesvirus replication-associated protein is mediated at both the transcriptional and posttranslational levels by binding to CCAAT/enhancer-binding protein α and p21(CIP-1). *J. Virol.* **2003**, *77*, 8893–8914.

145. Rossetto, C.; Yamboliev, I.; Pari, G.S. Kaposi's sarcoma-associated herpesvirus/human herpesvirus 8 K-bZIP modulates latency-associated nuclear protein-mediated suppression of lytic origin-dependent DNA synthesis. *J. Virol.* **2009**, *83*, 8492–8501.

146. Izumiya, Y.; Ellison, T.J.; Yeh, E.T.; Jung, J.U.; Luciw, P.A.; Kung, H.J. Kaposi's sarcoma-associated herpesvirus K-bZIP represses gene transcription via SUMO modification. *J. Virol.* **2005**, *79*, 9912–9925.

147. Lefort, S.; Soucy-Faulkner, A.; Grandvaux, N.; Flamand, L. Binding of Kaposi's sarcoma-associated herpesvirus K-bZIP to interferon-responsive factor 3 elements modulates antiviral gene expression. *J. Virol.* **2007**, *81*, 10950–10960.

148. Lefort, S.; Gravel, A.; Flamand, L. Repression of interferon-α stimulated genes expression by Kaposi's sarcoma-associated herpesvirus K-bZIP protein. *Virology* **2010**, *408*, 14–30.

149. Martinez, F.P.; Tang, Q. Leucine zipper domain is required for Kaposi sarcoma-associated herpesvirus (KSHV) K-bZIP protein to interact with histone deacetylase and is important for KSHV replication. *J. Biol. Chem.* **2012**, *287*, 15622–15634.
150. Izumiya, Y.; Izumiya, C.; van Geelen, A.; Wang, D.H.; Lam, K.S.; Luciw, P.A.; Kung, H.J. Kaposi's sarcoma-associated herpesvirus-encoded protein kinase and its interaction with K-bZIP. *J. Virol.* **2007**, *81*, 1072–1082.
151. Sinclair, A.J. bZIP proteins of human γ-herpesviruses. *J. Gen. Virol.* **2003**, *84*, 1941–1949.
152. Chen, Y.; Ciustea, M.; Ricciardi, R.P. Processivity factor of KSHV contains a nuclear localization signal and binding domains for transporting viral DNA polymerase into the nucleus. *Virology* **2005**, *340*, 183–191.
153. Xiao, Y.; Chen, J.; Liao, Q.; Wu, Y.; Peng, C.; Chen, X. Lytic infection of Kaposi's sarcoma-associated herpesvirus induces DNA double-strand breaks and impairs non-homologous end joining. *J. Gen. Virol.* **2013**, *94*, 1870–1875.
154. Nicholas, J.; Ruvolo, V.; Zong, J.; Ciufo, D.; Guo, H.G.; Reitz, M.S.; Hayward, G.S. A single 13-kilobase divergent locus in the Kaposi sarcoma-associated herpesvirus (human herpesvirus 8) genome contains nine open reading frames that are homologous to or related to cellular proteins. *J. Virol.* **1997**, *71*, 1963–1974.
155. Peng, C.; Chen, J.; Tang, W.; Liu, C.; Chen, X. Kaposi's sarcoma-associated herpesvirus ORF6 gene is essential in viral lytic replication. *PLoS One* **2014**, *9*, e99542.
156. Deutsch, E.; Cohen, A.; Kazimirsky, G.; Dovrat, S.; Rubinfeld, H.; Brodie, C.; Sarid, R. Role of protein kinase C delta in reactivation of Kaposi's sarcoma-associated herpesvirus. *J. Virol.* **2004**, *78*, 10187–10192.
157. Ford, P.W.; Bryan, B.A.; Dyson, O.F.; Weidner, D.A.; Chintalgattu, V.; Akula, S.M. Raf/MEK/ERK signalling triggers reactivation of Kaposi's sarcoma-associated herpesvirus latency. *J. Gen. Virol.* **2006**, *87*, 1139–1144.
158. Lan, K.; Choudhuri, T.; Murakami, M.; Kuppers, D.A.; Robertson, E.S. Intracellular activated Notch1 is critical for proliferation of Kaposi's sarcoma-associated herpesvirus-associated B-lymphoma cell lines *in vitro*. *J. Virol.* **2006**, *80*, 6411–6419.
159. Xie, J.; Ajibade, A.O.; Ye, F.; Kuhne, K.; Gao, S.J. Reactivation of Kaposi's sarcoma-associated herpesvirus from latency requires MEK/ERK, JNK and p38 multiple mitogen-activated protein kinase pathways. *Virology* **2008**, *371*, 139–154.
160. Cheng, F.; Weidner-Glunde, M.; Varjosalo, M.; Rainio, E.M.; Lehtonen, A.; Schulz, T.F.; Koskinen, P.J.; Taipale, J.; Ojala, P.M. KSHV reactivation from latency requires Pim-1 and Pim-3 kinases to inactivate the latency-associated nuclear antigen LANA. *PLoS Pathog.* **2009**, *5*, e1000324.
161. Qin, D.; Feng, N.; Fan, W.; Ma, X.; Yan, Q.; Lv, Z.; Zeng, Y.; Zhu, J.; Lu, C. Activation of PI3K/AKT and ERK MAPK signal pathways is required for the induction of lytic cycle replication of Kaposi's sarcoma-associated herpesvirus by herpes simplex virus type 1. *BMC Microbiol.* **2011**, *11*, 240.
162. Gregory, S.M.; West, J.A.; Dillon, P.J.; Hilscher, C.; Dittmer, D.P.; Damania, B. Toll-like receptor signaling controls reactivation of KSHV from latency. *Proc. Natl. Acad. Sci. USA* **2009**, *106*, 11725–11730.

163. Yu, F.; Harada, J.N.; Brown, H.J.; Deng, H.; Song, M.J.; Wu, T.T.; Kato-Stankiewicz, J.; Nelson, C.G.; Vieira, J.; Tamanoi, F.; *et al.* Systematic identification of cellular signals reactivating Kaposi sarcoma-associated herpesvirus. *PLoS Pathog.* **2007**, *3*, e44.
164. Li, D.J.; Verma, D.; Mosbruger, T.; Swaminathan, S. CTCF and Rad21 act as host cell restriction factors for Kaposi's sarcoma-associated herpesvirus (KSHV) lytic replication by modulating viral gene transcription. *PLoS Pathog.* **2014**, *10*, e1003880.
165. Lee, H.R.; Doganay, S.; Chung, B.; Toth, Z.; Brulois, K.; Lee, S.; Kanketayeva, Z.; Feng, P.; Ha, T.; Jung, J.U. Kaposi's sarcoma-associated herpesvirus viral interferon regulatory factor 4 (vIRF4) targets expression of cellular IRF4 and the Myc gene to facilitate lytic replication. *J. Virol.* **2014**, *88*, 2183–2194.
166. Bi, X.; Yang, L.; Mancl, M.E.; Barnes, B.J. Modulation of interferon regulatory factor 5 activities by the Kaposi sarcoma-associated herpesvirus-encoded viral interferon regulatory factor 3 contributes to immune evasion and lytic induction. *J. Interferon Cytokine Res.* **2011**, *31*, 373–382.
167. Kuang, E.; Fu, B.; Liang, Q.; Myoung, J.; Zhu, F. Phosphorylation of eukaryotic translation initiation factor 4B (EIF4B) by open reading frame 45/p90 ribosomal S6 kinase (ORF45/RSK) signaling axis facilitates protein translation during Kaposi sarcoma-associated herpesvirus (KSHV) lytic replication. *J. Biol. Chem.* **2011**, *286*, 41171–41182.
168. Zoeteweij, J.P.; Moses, A.V.; Rinderknecht, A.S.; Davis, D.A.; Overwijk, W.W.; Yarchoan, R.; Orenstein, J.M.; Blauvelt, A. Targeted inhibition of calcineurin signaling blocks calcium-dependent reactivation of Kaposi sarcoma-associated herpesvirus. *Blood* **2001**, *97*, 2374–2380.
169. Pan, H.; Xie, J.; Ye, F.; Gao, S.J. Modulation of Kaposi's sarcoma-associated herpesvirus infection and replication by MEK/ERK, JNK, and p38 multiple mitogen-activated protein kinase pathways during primary infection. *J. Virol.* **2006**, *80*, 5371–5382.
170. Sharma-Walia, N.; Krishnan, H.H.; Naranatt, P.P.; Zeng, L.; Smith, M.S.; Chandran, B. ERK1/2 and MEK1/2 induced by Kaposi's sarcoma-associated herpesvirus (human herpesvirus 8) early during infection of target cells are essential for expression of viral genes and for establishment of infection. *J. Virol.* **2005**, *79*, 10308–10329.
171. Hengge, U.R.; Ruzicka, T.; Tyring, S.K.; Stuschke, M.; Roggendorf, M.; Schwartz, R.A.; Seeber, S. Update on Kaposi's sarcoma and other HHV8 associated diseases. Part 2: Pathogenesis, Castleman's disease, and pleural effusion lymphoma. *Lancet Infect. Dis.* **2002**, *2*, 344–352.
172. Haas, D.A.; Bala, K.; Busche, G.; Weidner-Glunde, M.; Santag, S.; Kati, S.; Gramolelli, S.; Damas, M.; Dittrich-Breiholz, O.; Kracht, M.; *et al.* The inflammatory kinase MAP4K4 promotes reactivation of Kaposi's sarcoma herpesvirus and enhances the invasiveness of infected endothelial cells. *PLoS Pathog.* **2013**, *9*, e1003737.
173. Wang, S.E.; Wu, F.Y.; Chen, H.; Shamay, M.; Zheng, Q.; Hayward, G.S. Early activation of the Kaposi's sarcoma-associated herpesvirus RTA, RAP, and MTA promoters by the tetradecanoyl phorbol acetate-induced AP1 pathway. *J. Virol.* **2004**, *78*, 4248–4267.
174. Xie, J.; Pan, H.; Yoo, S.; Gao, S.J. Kaposi's sarcoma-associated herpesvirus induction of AP-1 and interleukin 6 during primary infection mediated by multiple mitogen-activated protein kinase pathways. *J. Virol.* **2005**, *79*, 15027–15037.
175. Cohen, A.; Brodie, C.; Sarid, R. An essential role of ERK signalling in TPA-induced reactivation of Kaposi's sarcoma-associated herpesvirus. *J. Gen. Virol.* **2006**, *87*, 795–802.

176. Emuss, V.; Lagos, D.; Pizzey, A.; Gratrix, F.; Henderson, S.R.; Boshoff, C. KSHV manipulates Notch signaling by DLL4 and JAG1 to alter cell cycle genes in lymphatic endothelia. *PLoS Pathog.* **2009**, *5*, e1000616.
177. Carroll, K.D.; Bu, W.; Palmeri, D.; Spadavecchia, S.; Lynch, S.J.; Marras, S.A.; Tyagi, S.; Lukac, D.M. Kaposi's Sarcoma-associated herpesvirus lytic switch protein stimulates DNA binding of RBP-Jk/CSL to activate the Notch pathway. *J. Virol.* **2006**, *80*, 9697–9709.
178. Cai, Q.; Murakami, M.; Si, H.; Robertson, E.S. A potential α-helix motif in the amino terminus of LANA encoded by Kaposi's sarcoma-associated herpesvirus is critical for nuclear accumulation of HIF-1α in normoxia. *J. Virol.* **2007**, *81*, 10413–10423.
179. Carroll, P.A.; Kenerson, H.L.; Yeung, R.S.; Lagunoff, M. Latent Kaposi's sarcoma-associated herpesvirus infection of endothelial cells activates hypoxia-induced factors. *J. Virol.* **2006**, *80*, 10802–10812.
180. Shin, Y.C.; Joo, C.H.; Gack, M.U.; Lee, H.R.; Jung, J.U. Kaposi's sarcoma-associated herpesvirus viral IFN regulatory factor 3 stabilizes hypoxia-inducible factor-1α to induce vascular endothelial growth factor expression. *Cancer Res.* **2008**, *68*, 1751–1759.
181. Gregory, S.M.; Damania, B. KSHV and the toll of innate immune activation. *Cell Cycle* **2009**, *8*, 3246–3247.
182. Yu, X.; Shahir, A.M.; Sha, J.; Feng, Z.; Eapen, B.; Nithianantham, S.; Das, B.; Karn, J.; Weinberg, A.; Bissada, N.F.; *et al.* Short-chain fatty acids from periodontal pathogens suppress histone deacetylases, EZH2, and SUV39H1 to promote Kaposi's sarcoma-associated herpesvirus replication. *J. Virol.* **2014**, *88*, 4466–4479.
183. Morris, V.A.; Punjabi, A.S.; Lagunoff, M. Activation of Akt through gp130 receptor signaling is required for Kaposi's sarcoma-associated herpesvirus-induced lymphatic reprogramming of endothelial cells. *J. Virol.* **2008**, *82*, 8771–8779.
184. Punjabi, A.S.; Carroll, P.A.; Chen, L.; Lagunoff, M. Persistent activation of STAT3 by latent Kaposi's sarcoma-associated herpesvirus infection of endothelial cells. *J. Virol.* **2007**, *81*, 2449–2458.
185. Burger, M.; Hartmann, T.; Burger, J.A.; Schraufstatter, I. KSHV-GPCR and CXCR2 transforming capacity and angiogenic responses are mediated through a JAK2-STAT3-dependent pathway. *Oncogene* **2005**, *24*, 2067–2075.
186. Muromoto, R.; Okabe, K.; Fujimuro, M.; Sugiyama, K.; Yokosawa, H.; Seya, T.; Matsuda, T. Physical and functional interactions between STAT3 and Kaposi's sarcoma-associated herpesvirus-encoded LANA. *FEBS Lett.* **2006**, *580*, 93–98.
187. Cai, Q.; Verma, S.C.; Choi, J.Y.; Ma, M.; Robertson, E.S. Kaposi's sarcoma-associated herpesvirus inhibits interleukin-4-mediated STAT6 phosphorylation to regulate apoptosis and maintain latency. *J. Virol.* **2010**, *84*, 11134–11144.
188. Ye, F.C.; Zhou, F.C.; Xie, J.P.; Kang, T.; Greene, W.; Kuhne, K.; Lei, X.F.; Li, Q.H.; Gao, S.J. Kaposi's sarcoma-associated herpesvirus latent gene vFLIP inhibits viral lytic replication through NF-κB-mediated suppression of the AP-1 pathway: A novel mechanism of virus control of latency. *J. Virol.* **2008**, *82*, 4235–4249.
189. Lei, X.; Bai, Z.; Ye, F.; Xie, J.; Kim, C.G.; Huang, Y.; Gao, S.J. Regulation of NF-κB inhibitor IκBα and viral replication by a KSHV microRNA. *Nat. Cell Biol.* **2010**, *12*, 193–199.

190. Izumiya, Y.; Izumiya, C.; Hsia, D.; Ellison, T.J.; Luciw, P.A.; Kung, H.J. NF-κB serves as a cellular sensor of Kaposi's sarcoma-associated herpesvirus latency and negatively regulates K-Rta by antagonizing the RBP-Jκ coactivator. *J. Virol.* **2009**, *83*, 4435–4446.
191. Brown, H.J.; Song, M.J.; Deng, H.; Wu, T.T.; Cheng, G.; Sun, R. NF-κB inhibits γ-herpesvirus lytic replication. *J. Virol.* **2003**, *77*, 8532–8540.
192. Grossmann, C.; Ganem, D. Effects of NF-κB activation on KSHV latency and lytic reactivation are complex and context-dependent. *Virology* **2008**, *375*, 94–102.
193. Gao, S.J.; Deng, J.H.; Zhou, F.C. Productive lytic replication of a recombinant Kaposi's sarcoma-associated herpesvirus in efficient primary infection of primary human endothelial cells. *J. Virol.* **2003**, *77*, 9738–9749.
194. Sadagopan, S.; Sharma-Walia, N.; Veettil, M.V.; Raghu, H.; Sivakumar, R.; Bottero, V.; Chandran, B. Kaposi's sarcoma-associated herpesvirus induces sustained NF-κB activation during *de novo* infection of primary human dermal microvascular endothelial cells that is essential for viral gene expression. *J. Virol.* **2007**, *81*, 3949–3968.
195. Peng, L.; Wu, T.T.; Tchieu, J.H.; Feng, J.; Brown, H.J.; Feng, J.; Li, X.; Qi, J.; Deng, H.; Vivanco, I.; *et al.* Inhibition of the phosphatidylinositol 3-kinase-Akt pathway enhances γ2-herpesvirus lytic replication and facilitates reactivation from latency. *J. Gen. Virol.* **2010**, *91*, 463–469.
196. Bhatt, A.P.; Damania, B. AKTivation of PI3K/AKT/mTOR signaling pathway by KSHV. *Front. Immunol.* **2012**, *3*, 401.
197. Lagunoff, M.; Majeti, R.; Weiss, A.; Ganem, D. Deregulated signal transduction by the K1 gene product of Kaposi's sarcoma-associated herpesvirus. *Proc. Natl. Acad. Sci. USA* **1999**, *96*, 5704–5709.
198. Tomlinson, C.C.; Damania, B. The K1 protein of Kaposi's sarcoma-associated herpesvirus activates the Akt signaling pathway. *J. Virol.* **2004**, *78*, 1918–1927.
199. Martin, D.; Galisteo, R.; Molinolo, A.A.; Wetzker, R.; Hirsch, E.; Gutkind, J.S. PI3Kγ mediates kaposi's sarcoma-associated herpesvirus vGPCR-induced sarcomagenesis. *Cancer Cell* **2011**, *19*, 805–813.
200. Sodhi, A.; Chaisuparat, R.; Hu, J.; Ramsdell, A.K.; Manning, B.D.; Sausville, E.A.; Sawai, E.T.; Molinolo, A.; Gutkind, J.S.; Montaner, S. The TSC2/mTOR pathway drives endothelial cell transformation induced by the Kaposi's sarcoma-associated herpesvirus G protein-coupled receptor. *Cancer Cell* **2006**, *10*, 133–143.
201. Hong, Y.K.; Foreman, K.; Shin, J.W.; Hirakawa, S.; Curry, C.L.; Sage, D.R.; Libermann, T.; Dezube, B.J.; Fingeroth, J.D.; Detmar, M. Lymphatic reprogramming of blood vascular endothelium by Kaposi sarcoma-associated herpesvirus. *Nat. Genet.* **2004**, *36*, 683–685.
202. Fu, B.; Kuang, E.; Li, W.; Avey, D.; Li, X.; Turpin, Z.; Valdes, A.; Brulois, K.; Myoung, J.; Zhu, F. Activation of p90 Ribosomal S6 Kinases (RSKs) by ORF45 of Kaposi Sarcoma-Associated Herpesvirus is Critical for Optimal Production of Infectious Viruses. *J. Virol.* **2015**, *89*, 195–207.
203. Wang, S.; Wang, S.; Maeng, H.; Young, D.P.; Prakash, O.; Fayad, L.E.; Younes, A.; Samaniego, F. K1 protein of human herpesvirus 8 suppresses lymphoma cell Fas-mediated apoptosis. *Blood* **2007**, *109*, 2174–2182.
204. Hideshima, T.; Chauhan, D.; Teoh, G.; Raje, N.; Treon, S.P.; Tai, Y.T.; Shima, Y.; Anderson, K.C. Characterization of signaling cascades triggered by human interleukin-6 *versus* Kaposi's sarcoma-associated herpes virus-encoded viral interleukin 6. *Clin. Cancer Res.* **2000**, *6*, 1180–1189.

205. Wang, Q.J.; Jenkins, F.J.; Jacobson, L.P.; Meng, Y.X.; Pellett, P.E.; Kingsley, L.A.; Kousoulas, K.G.; Baghian, A.; Rinaldo, C.R., Jr. CD8+ cytotoxic T lymphocyte responses to lytic proteins of human herpes virus 8 in human immunodeficiency virus type 1-infected and -uninfected individuals. *J. Infect. Dis.* **2000**, *182*, 928–932.
206. Lang, S.M.; Bynoe, M.O.; Karki, R.; Tartell, M.A.; Means, R.E. Kaposi's sarcoma-associated herpesvirus K3 and K5 proteins down regulate both DC-SIGN and DC-SIGNR. *PLoS One* **2013**, *8*, e58056.
207. Weber, K.S.; Grone, H.J.; Rocken, M.; Klier, C.; Gu, S.; Wank, R.; Proudfoot, A.E.; Nelson, P.J.; Weber, C. Selective recruitment of Th2-type cells and evasion from a cytotoxic immune response mediated by viral macrophage inhibitory protein-II. *Eur. J. Immunol.* **2001**, *31*, 2458–2466.
208. Kledal, T.N.; Rosenkilde, M.M.; Coulin, F.; Simmons, G.; Johnsen, A.H.; Alouani, S.; Power, C.A.; Luttichau, H.R.; Gerstoft, J.; Clapham, P.R.; *et al.* A broad-spectrum chemokine antagonist encoded by Kaposi's sarcoma-associated herpesvirus. *Science* **1997**, *277*, 1656–1659.
209. Feng, H.; Dong, X.; Negaard, A.; Feng, P. Kaposi's sarcoma-associated herpesvirus K7 induces viral G protein-coupled receptor degradation and reduces its tumorigenicity. *PLoS Pathog.* **2008**, *4*, e1000157.
210. Wang, H.W.; Sharp, T.V.; Koumi, A.; Koentges, G.; Boshoff, C. Characterization of an anti-apoptotic glycoprotein encoded by Kaposi's sarcoma-associated herpesvirus which resembles a spliced variant of human survivin. *EMBO J.* **2002**, *21*, 2602–2615.
211. Burysek, L.; Yeow, W.S.; Lubyova, B.; Kellum, M.; Schafer, S.L.; Huang, Y.Q.; Pitha, P.M. Functional analysis of human herpesvirus 8-encoded viral interferon regulatory factor 1 and its association with cellular interferon regulatory factors and p300. *J. Virol.* **1999**, *73*, 7334–7342.
212. Lin, R.; Genin, P.; Mamane, Y.; Sgarbanti, M.; Battistini, A.; Harrington, W.J., Jr.; Barber, G.N.; Hiscott, J. HHV-8 encoded vIRF-1 represses the interferon antiviral response by blocking IRF-3 recruitment of the CBP/p300 coactivators. *Oncogene* **2001**, *20*, 800–811.
213. Foster-Cuevas, M.; Wright, G.J.; Puklavec, M.J.; Brown, M.H.; Barclay, A.N. Human herpesvirus 8 K14 protein mimics CD200 in down-regulating macrophage activation through CD200 receptor. *J. Virol.* **2004**, *78*, 7667–7676.
214. Rezaee, S.A.; Gracie, J.A.; McInnes, I.B.; Blackbourn, D.J. Inhibition of neutrophil function by the Kaposi's sarcoma-associated herpesvirus vOX2 protein. *AIDS* **2005**, *19*, 1907–1910.
215. Brinkmann, M.M.; Pietrek, M.; Dittrich-Breiholz, O.; Kracht, M.; Schulz, T.F. Modulation of host gene expression by the K15 protein of Kaposi's sarcoma-associated herpesvirus. *J. Virol.* **2007**, *81*, 42–58.
216. Wang, L.; Brinkmann, M.M.; Pietrek, M.; Ottinger, M.; Dittrich-Breiholz, O.; Kracht, M.; Schulz, T.F. Functional characterization of the M-type K15-encoded membrane protein of Kaposi's sarcoma-associated herpesvirus. *J. Gen. Virol.* **2007**, *88*, 1698–1707.
217. Mullick, J.; Bernet, J.; Singh, A.K.; Lambris, J.D.; Sahu, A. Kaposi's sarcoma-associated herpesvirus (human herpesvirus 8) open reading frame 4 protein (kaposica) is a functional homolog of complement control proteins. *J. Virol.* **2003**, *77*, 3878–3881.
218. Mullick, J.; Singh, A.K.; Panse, Y.; Yadav, V.; Bernet, J.; Sahu, A. Identification of functional domains in kaposica, the complement control protein homolog of Kaposi's sarcoma-associated herpesvirus (human herpesvirus 8). *J. Virol.* **2005**, *79*, 5850–5856.

219. Spiller, O.B.; Robinson, M.; O'Donnell, E.; Milligan, S.; Morgan, B.P.; Davison, A.J.; Blackbourn, D.J. Complement regulation by Kaposi's sarcoma-associated herpesvirus ORF4 protein. *J. Virol.* **2003**, *77*, 592–599.
220. Zhu, F.X.; King, S.M.; Smith, E.J.; Levy, D.E.; Yuan, Y. A Kaposi's sarcoma-associated herpesviral protein inhibits virus-mediated induction of type I interferon by blocking IRF-7 phosphorylation and nuclear accumulation. *Proc. Natl. Acad. Sci. USA* **2002**, *99*, 5573–5578.
221. Sathish, N.; Zhu, F.X.; Golub, E.E.; Liang, Q.; Yuan, Y. Mechanisms of autoinhibition of IRF-7 and a probable model for inactivation of IRF-7 by Kaposi's sarcoma-associated herpesvirus protein ORF45. *J. Biol. Chem.* **2011**, *286*, 746–756.
222. Gregory, S.M.; Damania, B. Inhibition of the inflammasome response by a viral protein that interacts with NLRs. *Commun. Integr. Biol.* **2011**, *4*, 416–418.
223. Inn, K.S.; Lee, S.H.; Rathbun, J.Y.; Wong, L.Y.; Toth, Z.; Machida, K.; Ou, J.H.; Jung, J.U. Inhibition of RIG-I-mediated signaling by Kaposi's sarcoma-associated herpesvirus-encoded deubiquitinase ORF64. *J. Virol* **2011**, *85*, 10899–10904.
224. Cesarman, E.; Nador, R.G.; Bai, F.; Bohenzky, R.A.; Russo, J.J.; Moore, P.S.; Chang, Y.; Knowles, D.M. Kaposi's sarcoma-associated herpesvirus contains G protein-coupled receptor and cyclin D homologs which are expressed in Kaposi's sarcoma and malignant lymphoma. *J. Virol.* **1996**, *70*, 8218–8223.
225. Full, F.; Jungnickl, D.; Reuter, N.; Bogner, E.; Brulois, K.; Scholz, B.; Sturzl, M.; Myoung, J.; Jung, J.U.; Stamminger, T.; *et al.* Kaposi's sarcoma associated herpesvirus tegument protein ORF75 is essential for viral lytic replication and plays a critical role in the antagonization of ND10-instituted intrinsic immunity. *PLoS Pathog.* **2014**, *10*, e1003863.
226. Gunther, T.; Schreiner, S.; Dobner, T.; Tessmer, U.; Grundhoff, A. Influence of ND10 components on epigenetic determinants of early KSHV latency establishment. *PLoS Pathog.* **2014**, *10*, e1004274.
227. Borah, S.; Darricarrere, N.; Darnell, A.; Myoung, J.; Steitz, J.A. A viral nuclear noncoding RNA binds re-localized poly(A) binding protein and is required for late KSHV gene expression. *PLoS Pathog.* **2011**, *7*, e1002300.
228. Borah, S.; Nichols, L.A.; Hassman, L.M.; Kedes, D.H.; Steitz, J.A. Tracking expression and subcellular localization of RNA and protein species using high-throughput single cell imaging flow cytometry. *RNA* **2012**, *18*, 1573–1579.
229. Rossetto, C.C.; Tarrant-Elorza, M.; Verma, S.; Purushothaman, P.; Pari, G.S. Regulation of viral and cellular gene expression by Kaposi's sarcoma-associated herpesvirus polyadenylated nuclear RNA. *J. Virol.* **2013**, *87*, 5540–5553.
230. Rossetto, C.C.; Pari, G.S. Kaposi's sarcoma-associated herpesvirus noncoding polyadenylated nuclear RNA interacts with virus- and host cell-encoded proteins and suppresses expression of genes involved in immune modulation. *J. Virol.* **2011**, *85*, 13290–13297.
231. Monini, P.; Carlini, F.; Sturzl, M.; Rimessi, P.; Superti, F.; Franco, M.; Melucci-Vigo, G.; Cafaro, A.; Goletti, D.; Sgadari, C.; *et al.* α-interferon inhibits human herpesvirus 8 (HHV-8) reactivation in primary effusion lymphoma cells and reduces HHV-8 load in cultured peripheral blood mononuclear cells. *J. Virol.* **1999**, *73*, 4029–4041.
232. Jacobs, S.R.; Damania, B. The viral interferon regulatory factors of KSHV: Immunosuppressors or oncogenes? *Front. Immunol.* **2011**, *2*, 19.

233. Jacobs, S.R.; Gregory, S.M.; West, J.A.; Wollish, A.C.; Bennett, C.L.; Blackbourn, D.J.; Heise, M.T.; Damania, B. The viral interferon regulatory factors of kaposi's sarcoma-associated herpesvirus differ in their inhibition of interferon activation mediated by toll-like receptor 3. *J. Virol.* **2013**, *87*, 798–806.
234. Lee, H.R.; Choi, W.C.; Lee, S.; Hwang, J.; Hwang, E.; Guchhait, K.; Haas, J.; Toth, Z.; Jeon, Y.H.; Oh, T.K.; *et al.* Jung Bilateral inhibition of HAUSP deubiquitinase by a viral interferon regulatory factor protein. *Nat. Struct. Mol. Biol.* **2011**, *18*, 1336–1344.
235. Molden, J.; Chang, Y.; You, Y.; Moore, P.S.; Goldsmith, M.A. A Kaposi's sarcoma-associated herpesvirus-encoded cytokine homolog (vIL-6) activates signaling through the shared gp130 receptor subunit. *J. Biol. Chem.* **1997**, *272*, 19625–19631.
236. Arvanitakis, L.; Geras-Raaka, E.; Varma, A.; Gershengorn, M.C.; Cesarman, E. Human herpesvirus KSHV encodes a constitutively active G-protein-coupled receptor linked to cell proliferation. *Nature* **1997**, *385*, 347–350.
237. Montaner, S. Akt/TSC/mTOR activation by the KSHV G protein-coupled receptor: Emerging insights into the molecular oncogenesis and treatment of Kaposi's sarcoma. *Cell Cycle* **2007**, *6*, 438–443.
238. Matsuo, T.; Itami, M. Seropositivity of human herpesvirus-8 in patients with uveitis. *Ocul. Immunol. Inflamm.* **2002**, *10*, 197–199.
239. Lee, B.S.; Lee, S.H.; Feng, P.; Chang, H.; Cho, N.H.; Jung, J.U. Characterization of the Kaposi's sarcoma-associated herpesvirus K1 signalosome. *J. Virol.* **2005**, *79*, 12173–12184.
240. Prakash, O.; Swamy, O.R.; Peng, X.; Tang, Z.Y.; Li, L.; Larson, J.E.; Cohen, J.C.; Gill, J.; Farr, G.; Wang, S.; *et al.* Activation of Src kinase Lyn by the Kaposi sarcoma-associated herpesvirus K1 protein: Implications for lymphomagenesis. *Blood* **2005**, *105*, 3987–3994.
241. Feng, P.; Park, J.; Lee, B.S.; Lee, S.H.; Bram, R.J.; Jung, J.U. Kaposi's sarcoma-associated herpesvirus mitochondrial K7 protein targets a cellular calcium-modulating cyclophilin ligand to modulate intracellular calcium concentration and inhibit apoptosis. *J. Virol.* **2002**, *76*, 11491–11504.
242. Brulois, K.; Jung, J.U. Interplay between Kaposi's sarcoma-associated herpesvirus and the innate immune system. *Cytokine Growth Factor Rev.* **2014**, *25*, 597–609.
243. Coscoy, L.; Ganem, D. A viral protein that selectively downregulates ICAM-1 and B7–2 and modulates T cell costimulation. *J. Clin. Investig.* **2001**, *107*, 1599–1606.
244. Neipel, F.; Albrecht, J.C.; Ensser, A.; Huang, Y.Q.; Li, J.J.; Friedman-Kien, A.E.; Fleckenstein, B. Human herpesvirus 8 encodes a homolog of interleukin-6. *J. Virol.* **1997**, *71*, 839–842.
245. Russo, J.J.; Bohenzky, R.A.; Chien, M.C.; Chen, J.; Yan, M.; Maddalena, D.; Parry, J.P.; Peruzzi, D.; Edelman, I.S.; Chang, Y.; *et al.* Nucleotide sequence of the Kaposi sarcoma-associated herpesvirus (HHV8). *Proc. Natl. Acad. Sci. USA* **1996**, *93*, 14862–14867.
246. Liu, C.; Okruzhnov, Y.; Li, H.; Nicholas, J. Human herpesvirus 8 (HHV-8)-encoded cytokines induce expression of and autocrine signaling by vascular endothelial growth factor (VEGF) in HHV-8-infected primary-effusion lymphoma cell lines and mediate VEGF-independent antiapoptotic effects. *J. Virol.* **2001**, *75*, 10933–10940.
247. Stine, J.T.; Wood, C.; Hill, M.; Epp, A.; Raport, C.J.; Schweickart, V.L.; Endo, Y.; Sasaki, T.; Simmons, G.; Boshoff, C.; *et al.* KSHV-encoded CC chemokine vMIP-III is a CCR4 agonist, stimulates angiogenesis, and selectively chemoattracts TH2 cells. *Blood* **2000**, *95*, 1151–1157.

248. Shiratori, I.; Yamaguchi, M.; Suzukawa, M.; Yamamoto, K.; Lanier, L.L.; Saito, T.; Arase, H. Down-regulation of basophil function by human CD200 and human herpesvirus-8 CD200. *J. Immunol.* **2005**, *175*, 4441–4449.
249. Mark, L.; Lee, W.H.; Spiller, O.B.; Proctor, D.; Blackbourn, D.J.; Villoutreix, B.O.; Blom, A.M. The Kaposi's sarcoma-associated herpesvirus complement control protein mimics human molecular mechanisms for inhibition of the complement system. *J. Biol. Chem.* **2004**, *279*, 45093–45101.
250. Spiller, O.B.; Blackbourn, D.J.; Mark, L.; Proctor, D.G.; Blom, A.M. Functional activity of the complement regulator encoded by Kaposi's sarcoma-associated herpesvirus. *J. Biol. Chem.* **2003**, *278*, 9283–9289.
251. Spiller, O.B.; Mark, L.; Blue, C.E.; Proctor, D.G.; Aitken, J.A.; Blom, A.M.; Blackbourn, D.J. Dissecting the regions of virion-associated Kaposi's sarcoma-associated herpesvirus complement control protein required for complement regulation and cell binding. *J. Virol.* **2006**, *80*, 4068–4078.
252. Mark, L.; Proctor, D.G.; Blackbourn, D.J.; Blom, A.M.; Spiller, O.B. Separation of decay-accelerating and cofactor functional activities of Kaposi's sarcoma-associated herpesvirus complement control protein using monoclonal antibodies. *Immunology* **2008**, *123*, 228–238.
253. Zhu, F.X.; Cusano, T.; Yuan, Y. Identification of the immediate-early transcripts of Kaposi's sarcoma-associated herpesvirus. *J. Virol.* **1999**, *73*, 5556–5567.
254. Sathish, N.; Zhu, F.X.; Yuan, Y. Kaposi's sarcoma-associated herpesvirus ORF45 interacts with kinesin-2 transporting viral capsid-tegument complexes along microtubules. *PLoS Pathog.* **2009**, *5*, e1000332.
255. Brousset, P.; Meggetto, F.; Laharrague, P.; Attal, M.; Delsol, G. Kaposi's sarcoma-associated herpesvirus (KSHV) in bone marrow biopsy from patients with multiple myeloma: PCR amplification of orf26 but not orf72 and orf75 sequences. *Br. J. Haematol.* **2000**, *108*, 197–198.
256. Konrad, A.; Wies, E.; Thurau, M.; Marquardt, G.; Naschberger, E.; Hentschel, S.; Jochmann, R.; Schulz, T.F.; Erfle, H.; Brors, B.; *et al.* A systems biology approach to identify the combination effects of human herpesvirus 8 genes on NF-κB activation. *J. Virol.* **2009**, *83*, 2563–2574.
257. Gregory, S.M.; Davis, B.K.; West, J.A.; Taxman, D.J.; Matsuzawa, S.; Reed, J.C.; Ting, J.P.; Damania, B. Discovery of a viral NLR homolog that inhibits the inflammasome. *Science* **2011**, *331*, 330–334.
258. Rozen, R.; Sathish, N.; Li, Y.; Yuan, Y. Virion-wide protein interactions of Kaposi's sarcoma-associated herpesvirus. *J. Virol.* **2008**, *82*, 4742–4750.
259. Campbell, M.; Kim, K.Y.; Chang, P.C.; Huerta, S.; Shevchenko, B.; Wang, D.H.; Izumiya, C.; Kung, H.J.; Izumiya, Y. A lytic viral long noncoding RNA modulates the function of a latent protein. *J. Virol.* **2014**, *88*, 1843–1848.
260. Rossetto, C.C.; Pari, G. KSHV PAN RNA associates with demethylases UTX and JMJD3 to activate lytic replication through a physical interaction with the virus genome. *PLoS Pathog.* **2012**, *8*, e1002680.

9

Genotypic Analysis of Kaposi's Sarcoma-Associated Herpesvirus from Patients with Kaposi's Sarcoma in Xinjiang, China

Xinxing Ouyang [1,†], Yan Zeng [2,†], Bishi Fu [1], Xiaowu Wang [3], Wei Chen [3], Yuan Fang [3], Minhua Luo [1,*] and Linding Wang [3,4,*]

1 State Key Laboratory of Virology, Wuhan Institute of Virology, Chinese Academy of Sciences, Wuhan 430071, China; E-Mails: annlooy@gmail.com (X.O.); fubishi001@yeah.net (B.F.)

2 Laboratory of Xinjiang Endemic and Ethnic Diseases, Shihezi University, Shihezi 832002, China; E-Mail: yzeng910@163.com

3 Department of Microbiology, Anhui Medical University, Hefei 230032, China; E-Mails: Wangxiaowuxiaowu@sina.com (X.W.); chenwei1413@163.com (W.C.); yuanslience@163.com (Y.F.)

4 Central Laboratory of Molecular and Cellular Biology, School of Basic Medicine, Anhui Medical University, Hefei, Anhui 230032, China

† These authors contributed equally to this work.

* Authors to whom correspondence should be addressed; E-Mails: luomh@wh.iov.cn (M.L); wanglinding@ahmu.edu.cn (L.W.);

External Editor: Zhi-Ming Zheng

Abstract: Kaposi's sarcoma-associated herpesvirus (KSHV) is the causal agent of all forms of Kaposi's sarcoma (KS), including AIDS-KS, endemic KS, classic KS and iatrogenic KS. Based on Open reading frame (ORF) K1 sequence analysis, KSHV has been classified into seven major molecular subtypes (A, B, C, D, E, F and Z). The distribution of KSHV strains varies according to geography and ethnicity. Xinjiang is a unique region where the seroprevalence of KSHV is significantly higher than other parts of China. The genotyping of KSHV strains in this region has not been thoroughly studied. The present study aimed to evaluate the frequency of KSHV genotypes isolated from KS tissues in Classical KS and AIDS KS patients from Xinjiang, China. ORF-K1 of KSHV from tissue samples of 28 KS patients was amplified and sequenced. Two subtypes of KSHV were identified according to K1 genotyping. Twenty-three of them belonged to subtype A, while five of them were subtype C. More genotype A than genotype C strains were found in both Classical KS and AIDS KS. No significant difference was found in the prevalence of different genotype between Classical KS and AIDS KS.

Keywords: Kaposi's sarcoma-associated herpesvirus (KSHV); genotyping; K1 gene; Xinjiang

1. Introduction

Kaposi's sarcoma (KS) is a tumor, originally described by Moritz Kaposi in 1872 [1] and became more widely known as one of the AIDS defining illness in the 1980s [2]. The clincoepidemiologic forms of KS have been classified as AIDS-KS, classic KS, endemic KS and iatrogenic KS [3]. Classic KS was an indolent cutaneous disease entity primarily involving the lower extremities, affecting elderly men especially from the Mediterranean region or of Eastern European decent [4,5]. In contrast, AIDS-associated KS is clinically more aggressive and can occur in all organs with the exception of the central nervous system; it stimulated the greatest interest as one of the first illness associated with AIDS [6,7].

Kaposi's sarcoma-associated herpesvirus (KSHV), also known as human herpesvirus 8 (HHV-8), is a gamma-2-herpesvirus or rhadinovirus. KSHV was first identified in a biopsy tissue from a patient with AIDS-related Kaposi's sarcoma (AIDS-KS) by representational difference analysis in 1994 by Chang *et al.* [8]. KSHV is the causal agent of all forms of Kaposi's sarcoma, including AIDS-KS, endemic KS and iatrogenic KS (in transplant recipients receiving immunosuppressive therapy) [8]. KSHV is also associated with two other lymphoproliferative malignancies, including primarily effusion lymphoma and multicentric Castleman's disease (MCD) [9,10].

As a large double-stranded DNA virus, KSHV has approximately 90 identified open reading frames, of which over 60 show homology with other rhadinoviruses and 15, designated K1–K15, were unique to KSHV when its genome was first sequenced [11]. ORF-K1, at the left end of KSHV genome, encodes an early lytic transmembrane glycoprotein of 289 amino acids (aa). The amino acid sequence of K1 varies from 0.4% to 44% between different KSHV isolates, with the variations concentrated in two hyper-variable regions, VR1 and VR2. Current genotyping method of KSHV is based primarily on the sequence variations of the ORF-K1 gene. Based on K1 sequence analysis, KSHV has been classified into seven major molecular subtypes (A, B, C, D, E, F and Z) [12–17]. The distribution of KSHV strains varies according to geography and ethnicity, which appears to be attributable to human migrations. Subtype A and C are found in Europe, the USA, Middle East and Northern Asia; Subtype B is characteristic for Africa; Subtype D was found in individuals from the pacific Islands; Subtype E was found in Brazilian Amerindians; Subtype Z has been found in a small cohort of Zambian children; A new subtype F has been recently identified in Uganda.

Xinjiang Uygur Autonomous region is the largest province in northwestern China, and located on the ancient Silk Road as an important staging post over a thousand years ago. Xinjiang borders on

Russia, Kazakhstan, Kyrgyzstan, Tajikistan, Pakistan, Mongolia, India and Afghanistan. Ethnic groups in Xinjiang are diverse and distinct, the main ethnic groups are the Uygur (45.7%) and the Han (39.7%), other ethnic minorities include Kazakh, Mongolians, Hui, Kirgiz, Manchu, and Xibo. Classic KS are rarely seen in the Han Chinese, but are seen more frequently in the Uygur ethnic group, more than one patient has been diagnosed histopathologically as having KS every year at the Affiliated Tumor Hospital of Xinjiang Medical University in Urumqi, the capital city of Xinjiang Uygur Autonomous Region. We have previously shown that the overall seroprevalence of KSHV was 19.2% in the general population in Xinjiang, which was substantially higher than the 9.5% seroprevalence of KSHV in the control subjects from the general population in Han Chinese in Hubei Province [18]. Our data indicated that Xinjiang is a unique region where the seroprevalence of KSHV is significantly higher than other parts of China. This high seroprevalence of KSHV is consistent with the high incidence of KS in this region.

Dilnur *et al.* (2001) had reported that KSHV strains from seven patients with classical KS in Xinjiang were classified as subtype C [19]. Zhang reported that, on the basis of the K1/VR1 amino acid sequence, that the majority of these KS patients were infected by subtype C ($n = 18$), and several by subtype A ($n = 4$) [20]. In this study, we collected 28 samples of KS patients in Xinjiang and examined the infection of KSHV by nested PCR and characterized the ORF-K1 genotypes.

2. Materials and Methods

2.1. Tissue Specimens

Twenty-eight KS classical tissue examples were collected from Xinjiang Uygur Autonomous Region, China. Seven of them were peripheral blood mononuclear cells examples and the rest were formalin-fixed paraffin-embedded tissues (FFPET). All the specimens were collected between 1997 and 2009. Eighteen of them were from Uygur patients, eight of them were from Kazakh Patients, one from a Hui patient and one from a Han patient. Permission to conduct the study and informed consent was obtained in accordance with a protocol approved by the Ethics Committee of Wuhan Institute of Virology, Chinese Academy of Sciences.

2.2. Preparation of DNA

DNA from peripheral blood mononuclear cells of KS patients was extracted with a Blood & Cell Culture DNA Mini Kit according to manufacturer's instructions (Qiagen, Valencia, CA, USA).

DNA from FFPET was extracted as the following steps. Tissues were incubated at 55 °C with 1 mL xylene for 10 min, and then centrifuged at 8000 rpm for 5 min, supernatant was collected. The supernatant was further centrifuged at 8000 rpm for 5 min to totally remove the paraffin. The remaining dimethylbenzene was removed by using anhydrous ethanol. Then the supernatant was added with 250 μL cell lysis buffer (0.02 M Tris-HCl pH 8.0, 0.01 M EDTA pH 8.0, 2% SDS and 50 μL proteinase K (20 mg/mL) and kept at 55 °C over night. Equal volumes of phenol, chloroform, and isoamyl alcohol (25:24:1) were added to the supernatant. The supernatant was clarified by centrifugation at 8000 rpm for 2 min at 4 °C. Then 2.5 times the volume of pre-cold anhydrous alcohol was added to precipitate DNA. 30 min later, the mixture was centrifuged at 13,000 rpm for 10 min at 4 °C. The supernatant was discarded. After slightly drying, 20 μL TE (pH8.0) was added to dissolve DNA, and kept in −20 °C.

2.3. PCR Amplification

Fragments (363 bp) of the VR1 region of KSHV ORF-K1 were amplified by nested-PCR. Primers were designed as described [16]. Briefly, the forward and reverse primers for first round of PCR were 5'-GACCTTGTTGGACATCCTGTA-3' and 5'-GAGTTTCTGGAGTTATATTG-3'. Primers for the second round of PCR were 5'-TTGTGCCCTGGAGTGATT-3' and 5'-CAGCGTAAAATTATAGTA-3'. The PCR program was set as 35 cycles 95 °C for 1 min, 53 °C for 45 s, and 72 °C for 1 min in the first reaction; and 94 °C for 1 min, 48 °C for 45 s, and 72 °C for 1 min in the second reaction. PCR products were evaluated in 1.5% agarose gel electrophoresis and stained with ethidium bromide. Afterward, the products were purified using a Gel Extraction Kit (Omega Bio-Tek, Winooski, VT, USA) according to instructions of manufacturer.

2.4. DNA Sequencing

The purified K1 gene fragments were cloned into pGEM-T vector for constructing recombinant plasmid pGEM-T-K1 and transformed into DH5α competent cells. Then, the recombinant plasmid was extracted from DH5α cells using TIANprep Mini Plasmid Kit (TIANGEN BIOTECH CO., LTD, Beijing, China) and sent for sequencing by Shanghai Majorbio Biotech Co., Ltd. (Shanghai, China)

2.5. Phylogenetic Tree Analysis

Twenty-eight of the unique sequences (GenBank Accession numbers: KM582680–KM582707) and 19 other KSHV strains obtained from GenBank were used to construct the phylogenetic tree and were analyzed with a DNAStar package and aligned with Clustal W in Bioedit (version 7.0.0). Phylogenetic trees were constructed by neighbor-joining analysis by Phylip (version 3.68) and MEGA (version 4.0.2), respectively. The statistical reliability of the NJ tree was evaluated using 1000 bootstrap samples. TreeView (version 1.6.6) was used to see the trees constructed by Phylip. The sequences of 19 other KSHV strains consisted of 1 strain of subtype A1 XJ-27 (FJ853386); 1 strain of subtype A2: Ema-7 (AF130305); 1 strain of subtype A3: IT-268 (GU097421), 1 strain of subtype A4: BCBL-B (AF133039),US114 (GU097431) ; 1 strain of subtype A5: KE229 (GU097433); 3 strains of subtype B: UG-65 (FJ884618), UG-85 (FJ884620), MP10 (AF387367); 2 strains of subtype C2: XJ-6 (FJ853368), XJ30 (FJ853388); 1 strain of subtype C3: XJ-20 (FJ853379), AF17170531; 2 strains of subtype D: TKS10 (AF133043), ZKS3 (AF133044), 2 strains of subtype E: Sio1 (AY329025), Tupi1 (AF220292); and 2 strains of subtype F: AF178810, FJ884616.

2.6. Statistical Analysis

All the information of cases was analyzed by descriptive statistics. MS Excel was used to calculate the mean, standard deviation, coefficient of variation, and median of data. Fisher's exact test was used to test the associations of sex ratio and KSHV genotype between two groups and the associations of

KS subtypes and KSHV subtypes. One-way analysis of variance (ANOVA) was used for statistical evaluation of differences between the two groups.

3. Results

3.1. KS Cases Characterization

Twenty-eight KS cases were evaluated for ORF-K1 genotype analysis. The number of male patients was higher than that of female patients. Male patients were 24/28 (85.71%), while female patients were 4/28 (14.29%). Classic KS cases were twenty-three, while AIDS-associated KS were five. Classic KS has a male/female ratio of 6.67 (20/3), and AIDS-associated KS has Male/Female Ratio of 4.0 (4/1). Compared to AIDS KS, the male/female ratio in Classic KS is higher but without significant difference by Fisher's exact test. The mean ages for Classic KS and AIDS-KS were 54.39 and 23.80 years old, respectively. AIDS associated KS patients were younger than Classic KS ($p < 0.001$; ANOVA). (Table 1).

Table 1. Sex and age for KS patients studied.

KS Groups	M/F Ratio	Age						
		Mean (Years)	SD (Years)	CV (%)	P_{25}	P_{75}	Median (Years)	Range (Years)
Classic KS (23)	6.67 *	54.39	13.94	25.64	40.00	65.50	53.00 §	32–73
AIDS-associated KS (5)	4.00 *	23.80	9.65	40.56	18.00	29.00	28.00 §	10–34
Total (28)	6.00	48.93	17.74	36.25	37.00	63.00	50.50	10–73

M/F: male/female; SD: standard deviation; CV: coefficient of variation; P_{25}: 25th percentile; P_{75}: 75th percentile;
* M/F ratios of two kind of KS has no significant difference (Fisher's exact test); § $p < 0.001$ (ANOVA).

3.2. ORF-K1 Amplification

Total DNA were extracted from FFPET and plasmid samples. 363 bp of ORF-K1 VR1 region of KSHV were amplified by nested PCR from each sample. PCR products were detected by 1.5% agarose gel electrophoresis (see Supplementary File). DNA extracted from gel was cloned into pGEM-T vector, and then sent for sequencing.

3.3. Phylogenetic Analysis

The 363 bp K1 sequences of the 28 KSHV isolates from Xinjiang Uygur Autonomous Region, China were aligned with 19 previously reported sequences in the database of NCBI using Bioedit (version 7.0.0). Phylogenetic trees were constructed by neighbor-joining analysis by MEGA (version 4.0.2) and Phylip (version 3.68), respectively. From the MEGA phylogenetic trees (Figure 1), the Xinjiang cases were categorized into subtypes A and C. Twenty-three cases (82.14%) were categorized into subtype A and five cases (17.86%) categorized into subtype C in total. Of classic KS, 19 subjects (82.61%) were infected with subtype A and four subjects (13.91%) were infected with subtype C, while of AIDS-associated KS, four subjects (80.00%) were infected with subtype A and one subject (20.00%) was infected with subtype C. Subtype C2 and C3 and all known subtype A, except A5, were found in classic KS. In AIDS-KS, subtype A1, A2 and C3 were identified (Table 2). In this study, subtype A/C ratios in Classic KS and AIDS-associated KS were not significantly different (Fisher's exact test); Subtype A/C ratios in Uygur patients and Kazakh patients were also not significant different (Fisher's exact test, Table 3). By using neighbor-joining analysis with Phylip, we saw a similar topology (see Supplementary File).

Table 2. KSHV K1 genotypes between classic KS and AIDS KS.

KS Subtype	**K1 Subtype ***		
Classic KS			
	A (19)	A1	2
		A2	11
		A3	4
		A4	2
	C (4)	C2	2
		C3	2
AIDS-associated KS			
	A (4)	A1	1
		A2	3
	C (1)	C3	1

* KSHV subtype A/C ratios of classic and AIDS-KS were not significantly different (Fisher's exact test).

Table 3. KSHV K1 genotypes in different ethnic patients.

KSHV Isolates	**GenBank Accession Numbers**	**Ethnicity**	**Subtypes**
XUAR1	KM582680	Uygur	A2
XUAR2	KM582681	Kazakh	A3
XUAR3	KM582682	Uygur	A3
XUAR4	KM582683	Uygur	A1
XUAR5	KM582684	Uygur	A2
XUAR 6	KM582685	Uygur	A3
XUAR7	KM582687	Uygur	C3
XUAR8	KM582688	Kazakh	A4
XUAR9	KM582689	Uygur	C3
XUAR10	KM582690	Uygur	C2
XUAR11	KM582691	Kazakh	C2
XUAR12	KM582692	Kazakh	A1
XUAR13	KM582693	Uygur	A2
XUAR14	KM582694	Hui	A2
XUAR15	KM582695	Uygur	A2
XUAR16	KM582696	Han	A4
XUAR17	KM582697	Uygur	A2
XUAR18	KM582698	Kazakh	A3
XUAR19	KM582699	Uygur	A2
XUAR20	KM582700	Kazakh	A2
XUAR21	KM582701	Kazakh	A2
XUAR22	KM582702	Uygur	A2
XUAR23	KM582703	Kazakh	A2
XUAR24	KM582704	Uygur	C3

Table 3. *Cont.*

KSHV Isolates	GenBank Accession Numbers	Ethnicity	Subtypes
XUAR25	KM582705	Uygur	A2
XUAR26	KM582706	Uygur	A2
XUAR27	KM582707	Uygur	A1
XUAR28	KM582708	Uygur	C3

KSHV subtype A/C ratios in different ethnic patients were not significantly different (Fisher's exact test).

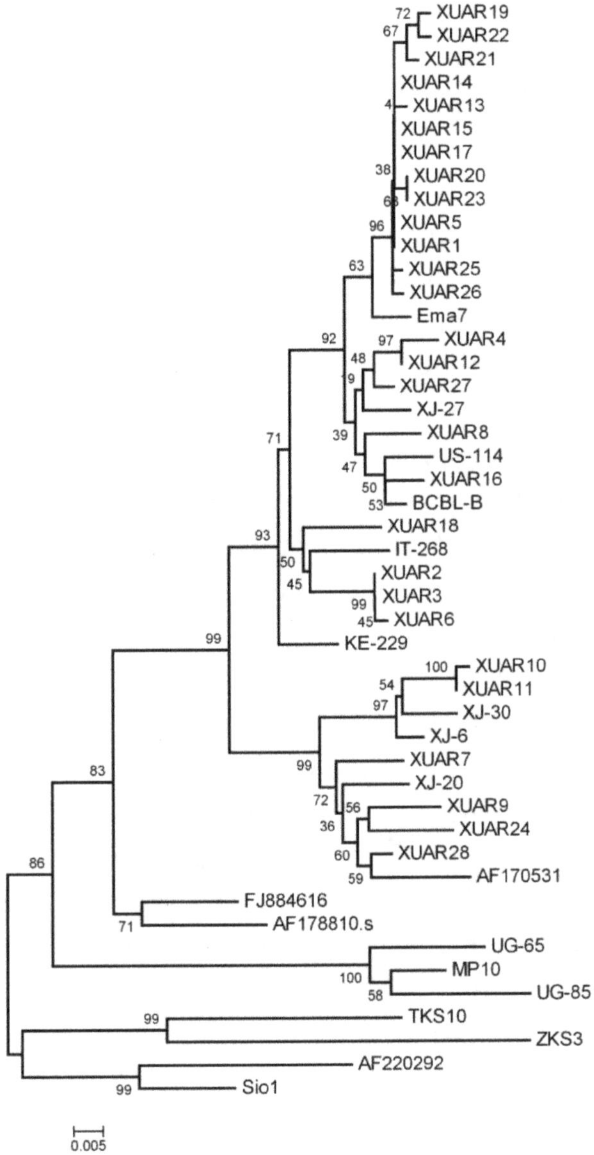

Figure 1. Phylogenetic tree of KSHV ORF-K1 DNA sequences constructed by MEGA (version 4.0.2) using the neighbor-joining algorithm. Relationships of KSHV isolates in present study and isolates in the literature are shown. KSHV isolates in present study were named by XUAR. XUAR1-XUAR23 were isolated from classic KS tissues, XUAR24-XUAR28 were isolated from AIDS-KS tissues. Ema7, US-114, BCBL-B, IT-268, KE229, XJ6, XJ20, XJ-27, XJ-30, AF170531, AF178810, AF220292, FJ884616, UG-65, UG-85, MP10, TKS10, ZKS3, and Sio1 were downloaded from NCBI.

4. Discussion

Traditionally, herpesviruses are considered to be co-evolved with their hosts throughout vertebrate evolution. KSHV is an ancient human virus and spread together with its migrating human hosts [21]. Many studies have found seven subtypes of KSHV (A, B, C, D, E, F and Z) and identified that KSHV subtypes have close associations with the geographic and ethnic background of patients. Subtypes A and C have the largest prevalence area, including Africa, Europe, Middle East, Asia [12,22], America and Australia. Subtype D has been reported from the Pacific Islands, like Japan [12,16], Chinese Taiwan [23], and Australia. Subtype B is found from patients of African origin [24], only one subtype B isolates separated from a Mexican female AIDS patient in Miami [14]. Subtype E was only found in Amerindians [15,25]. Subtype Z has been found in Zambian [17]. Subtype F is rare and was only found in Africa recently [13].

In China, KS cases are rare, only reported in Xinjiang Region [20] and Taiwan Island [26]. However, KSHV subtypes in these regions are quite different from each other. Previous research and this study have found the KSHV subtypes in Xinjiang region are subtypes A and C, while in Taiwan it is subtypes C and D [23]. This difference may have geographic reasons. Taiwan is an island in the Pacific, while Xinjiang located in central Asia. Xinjiang is home to a number of different ethnic groups, including the Uyghur, Han, Kazakh, Hui, Kyrgyz, Xibo and Mongolians. Historically, the population in Xinjiang consisted of various nomadic tribes, like Yuezhi, Xiongnu, and Wusun. Well-preserved Tarim mummies with Caucasoid features, often with reddish or blond hair, which dated from 1800 BCE to 200 CE, which were different from the Han Chinese, have been found in Xinjiang area. Today the minorities, including Uyghur, Kazakh and Hui, still have significant difference from the Han. The epidemiology study has found that the seroprevalence of KSHV in the general population was higher than the other provinces in China, and the minorities (Uyghur, Kazakh, Hui, and Xibo) have higher KSHV-seropositive ratio than the Han [18]. Previous researches have suggested that KSHV has been introduced to Xinjiang Region along the Silk Road. The Silk Road was a path for cultural, commercial and technological exchange between Ancient China, India, the Persian Empire and Mediterranean countries for nearly 3000 years. The Silk Road was extended to the center of China, and Xinjiang was a section of the Silk Road. In our study, two subtypes of KSHV were identified, 23 of them belonged to subtype A, while five of them were subtype C. More genotype A was found than genotype C, both in Classical KS and AIDS KS. No significant difference was found in the predominance of different genotype between Classical KS and AIDS KS.

Co-migration of viruses with the human population is common. KSHV subtypes A and C are predominant in the Mediterranean, Middle Eastern and Asian regions [12,22], which were parts of the ancient Great Silk Road that went through the Xinjiang region. KSHV subtype C has been found in Xinjiang in 2001 [19]. KSHV subtype A has also been found in Xinjiang by our study in 2009 [18]. In this study, both KSHV subtype A and C were identified. Thus, it is tempting to speculate that KSHV might have been spread along the Silk Road with the human population.

Classic KS was only found in Xinjiang Minority, but not often seen in Han population. More epidemiologic study should be done to investigate the risk factors, including environmental and human genetic factors.

Acknowledgments

This work was supported by National Science Foundation of China (Grant No. 81271837) and Natural Science Foundation Key project of Anhui Education Department (KJ2012A161).

Author Contributions

XO and YZ carried out the experiment and drafted the manuscript. BF, XW, WC and YF participated in design and coordination of study, data and sample collection. ML and LW conceived study, provided guidance to all aspects of study, and performed quality assessment of data, data analysis, data preparation, and drafted manuscript. All authors read and approved final manuscript.

References and Notes

1. Kaposi, M. Idiopathisches multiples pigmentsarkom der haut. *Arch. Dermatol. Syph.* **1872**, *4*, 265–273.
2. Beral, V.; Peterman, T.A.; Berkelman, R.L.; Jaffe, H.W. Kaposi's sarcoma among persons with aids: A sexually transmitted infection? *Lancet* **1990**, *335*, 123–128.
3. Antman, K.; Chang, Y. Kaposi's sarcoma. *N. Engl. J. Med.* **2000**, *342*, 1027–1038.
4. Friedman-Kien, A.E.; Saltzman, B.R. Clinical manifestations of classical, endemic african, and epidemic aids-associated kaposi's sarcoma. *J. Am. Acad. Dermatol.* **1990**, *22*, 1237–1250.
5. Iscovich, J.; Boffetta, P.; Winkelmann, R.; Brennan, P.; Azizi, E. Classic kaposi's sarcoma in jews living in israel, 1961–1989: A population-based incidence study. *Aids* **1998**, *12*, 2067–2072.
6. Guo, W.X.; Antakly, T. Aids-related kaposi's sarcoma: Evidence for direct stimulatory effect of glucocorticoid on cell proliferation. *Am. J. Pathol.* **1995**, *146*, 727–734.
7. Fatahzadeh, M. Kaposi sarcoma: Review and medical management update. *Oral Surg. Oral Med. Oral Pathol. Oral Radiol.* **2012**, *113*, 2–16.
8. Chang, Y.; Cesarman, E.; Pessin, M.S.; Lee, F.; Culpepper, J.; Knowles, D.M.; Moore, P.S. Identification of herpesvirus-like DNA sequences in aids-associated kaposi's sarcoma. *Science* **1994**, *266*, 1865–1869.
9. Cesarman, E.; Chang, Y.; Moore, P.S.; Said, J.W.; Knowles, D.M. Kaposi's sarcoma-associated herpesvirus-like DNA sequences in aids-related body-cavity-based lymphomas. *N. Engl. J. Med.* **1995**, *332*, 1186–1191.
10. Soulier, J.; Grollet, L.; Oksenhendler, E.; Cacoub, P.; Cazals-Hatem, D.; Babinet, P.; d'Agay, M.F.; Clauvel, J.P.; Raphael, M.; Degos, L.; *et al.* Kaposi's sarcoma-associated herpesvirus-like DNA sequences in multicentric castleman's disease. *Blood* **1995**, *86*, 1276–1280.
11. Russo, J.J.; Bohenzky, R.A.; Chien, M.C.; Chen, J.; Yan, M.; Maddalena, D.; Parry, J.P.; Peruzzi, D.; Edelman, I.S.; Chang, Y.; *et al.* Nucleotide sequence of the kaposi sarcoma-associated herpesvirus (hhv8). *Proc. Natl. Acad. Sci. USA* **1996**, *93*, 14862–14867.

12. Kanno, T.; Sato, Y.; Nakamura, T.; Sakamoto, K.; Sata, T.; Katano, H. Genotypic and clinicopathological characterization of kaposi's sarcoma-associated herpesvirus infection in japan. *J. Med. Virol.* **2010**, *82*, 400–406.
13. Kajumbula, H.; Wallace, R.G.; Zong, J.C.; Hokello, J.; Sussman, N.; Simms, S.; Rockwell, R.F.; Pozos, R.; Hayward, G.S.; Boto, W. Ugandan kaposi's sarcoma-associated herpesvirus phylogeny: Evidence for cross-ethnic transmission of viral subtypes. *Intervirology* **2006**, *49*, 133–143.
14. Hayward, G.S.; Zong, J.C. Modern evolutionary history of the human kshv genome. *Curr. Top. Microbiol. Immunol.* **2007**, *312*, 1–42.
15. Biggar, R.J.; Whitby, D.; Marshall, V.; Linhares, A.C.; Black, F. Human herpesvirus 8 in brazilian amerindians: A hyperendemic population with a new subtype. *J. Infect. Dis.* **2000**, *181*, 1562–1568.
16. Meng, Y.X.; Sata, T.; Stamey, F.R.; Voevodin, A.; Katano, H.; Koizumi, H.; Deleon, M.; de Cristofano, M.A.; Galimberti, R.; Pellett, P.E. Molecular characterization of strains of human herpesvirus 8 from japan, argentina and kuwait. *J. Gen. Virol.* **2001**, *82*, 499–506.
17. Kasolo, F.C.; Monze, M.; Obel, N.; Anderson, R.A.; French, C.; Gompels, U.A. Sequence analyses of human herpesvirus-8 strains from both african human immunodeficiency virus-negative and -positive childhood endemic kaposi's sarcoma show a close relationship with strains identified in febrile children and high variation in the k1 glycoprotein. *J. Gen. Virol.* **1998**, *79*, 3055–3065.
18. Fu, B.; Sun, F.; Li, B.; Yang, L.; Zeng, Y.; Sun, X.; Xu, F.; Rayner, S.; Guadalupe, M.; Gao, S.J.; *et al.* Seroprevalence of kaposi's sarcoma-associated herpesvirus and risk factors in xinjiang, china. *J. Med. Virol.* **2009**, *81*, 1422–1431.
19. Dilnur, P.; Katano, H.; Wang, Z.H.; Osakabe, Y.; Kudo, M.; Sata, T.; Ebihara, Y. Classic type of kaposi's sarcoma and human herpesvirus 8 infection in xinjiang, china. *Pathol. Int.* **2001**, *51*, 845–852.
20. Zhang, D.Z.; Pu, X.M.; Wu, W.D.; Jin, Y.; Juhear, M.; Wu, X.J. Genotypic analysis on the orf-k1 gene of human herpesvirus 8 from patients with kaposi's sarcoma in xinjiang, china. *J. Genet. Genomics* **2008**, *35*, 657–663.
21. Hayward, G.S. Kshv strains: The origins and global spread of the virus. *Semin. Cancer Biol.* **1999**, *9*, 187–199.
22. Cook, P.M.; Whitby, D.; Calabro, M.L.; Luppi, M.; Kakoola, D.N.; Hjalgrim, H.; Ariyoshi, K.; Ensoli, B.; Davison, A.J.; Schulz, T.F. Variability and evolution of kaposi's sarcoma-associated herpesvirus in europe and africa. International collaborative group. *Aids* **1999**, *13*, 1165–1176.
23. Zong, J.C.; Ciufo, D.M.; Alcendor, D.J.; Wan, X.; Nicholas, J.; Browning, P.J.; Rady, P.L.; Tyring, S.K.; Orenstein, J.M.; Rabkin, C.S.; *et al.* High-level variability in the orf-k1 membrane protein gene at the left end of the kaposi's sarcoma-associated herpesvirus genome defines four major virus subtypes and multiple variants or clades in different human populations. *J. Virol.* **1999**, *73*, 4156–4170.
24. White, T.; Hagen, M.; Gudza, I.; White, I.E.; Ndemera, B.; Gwanzura, L.; Borok, M.; Campbell, T.B. Genetic diversity of the kaposi's sarcoma herpesvirus k1 protein in aids-ks in zimbabwe. *J. Clin. Virol.* **2008**, *42*, 165–171.

25. Whitby, D.; Marshall, V.A.; Bagni, R.K.; Wang, C.D.; Gamache, C.J.; Guzman, J.R.; Kron, M.; Ebbesen, P.; Biggar, R.J. Genotypic characterization of kaposi's sarcoma-associated herpesvirus in asymptomatic infected subjects from isolated populations. *J. Gen. Virol.* **2004**, *85*, 155–163.
26. Su, I.J.; Hsu, Y.S.; Chang, Y.C.; Wang, I.W. Herpesvirus-like DNA sequence in kaposi's sarcoma from aids and non-aids patients in taiwan. *Lancet* **1995**, *345*, 722–723.

10

Epidemiology and Transmission of Kaposi's Sarcoma-Associated Herpesvirus

Veenu Minhas † and Charles Wood *

Nebraska Center for Virology, School of Biological Sciences, University of Nebraska, Morrison Center, 4240 Fair Street, Lincoln, NE 68583, USA; E-Mail: vminhas2@unl.edu

† Present Address: Department of Epidemiology, College of Public Health, University of Nebraska Medical Center, Omaha, NE 68198, USA.

* Author to whom correspondence should be addressed; E-Mail: cwood1@unl.edu;

External Editor: Zhi-Ming Zheng

Abstract: This review summarizes the current knowledge pertaining to Kaposi sarcoma-associated herpesvirus (KSHV) epidemiology and transmission. Since the identification of KSHV twenty years ago, it is now known to be associated with Kaposi's sarcoma (KS), primary effusion lymphoma, and multicentric Castleman's disease. Many studies have been conducted to understand its epidemiology and pathogenesis and their results clearly show that the worldwide distribution of KSHV is uneven. Some geographical areas, such as sub-Saharan Africa, the Mediterranean region and the Xinjiang region of China, are endemic areas, but Western Europe and United States have a low prevalence in the general population. This makes it imperative to understand the risk factors associated with acquisition of infection. KSHV can be transmitted via sexual contact and non-sexual routes, such as transfusion of contaminated blood and tissues transplants, or via saliva contact. There is now a general consensus that salivary transmission is the main route of transmission, especially in children residing in endemic areas. Therefore, there is a need to better understand the sources of transmission to young children. Additionally, lack of animal models to study transmission, gold standard serological assay and the lack of emphasis on endemic KS research has hampered the efforts to further delineate KSHV transmission in order to design effective prevention strategies.

Keywords: KSHV; HHV-8; epidemiology; transmission

1. Introduction

Human herpesviruses are large double stranded DNA viruses that are ubiquitous in nature. Of the eight known human herpesviruses two have been implicated to be the etiologic agents for a number of cancers, Epstein-Barr virus, and the Kaposi sarcoma-associated herpesvirus (KSHV). Since its identification in 1994, KSHV is known to be associated with all forms of Kaposi's sarcoma (KS) primary effusion lymphoma and multicentric Castleman's disease [1–4]. KS is an AIDS-defining illness and is the most common malignancy present in HIV-1 infected patients [5]. One of the earliest indicators of the upcoming HIV/AIDS epidemic was the sudden appearance of KS and high-grade non-Hodgkin's lymphoma in young men having sex with men (MSM) [6].

Since KSHV DNA cannot be detected in all infected individuals, a serological assay to detect the presence of antibodies against KSHV is the method of choice to investigate KSHV prevalence in a population. Unlike EBV, global seroprevalence of KSHV varies greatly and is generally high in areas where non-HIV associated forms of KS (classic or endemic forms) have been common [7]. These areas include African and Mediterranean regions, where KSHV seroprevalence ranges from 20% to 80% in the adult populations, whereas seroprevalence in the United States and Northern Europe is generally low (<10%) [8–14]. In South America, the Amerindians have also been identified as a hyperendemic population for KSHV infection [15]. Interestingly, the prevalence in the non-Amerindians is significantly lower. de Souza *et al.* have recently reported that >70% of children (four to nine years) were positive for KSHV and the prevalence increased with age (>90% in >40 year age group) [16]. This pattern is strikingly similar to Africa, where infection seems to occur in early childhood. In contrast, KS incidence is uncommon and KSHV prevalence is low in most Asian countries [17]. However, in China, KSHV prevalence varies considerably between different regions of the country and is between 7.3% percent and 16.1 percent in adults in different provinces [18–21]. In particular, the prevalence was found to be high in the Xinjiang province, which is located in the northwestern region of China. Interestingly, cases of both classic KS and AIDS-associated KS are found mostly in the Uyghurs and the Kazakh ethnicities [20,21]. The reasons for such geographical and population variations of KS and KSHV distribution remain to be addressed.

Our laboratory has conducted extensive cohort studies in Zambia, a sub-Saharan African country. Zambia has been considered as part of the "KS belt", where endemic KS was prevalent and where significant increase in KS incidence in adults and children has coincided with the emergence of the HIV-1 epidemic [22–24]. This increase in KS incidence is significant because, by 1992, KS accounted for approximately 25% of all childhood cancers diagnosed in Lusaka, the capital of Zambia [25]. This review summarizes some of the results of our cohort studies in Zambia, together with those from other significant studies that have contributed to further our understanding of KSHV transmission and pathogenesis, and, more importantly, explore some of the questions that still remain unanswered and warrant further research.

2. Modes of Transmission

Primary KSHV infection can occur during childhood and as an adult and can be transmitted via both sexual and non-sexual routes. KSHV can be found in the peripheral blood mononuclear cells (PBMCs), saliva, oropharyngeal mucosa, semen, cervico-vaginal secretions, and prostate glands, which may represent the source of both vertical and horizontal transmission [26–29]. The modes of transmission of KSHV may vary in different parts of the world, depending on the endemicity of that region. In non-endemic areas, especially in adults, sexual transmission (homosexual and heterosexual) of KSHV may be the route of transmission [30–32]. However, there is limited information about this route and further extensive studies are needed. Studies have shown that sexual transmission, particularly among MSMs, may play a major role in transmission in non-endemic areas, such as the United States and Western Europe, because the seroprevalence among MSMs is significantly higher as compared to the general population. In the San Francisco Mens Health Study, the prevalence of KSHV infection was reported to be high among MSMs (37.6%) when compared to heterosexual men (no prevalence in this group) [31]. They also reported an association of KSHV prevalence with the number of sexual partners. Other studies have reported that besides the number of sex partners, syphilis infection and hepatitis B infection are also risk factors for acquisition of infection [27,31,33,34]. However, the possibility of sexual transmission among heterosexual individuals remains unclear. Different studies have been inconsistent about the evidence of transmission through heterosexual contact. This can be explained in part by the populations that were investigated. Two African studies have been conducted in commercial sex workers in Nigeria and Kenya, and clearly suggest sexual transmission occurring in the adult population [35,36]. Another study conducted in truckers in Kenya also suggest that there is ongoing sexual transmission during adulthood [37]. However, there are other studies that do not observe any evidence of sexual transmission [38,39]. One of them is a study conducted by Malope *et al.*, in South Africa. This is an interesting study because it had a large sample size and sampling was conducted from miners, sex workers, and other residents of the city. This study did not find any association of KSHV prevalence with sexual behavior. In addition, our study conducted in a group of female sex workers in China showed KSHV prevalence of 10%, which is similar to the general women population, even though the prevalence of other sexually transmitted infections (STI), such as HSV-2 and syphilis, were higher than the general women population. Our study suggested that heterosexual transmission may not be a dominant route for KSHV, at least among Chinese women [40]. Another important issue while investigating sexual transmission is that both kissing and salivary exchange during hetero and homosexual contact is common. This makes it harder to delineate whether the transmission was sexual or oral through saliva. Additionally, it is important to consider that all these studies were cross-sectional in design. Longitudinal cohort studies that can follow individuals for seroconversion and incident infection are needed to truly understand the impact of this route of transmission.

Besides sexual transmission, another possible horizontal transmission route is via KSHV contaminated blood transfusion because viral DNA may be detected in 10–15 percent of PBMCs of healthy KSHV seropositive individuals [41,42]. Transmission of KSHV via this route is likely to be inefficient because of the cell associated nature of the virus, and viremia is uncommon, even in KSHV infected individuals. However, blood transfusion has recently been investigated in an elegant study by

Hladik *et al.*, where they provided strong evidence that KSHV can be transmitted by blood transfusion [43]. This study was conducted in Uganda, where 991 KSHV seronegative recipients were followed after transfusion. Among them, 425 patients received KSHV seropositive blood and 566 patients received KSHV seronegative blood. They observed that KSHV seroconversion occurred in 41 of the 991 recipients. The excess risk of KSHV seroconversion after transfusion with KSHV seropositive blood (during the observed 24-week follow-up) was 2.8%. This suggests that 12 patients (of the 425 patients exposed to KSHV seropositive blood) were infected by transfusion. Another interesting result of this study was that an excess risk of 4.2% was observed among patients who received blood stored for less than four days as compared to those who received blood stored for more than four days. The results from this study led them to further investigate whether transfusion of KSHV positive blood led to an increase in mortality. Their results show that transfusion of short stored blood was associated with increased risk of death (adjusted hazard ratio – 1.79) [44]. Whether the increased risk was due to KSHV or other factors in short stored blood will need to be further investigated.

After KSHV was identified, several observations and published studies led researchers to hypothesize that KSHV can be transmitted by another non-sexual route, specifically from mother to child vertically. The high seroprevalence level in children, especially in endemic areas was one of the rationales. Additionally, KSHV has been found in cervico-vaginal secretions of HIV-1 and KSHV co-infected women, suggesting that KSHV viral load in the female genital tract might influence the vertical transmission of KSHV [45]. Moreover, development of KS in children less than one year of age was indicative of perinatal transmission [46]. These observations also formed the basis of our cohort study conducted between 1998 and 2004. Pregnant mothers in early stages of labor were recruited in the cohort and both the mother and infant were enrolled and followed after delivery. Vertical transmission was investigated in 89 mother-infant pairs. All of these 89 children were born to KSHV seropositive mothers whose serostatus was tested at the time of delivery [28]. We have reported that KSHV DNA was detected in PBMCs in 2/89 children within 24 h after birth. More recently, Lisco *et al.* have addressed this question from a different perspective [45]. As herpesvirus reactivation occurs during pregnancy, they investigated the KSHV presence and viral load in PBMCs and cervicovaginal secretions (CVS) from 15 pregnant Italian women. They have reported viral reactivation in 5/15 women in PBMCs and CVS. One infant also had the same viral subtype as the mother at two months after birth. Together, these reports strongly indicate that *in utero* or intrapartum KSHV infection might, albeit rarely, occur in countries where KSHV is endemic. More studies, with larger sample sizes are required to further delineate this route of transmission.

Breast milk transmission has also been hypothesized to be contributing to KSHV transmission from mother to infants. Breast milk has been reported to contain herpesviruses, such as CMV, EBV, and HHV-7 [47–51]. Both CMV and KSHV are known to infect monocytes, macrophages, and epithelial cells. These cells are found in the cellular components of mature milk suggesting that KSHV might also be found in breast milk. Indeed, a study by Dedicoat *et al.* has shown that KSHV DNA could be detected in 12/43 breast milk samples from South African mothers [52]. Our laboratory has investigated the presence of KSHV DNA both in breast milk cells and in the liquid portion of breast milk (supernatant) [53]. However, we failed to detect any viral DNA in breast milk, whereas EBV DNA was readily detectable. Since then, there has been no recent study that has investigated the potential of breast milk transmission. Therefore, the current evidence does not point towards breast milk transmission being an important route of horizontal KSHV transmission from the mothers to their infants, at least in Zambia.

The high prevalence of KSHV infection reported by many studies, especially those conducted in African children cannot be explained by transmission routes discussed above; vertical mother to child transmission, breast milk transmission or via blood contamination. Therefore, another major horizontal route of transmission that needs to be investigated is salivary transmission. Indeed, several interesting studies indicated that saliva as the likely candidate, which mediates childhood transmission. Soon after its discovery, there were studies that reported the presence of KSHV DNA sequences in saliva [54–56]. In 2000, Pauk *et al.* reported that oral exposure to infectious saliva was a potential risk factor for the acquisition of KSHV among men who have sex with men. Since then, our group and others have conducted several studies to show that KSHV is shed in saliva of infected individuals regardless of their HIV-1 status [53,57]. We have reported that the group of mothers who were not shedding KSHV in breast milk, did shed KSHV in saliva (19/65 samples were positive). Other studies have also investigated the frequency of shedding over a period [57,58]. The results showed that not all KSHV seropositive individuals shed the virus in saliva, and individuals in whom KSHV can be detected can range from occasional shedders to daily shedders. All the above studies provided strong evidence that horizontal transmission, predominantly via saliva, is the major route of transmission especially in endemic countries.

3. Sources and Risk Factors Associated with Transmission

Soon after the discovery of KSHV, a majority of the studies conducted in Africa, even though cross-sectional in design, indicated that KSHV infection was prevalent in young children [11,59,60]. Our cohort studies first started in 1998 with the aim to study the epidemiology of KSHV in Zambia. Our group has reported that more than 10 percent of Zambian children may be infected by 12 months of age [61]. Other studies, though cross-sectional in design, have also shown that especially in endemic regions, primary infection likely occurs in early childhood leading to accumulation of infection. This is evident from the observed high seroprevalence during childhood (Table 1). Therefore, it is important to understand the source of the transmission of virus to children and the risk factors associated with acquisition of infection. The exact risk factors that predispose children to acquisition of KSHV infection are still not fully understood but immunosuppression, especially due to HIV-1 infection is a major risk factor. The route(s) of transmission most likely varies among populations, and given the increasing availability of anti-retroviral therapy (ART) in the high HIV-1 prevalence regions, such as sub-Saharan Africa, there is a need to study the effect of ART on transmission to HIV-1 infected individuals including children.

It has been reported that there is an increased prevalence of KSHV antibodies in children of mothers shedding high number of viral DNA copies/ml of saliva, suggesting that KSHV in maternal saliva may be associated with transmission to the child [62]. As mentioned above, we have observed KSHV shedding in maternal saliva. Therefore, it will be important to determine whether the maternal KSHV shedding in saliva is important for transmission to children and the specific common practices that are associated with transmission of infection. The specific practices that expose individuals, especially young children, to salivary transmission still remain to be explored and are a subject of our ongoing research study in Zambia. Other studies also indicate that, in Africa, transmission from mother to child and between siblings accounts for a substantial proportion of childhood infections [29,60]. Preliminary analysis of our ongoing cohort studies in Zambia reveal that transmission to children also occurs in households where no household member is KSHV seropositive, who could serve as a source of infection to the child. This indicates that transmission may occur from sources outside the family unit. In fact, we found that children from several families have KSHV genotype that differed from those of their mothers and other family members, suggesting that they have acquired KSHV infection from sources outside the household [63]. Other than these studies from Africa, very little is known about transmission of KSHV in children in developed countries, including the United States, since the prevalence of childhood infection is low, and the results of reports from these countries are conflicting [11].

Table 1. Selected recent reports summarizing the observed percent seroprevalence of KSHV in early childhood.

Reference (Year)	Country	Age (Years)	Percent Prevalence
Cao *et al.* (2014) [21]	Xinjiang, China	0.5–5	48.3
Wakeham *et al.* (2013) [64]	Entebbe, Uganda	1	4
		2	7
		3	10
		4	13
		5	14
Butler *et al.* (2011) [65]	Buziika B Parish, Uganda	1.5–2	15.5
		3–5	22.7
		6–9	31.6
		10–13	32.0
Pfeiffer *et al.* (2010) [66]	Kampala, Uganda	0–4	32.1
		5–9	37.4
		10–19	30.6
	Shirati District, Tanzania	0–4	33.9
		5–9	33.7
		10–19	32.4
	Lagos, Nigeria	0–4	14.6
		5–9	2.3
		10–19	83.1
Butler *et al.* (2009) [38]	Kampala, Uganda	2–8	10–30.6

The current knowledge about source and route of infection, and risk factors associated with acquisition of infection has been summarized in Figure 1. As described in detail above, for young children, especially in endemic areas where children can be infected by members from within and outside the family mainly via saliva, other routes are possible but may not be the major routes of transmission. In addition, factors present both within and outside the household or in the environment may also increase the risk of acquisition of infection by the child. In fact it has been reported that environmental or ecological factors may be associated with the development of KS. This includes geographical areas with volcanic soils, chronic schistosome or other parasite infections, biting flies or contact with phorbol esters or other constituents of plants [67–70]. Whether these factors also increase the risk of acquisition of KSHV infection is not known.

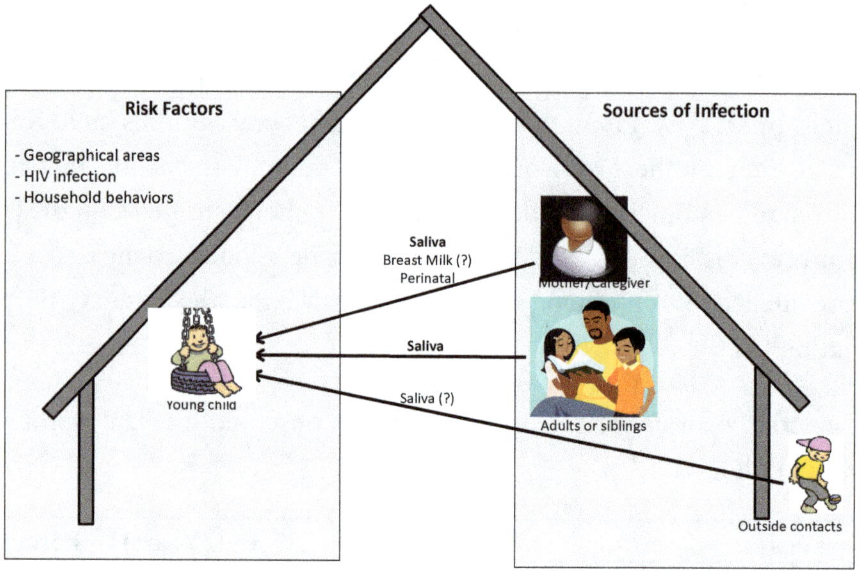

Figure 1. Pictorial representation of possible sources and risk factors associated with acquisition of KSHV infection by children in endemic areas.

4. Future Research

The year 2014, marks the 20th year since the discovery of KSHV. This is a time to reflect on the great strides achieved towards understanding the biology and epidemiology of KSHV, and look towards the future to explore other critical questions that need to be investigated. KSHV pathogenesis is an area of active research and animal models have been used to study several aspects of KSHV infection and pathogenesis. However, there was no ideal model available to understand KSHV transmission, routes of transmission, early events following primary viral infection and its interaction with the host particularly regarding disease development. One of the main hurdles was the lack of an effective model that supported natural routes of viral infection followed by a long-term sustainable infection involving both latent and lytic viral gene expression, and leading to the infection of target cells and tissues. There have been several studies that have developed animal models to study KSHV infection; they include the non-human primate model. However, primate models are difficult to work with and not all reported models were able to establish persistent infection [71,72]. Therefore, there is still a great need to develop small animal models, such as a rodent model, which can be used to study KSHV infection and pathogenesis efficiently. Such a model will need to have relatively short generation time, inexpensive, and easy to manipulate. Additionally, this model needs to support natural route of viral infection, a long term sustainable infection involving both latent and lytic viral gene expression, and the infection of target tissues and cells that reflect those of human infection. Recently, a new generation of humanized mouse, the BLT (bone marrow, liver and thymus) mouse (hu-BLT) has been shown to be an excellent model for studying human viral infections [73]. This is the only mouse model which can generate the human mucosal immune systems, and a human HLA restricted antigen specific humoral and cellular responses [74]. To this effect, our laboratory has recently used this mouse model to study KSHV infection [75]. Our study demonstrated that both latent and lytic viral transcripts, as well as viral protein expression in various tissues, including spleen and skin. Interestingly, mice could be infected via several routes, including the oral mucosal route. Furthermore, we found that KSHV can establish infection in human B cells and macrophages in this model. Our study represents the first successful effort to recapitulate KSHV infection in a small animal model via a natural route of infection. The hu-BLT mouse could potentially be further developed as a model not only for studying the pathogenesis of KSHV *in vivo*, but can also be applied to study routes and mechanisms of KSHV transmission and systemic dissemination.

As discussed earlier by a number of reports, there is still a lack of a gold standard serological assay that is equally effective in detecting anti-KSHV antibodies in KS patients (generally having high antibody titers) and in KSHV infected but non-KS individuals (generally having lower antibody titers). Lack of such an assay has made it difficult to develop a clearer global picture about KSHV epidemiology; since the performance of various assays differs, and it has been difficult to compare across studies that also differ in the characteristics of the population and study design. Currently, there now are two general platforms of KSHV serologic assays, the ELISA based assay and the immunofluorescence-based assay. The ELISA based assays in general involve the use of recombinant KSHV structural proteins. Such an assay is objective and of high throughout, readily screening through a larger number of samples. However, the common format involves the use of only two or three viral proteins, and it is known to underestimate the prevalence of infection in a study population. In addition, setting the cutoff is critical because the titers of KSHV antibodies vary greatly between different study populations and between geographical locations. Recently, Labo *et al.* have reported an important study about the development of a bead-based multiplex assay that detects antibodies to six KSHV antigens [76]. Interestingly, the authors conducted a systematic antigenic analysis of the entire KSHV proteome by screening 72 KSHV proteins to understand the seroreactivity pattern. Further evaluation of this assay on a larger sample size composed of different study populations may prove this approach to be very useful for KSHV serodiagnosis.

The second KSHV serology platform is immunofluorescence assay (IFA). This assay in general utilizes KSHV chronically infected B lymphoma cell lines such BC3 cells, which are latently infected with KSHV. These cells are then stimulated by TPA to undergo lytic reactivation so that both lytic and latent viral proteins can be expressed effective, and then used for the detection of anti-KSHV antibodies in study population using indirect immunofluorescence [77]. There are a number of laboratories, including ours, that have used this assay format because of the sensitivity of the assay and its ability to detect all viral antigens expressed. Our laboratory has used this assay to conduct a number epidemiology studies in different study populations, both in Zambia and China [40,61]. However, such an assay is subjective, labor intensive, has low throughput and also required highly experienced laboratory personnel, and IFA is not a high throughput assay when compared to ELISAs. Using such assays in Zambia, we have observed that children generally have lower titers as compared to adults. Development of a more efficient highly sensitive and standardized assay will be important to further confirm the seroprevalence of KSHV in different parts of world, especially in non-KS individuals. Use of such an assay will also help to confirm whether the currently reported seroreversion in infected individuals could be due to a true drop in antibody titers to below detection limit of our current assays, or whether there is a true loss of KSHV antibody with time due to the viral latency [41,61].

KSHV antibodies, for a large part, has been used to investigate KSHV epidemiology, but there is still a need to understand the role of the immune response in preventing KSHV infection, in controlling viral infection and the development of the KS in infected individuals. There is also a need to further study not only the role of the humoral immune response but the cellular immune response as well. For the humoral response, an in-depth understanding of the role broadly protective immune response on the control of KSHV infection and development of KS will be important. Two studies have quantified the prevalence and titers of neutralizing antibodies (nAb) in KS patients or in KSHV infected asymptomatic controls. Both reports have focused on a small number of KS patients from the US. One suggested that KS patients had lower titers of neutralizing antibodies compared to asymptomatic individuals irrespective of their HIV status [78], while the other study found no significant difference between the two groups [79]. In addition, the overall prevalence of nAb in KS patients or asymptomatic subjects was found to be low and comparable between KS and asymptomatic individuals [79]. Our laboratory has investigated the prevalence and titer of KSHV nAb in a cross-sectional cohort of Zambian patients [80]. HIV-1-associated KS patients had a significantly higher prevalence of neutralizing antibody response when compared to non-KS but KSHV infected individuals. This suggested that viral lytic replication and the resulting antigenemia, which occur during KS development, might be essential contributors to development of neutralizing responses. However, whether these responses are protective if passively introduced prior to primary KSHV infection is currently not known. In addition to nAb, further studies are required to decipher the role of cell-mediated immunity in protecting against KSHV infection and preventing the development of KS. A major hurdle in this area of research has been a lack of a robust T cell response against KSHV antigens. A weak adaptive immune response following primary infection has proved to be the major obstacle in this area. However, recent studies have provided some interesting data. One of these studies by Lepone *et al.* is very interesting that has reported on the identification of five novel HLA A*0201-restricted $CD8^+$ T cell epitopes present in LANA-1, K12, gB, and K8.1 proteins [81]. These epitopes induced both single and multiple immune mediators in $CD8^+$ T cells from healthy KSHV seropositive individuals. A review by Robey *et al.* has provided a list of identified epitopes to known KSHV immunogenic open reading frames [82]. These are critical steps towards understanding the role of T cells in the pathogenesis of KSHV infection though further research in identification of other immunodominant epitopes is needed. Further studies are also required to understand the role of CD8+ and CD4+ T cell responses to primary infection and disease development. There is a gap in our understanding of broad range of antigenic epitopes that can be restricted by other MHC class I and II haplotypes.

In addition to the immune response against KSHV, there needs to be a better understanding of the routes of transmission in different populations, the risk factors associated with incident infection and the potential source(s) of transmission. This information is pertinent in the public health disease management for the design of strategies to prevent KSHV infection, especially in endemic regions. Prevention of transmission to high-risk individuals in Western countries and to children in endemic areas, especially in Africa, may be the best approach to reduce the burden of infection and the subsequent development of KS. We believe that further work in understanding local risk factors and sources of transmission will lead to the development of strategies to prevent transmission of infection to susceptible individuals.

In parallel to the development of public health policy to prevent KSHV infection, there is also a need towards the development of protective vaccines since vaccination is the most efficient method to eliminating viral infectious agents and to curb the epidemic. However, there are ongoing discussions among researchers and policy makers whether such a vaccine is needed for KSHV since the KS incidence is decreasing because of the effectiveness of the ART, at least in the developed countries [83,84]. However, due to the high prevalence of KSHV in the African countries and in other KS endemic regions, it is questionable whether ART and other public health prevention strategies will be effective in preventing KSHV infection and the subsequent risk in developing KS. Therefore, there is a still a continuous need to better understand the viral antigens and the generation of protective immune response against KSHV antigens. Such an understanding will be a first step towards developing a protective KSHV vaccine.

In addition to the need to prevent epidemic KS due to its sudden rise in concordance to the AIDS epidemic, there is a severe lack of information about the endemic KS. This form of KS is still prevalent in many parts of Africa and its epidemiology and pathogenesis needs to be better characterized. It is most likely that the endemic KS will continue to persist in these regions due to the high prevalence of KSHV infection, even in the HIV-1 uninfected population. We have recently conducted a study at the Dermatology and Venereology clinic at University Teaching Hospital, Lusaka, Zambia, which is a national referral clinic for all suspected KS cases. From 2008 to 2013, we document 726 pathologist confirmed KS cases at this clinic [85]. Of these, one third of cases were diagnosed in HIV negative patients (endemic KS). This underscores the need to understand the epidemiology of endemic form of KS in this region. Currently, there is a gap in the literature regarding the incidence of endemic KS, whether prevalence is stable or increasing, risk factors associated with development of endemic form of KS. Up to now endemic KS has received little attention because it is overshadowed by the recent epidemic KS, especially when a number of studies have reported a decrease in epidemic KS incidence in anti-retroviral treated HIV positive patients, even though they are mostly occurring in the developed countries [83,85,86]. Our study in Zambia clearly shows that the endemic KS is still prevalent and it is likely that antiretroviral therapy will have little or no impact on its incidence. Therefore, it is likely that endemic areas will continue to bear the burden of endemic KS. Furthermore, while great strides have been taken in increasing the coverage of ART in HIV-1 positive patients, the degree of coverage in different African countries varies. Therefore, to further reduce epidemic KS in areas that are endemic for KS (such as the "KS belt" in Africa) without complete ART coverage may even prove difficult in the near future. Our study in Zambia and other studies have clearly shown that classic and endemic forms of KS are more prevalent in males as compared to females [85,87]. These observations also raise the question that genetic factors linked to gender may have a role in progression to KS.

While conducting our cohort studies in Zambia, we have observed that laboratory research findings have yet to be translated into public health practices and programs to reduce transmission of KSHV. Health care professionals are often not aware of KSHV and its link to KS, and current research in this field and better programs similar to human papilloma virus, hepatitis B virus and HIV-1 prevention need to be developed. Educational programs focusing on prevention of transmission are especially needed in endemic regions with limited resources for treatment of KS. It is imperative that results of laboratory research be translated into effective health behavior interventions for prevention of KSHV infection and for overall community health promotion.

Acknowledgments

This work was supported in part by the Public Health Service grants CA-75903 and P30 GM103509 from the National Institutes of Health.

Author Contributions

Both VM and CW have participated in literature search, evaluation of current data and manuscript writing.

References and Notes

1. Cesarman, E.; Chang, Y.; Moore, P.S.; Said, J.W.; Knowles, D.M. Kaposi's sarcoma-associated herpesvirus-like DNA sequences in aids-related body-cavity-based lymphomas. *N. Engl. J. Med.* **1995**, *332*, 1186–1191.
2. Chang, Y.; Cesarman, E.; Pessin, M.S.; Lee, F.; Culpepper, J.; Knowles, D.M.; Moore, P.S. Identification of herpesvirus-like DNA sequences in aids-associated kaposi's sarcoma. *Science* **1994**, *266*, 1865–1869.
3. Schulz, T.F. Kaposi's sarcoma-associated herpesvirus (human herpesvirus-8). *J. Gen. Virol.* **1998**, *79*, 1573–1591.
4. Soulier, J.; Grollet, L.; Oksenhendler, E.; Cacoub, P.; Cazals-Hatem, D.; Babinet, P.; d'Agay, M.F.; Clauvel, J.P.; Raphael, M.; Degos, L.; *et al.* Kaposi's sarcoma-associated herpesvirus-like DNA sequences in multicentric castleman's disease. *Blood* **1995**, *86*, 1276–1280.
5. Sunil, M.; Reid, E.; Lechowicz, M.J. Update on hhv-8-associated malignancies. *Curr. Infect. Dis. Rep.* **2010**, *12*, 147–154.
6. Ziegler, J.L.; Bragg, K.; Abrams, D.; Beckstead, J.; Cogan, M.; Volberding, P.; Baer, D.; Wilkinson, L.; Rosenbaum, E.; Grant, K.; *et al.* High-grade non-hodgkin's lymphoma in patients with aids. *Ann. N. Y. Acad. Sci.* **1984**, *437*, 412–419.
7. Cook-Mozaffari, P.; Newton, R.; Beral, V.; Burkitt, D.P. The geographical distribution of kaposi's sarcoma and of lymphomas in africa before the aids epidemic. *Br. J. Cancer* **1998**, *78*, 1521–1528.
8. Blauvelt, A.; Sei, S.; Cook, P.M.; Schulz, T.F.; Jeang, K.T. Human herpesvirus 8 infection occurs following adolescence in the united states. *J. Infect. Dis.* **1997**, *176*, 771–774.
9. Dedicoat, M.; Newton, R. Review of the distribution of kaposi's sarcoma-associated herpesvirus (kshv) in africa in relation to the incidence of kaposi's sarcoma. *Br. J. Cancer* **2003**, *88*, 1–3.

10. Gao, S.J.; Kingsley, L.; Li, M.; Zheng, W.; Parravicini, C.; Ziegler, J.; Newton, R.; Rinaldo, C.R.; Saah, A.; Phair, J.; *et al.* Kshv antibodies among americans, italians and ugandans with and without kaposi's sarcoma. *Nat. Med.* **1996**, *2*, 925–928.
11. Martro, E.; Bulterys, M.; Stewart, J.A.; Spira, T.J.; Cannon, M.J.; Thacher, T.D.; Bruns, R.; Pellett, P.E.; Dollard, S.C. Comparison of human herpesvirus 8 and epstein-barr virus seropositivity among children in areas endemic and non-endemic for kaposi's sarcoma. *J. Med. Virol.* **2004**, *72*, 126–131.
12. Pellett, P.E.; Wright, D.J.; Engels, E.A.; Ablashi, D.V.; Dollard, S.C.; Forghani, B.; Glynn, S.A.; Goedert, J.J.; Jenkins, F.J.; Lee, T.H.; *et al.* Multicenter comparison of serologic assays and estimation of human herpesvirus 8 seroprevalence among us blood donors. *Transfusion* **2003**, *43*, 1260–1268.
13. Weiss, R.A.; Whitby, D.; Talbot, S.; Kellam, P.; Boshoff, C. Human herpesvirus type 8 and kaposi's sarcoma. *J. Natl. Cancer Inst. Monogr.* **1998**, *23*, 51–54.
14. Stolka, K.; Ndom, P.; Hemingway-Foday, J.; Iriondo-Perez, J.; Miley, W.; Labo, N.; Stella, J.; Abassora, M.; Woelk, G.; Ryder, R.; *et al.* Risk factors for kaposi's sarcoma among hiv-positive individuals in a case control study in cameroon. *Cancer Epidemiol.* **2014**, *38*, 137–143.
15. Biggar, R.J.; Whitby, D.; Marshall, V.; Linhares, A.C.; Black, F. Human herpesvirus 8 in brazilian amerindians: A hyperendemic population with a new subtype. *J. Infect. Dis.* **2000**, *181*, 1562–1568.
16. de Souza, V.A.; Sumita, L.M.; Nascimento, M.C.; Oliveira, J.; Mascheretti, M.; Quiroga, M.; Freire, W.S.; Tateno, A.; Boulos, M.; Mayaud, P.; *et al.* Human Herpesvirus-8 infection and oral shedding in Amerindian and non-Amerindian populations in the Brazilian amazon region. *J. Infect. Dis.* **2007**, *196*, 844–852.
17. De Sanjose, S.; Mbisa, G.; Perez-Alvarez, S.; Benavente, Y.; Sukvirach, S.; Hieu, N.T.; Shin, H.R.; Anh, P.T.; Thomas, J.; Lazcano, E.; *et al.* Geographic variation in the prevalence of kaposi sarcoma-associated herpesvirus and risk factors for transmission. *J. Infect. Dis.* **2009**, *199*, 1449–1456.
18. Fu, B.; Sun, F.; Li, B.; Yang, L.; Zeng, Y.; Sun, X.; Xu, F.; Rayner, S.; Guadalupe, M.; Gao, S.J.; *et al.* Seroprevalence of kaposi's sarcoma-associated herpesvirus and risk factors in xinjiang, china. *J. Med. Virol.* **2009**, *81*, 1422–1431.
19. Mei, Q.; Ming, Z.W.; Ping, Y.X.; Hui, J.J.; Bin, Z.Y.; Hong, W.; Juan, L.; Zhe, C.Y.; Wei, T.; Han, Y. Hhv-8 seroprevalence in blood donors and hiv-positive individuals in shandong area, china. *J. Infect.* **2007**, *55*, 89–90.
20. Wang, H.; Liu, J.; Dilimulati; Li, L.; Ren, Z.; Wen, H.; Wang, X. Seroprevalence and risk factors of kaposi's sarcoma-associated herpesvirus infection among the general uygur population from south and north region of xinjiang, china. *Virol. J.* **2011**, *8*, 539.
21. Cao, Y.; Minhas, V.; Tan, X.; Huang, J.; Wang, B.; Zhu, M.; Gao, Y.; Zhao, T.; Yang, L.; Wood, C. High prevalence of early childhood infection by kaposi's sarcoma-associated herpesvirus in a minority population in china. *Clin. Microbiol. Infect.* **2014**, *20*, 475–481.
22. Bayley, A.C. Occurrence, clinical behaviour and management of kaposi's sarcoma in zambia. *Cancer Surv.* **1991**, *10*, 53–71.

23. Patil, P.; Elem, B.; Zumla, A. Pattern of adult malignancies in zambia (1980–1989) in light of the human immunodeficiency virus type 1 epidemic. *J. Trop. Med. Hyg.* **1995**, *98*, 281–284.
24. Patil, P.S.; Elem, B.; Gwavava, N.J.; Urban, M.I. The pattern of paediatric malignancy in zambia (1980–1989): A hospital-based histopathological study. *J. Trop. Med. Hyg.* **1992**, *95*, 124–127.
25. Chintu, C.; Athale, U.H.; Patil, P.S. Childhood cancers in zambia before and after the hiv epidemic. *Arch. Dis. Child* **1995**, *73*, 100–104.
26. Chen, T.; Hudnall, S.D. Anatomical mapping of human herpesvirus reservoirs of infection. *Mod. Pathol.* **2006**, *19*, 726–737.
27. Kedes, D.H.; Operskalski, E.; Busch, M.; Kohn, R.; Flood, J.; Ganem, D. The seroepidemiology of human herpesvirus 8 (kaposi's sarcoma-associated herpesvirus): Distribution of infection in ks risk groups and evidence for sexual transmission. *Nat. Med.* **1996**, *2*, 918–924.
28. Mantina, H.; Kankasa, C.; Klaskala, W.; Brayfield, B.; Campbell, J.; Du, Q.; Bhat, G.; Kasolo, F.; Mitchell, C.; Wood, C. Vertical transmission of kaposi's sarcoma-associated herpesvirus. *Int. J. Cancer* **2001**, *94*, 749–752.
29. Mbulaiteye, S.M.; Pfeiffer, R.M.; Whitby, D.; Brubaker, G.R.; Shao, J.; Biggar, R.J. Human herpesvirus 8 infection within families in rural tanzania. *J. Infect. Dis.* **2003**, *187*, 1780–1785.
30. Cannon, M.J.; Dollard, S.C.; Smith, D.K.; Klein, R.S.; Schuman, P.; Rich, J.D.; Vlahov, D.; Pellett, P.E. Blood-borne and sexual transmission of human herpesvirus 8 in women with or at risk for human immunodeficiency virus infection. *N. Engl. J. Med.* **2001**, *344*, 637–643.
31. Martin, J.N.; Ganem, D.E.; Osmond, D.H.; Page-Shafer, K.A.; Macrae, D.; Kedes, D.H. Sexual transmission and the natural history of human herpesvirus 8 infection. *N. Engl. J. Med.* **1998**, *338*, 948–954.
32. Melbye, M.; Cook, P.M.; Hjalgrim, H.; Begtrup, K.; Simpson, G.R.; Biggar, R.J.; Ebbesen, P.; Schulz, T.F. Risk factors for kaposi's-sarcoma-associated herpesvirus (kshv/hhv-8) seropositivity in a cohort of homosexual men, 1981–1996. *Int. J. Cancer* **1998**, *77*, 543–548.
33. Dukers, N.H.; Renwick, N.; Prins, M.; Geskus, R.B.; Schulz, T.F.; Weverling, G.J.; Coutinho, R.A.; Goudsmit, J. Risk factors for human herpesvirus 8 seropositivity and seroconversion in a cohort of homosexual men. *Am. J. Epidemiol.* **2000**, *151*, 213–224.
34. Smith, N.A.; Sabin, C.A.; Gopal, R.; Bourboulia, D.; Labbet, W.; Boshoff, C.; Barlow, D.; Band, B.; Peters, B.S.; de Ruiter, A.; *et al.* Serologic evidence of human herpesvirus 8 transmission by homosexual but not heterosexual sex. *J. Infect. Dis.* **1999**, *180*, 600–606.
35. Eltom, M.A.; Mbulaiteye, S.M.; Dada, A.J.; Whitby, D.; Biggar, R.J. Transmission of human herpesvirus 8 by sexual activity among adults in lagos, nigeria. *AIDS* **2002**, *16*, 2473–2478.
36. Lavreys, L.; Chohan, B.; Ashley, R.; Richardson, B.A.; Corey, L.; Mandaliya, K.; Ndinya-Achola, J.O.; Kreiss, J.K. Human herpesvirus 8: Seroprevalence and correlates in prostitutes in mombasa, kenya. *J. Infect. Dis.* **2003**, *187*, 359–363.
37. Baeten, J.M.; Chohan, B.H.; Lavreys, L.; Rakwar, J.P.; Ashley, R.; Richardson, B.A.; Mandaliya, K.; Bwayo, J.J.; Kreiss, J.K. Correlates of human herpesvirus 8 seropositivity among heterosexual men in kenya. *AIDS* **2002**, *16*, 2073–2078.
38. Butler, L.M.; Dorsey, G.; Hladik, W.; Rosenthal, P.J.; Brander, C.; Neilands, T.B.; Mbisa, G.; Whitby, D.; Kiepiela, P.; Mosam, A.; *et al.* Kaposi sarcoma-associated herpesvirus (kshv)

seroprevalence in population-based samples of african children: Evidence for at least 2 patterns of kshv transmission. *J. Infect. Dis.* **2009**, *200*, 430–438.

39. Malope, B.I.; MacPhail, P.; Mbisa, G.; MacPhail, C.; Stein, L.; Ratshikhopha, E.M.; Ndhlovu, L.; Sitas, F.; Whitby, D. No evidence of sexual transmission of kaposi's sarcoma herpes virus in a heterosexual south african population. *Aids* **2008**, *22*, 519–526.
40. Zhang, T.; Yang, Y.; Yu, F.; Zhao, Y.; Lin, F.; Minhas, V.; Wood, C.; He, N. Kaposi`s sarcoma associated herpesvirus infection among female sex workers and general population women in shanghai, china: A cross-sectional study. *BMC Infect. Dis.* **2014**, *14*, 58.
41. Zavitsanou, A.; Sypsa, V.; Petrodaskalaki, M.; Psichogiou, M.; Katsoulidou, A.; Boletis, J.; Hadjiconstantinou, V.; Karalis, D.; Kalapothaki, V.; Hatzakis, A. Human herpesvirus 8 infection in hemodialysis patients. *Am. J. Kidney Dis.* **2006**, *47*, 167–170.
42. Whitby, D.; Howard, M.R.; Tenant-Flowers, M.; Brink, N.S.; Copas, A.; Boshoff, C.; Hatzioannou, T.; Suggett, F.E.; Aldam, D.M.; Denton, A.S.; *et al.* Detection of kaposi sarcoma associated herpesvirus in peripheral blood of hiv-infected individuals and progression to kaposi's sarcoma. *Lancet* **1995**, *346*, 799–802.
43. Hladik, W.; Dollard, S.C.; Mermin, J.; Fowlkes, A.L.; Downing, R.; Amin, M.M.; Banage, F.; Nzaro, E.; Kataaha, P.; Dondero, T.J.; *et al.* Transmission of human herpesvirus 8 by blood transfusion. *N. Engl. J. Med.* **2006**, *355*, 1331–1338.
44. Hladik, W.; Pellett, P.E.; Hancock, J.; Downing, R.; Gao, H.; Packel, L.; Mimbe, D.; Nzaro, E.; Mermin, J. Association between transfusion with human herpesvirus 8 antibody-positive blood and subsequent mortality. *J. Infect. Dis.* **2012**, *206*, 1497–1503.
45. Lisco, A.; Barbierato, M.; Fiore, J.R.; Gasperini, P.; Favia, A.; Volpe, A.; Chironna, M.; Pastore, G.; Chieco-Bianchi, L.; Calabro, M.L. Pregnancy and human herpesvirus 8 reactivation in human immunodeficiency virus type 1-infected women. *J. Clin. Microbiol.* **2006**, *44*, 3863–3871.
46. Gutierrez-Ortega, P.; Hierro-Orozco, S.; Sanchez-Cisneros, R.; Montano, L.F. Kaposi's sarcoma in a 6-day-old infant with human immunodeficiency virus. *Arch. Dermatol.* **1989**, *125*, 432–433.
47. Ahlfors, K.; Ivarsson, S.A. Cytomegalovirus in breast milk of swedish milk donors. *Scand. J. Infect. Dis.* **1985**, *17*, 11–13.
48. Dworsky, M.; Yow, M.; Stagno, S.; Pass, R.F.; Alford, C. Cytomegalovirus infection of breast milk and transmission in infancy. *Pediatrics* **1983**, *72*, 295–299.
49. Fujisaki, H.; Tanaka-Taya, K.; Tanabe, H.; Hara, T.; Miyoshi, H.; Okada, S.; Yamanishi, K. Detection of human herpesvirus 7 (hhv-7) DNA in breast milk by polymerase chain reaction and prevalence of hhv-7 antibody in breast-fed and bottle-fed children. *J. Med. Virol.* **1998**, *56*, 275–279.
50. Hamprecht, K.; Vochem, M.; Baumeister, A.; Boniek, M.; Speer, C.P.; Jahn, G. Detection of cytomegaloviral DNA in human milk cells and cell free milk whey by nested pcr. *J. Virol. Methods* **1998**, *70*, 167–176.
51. Junker, A.K.; Thomas, E.E.; Radcliffe, A.; Forsyth, R.B.; Davidson, A.G.; Rymo, L. Epstein-barr virus shedding in breast milk. *Am. J. Med. Sci.* **1991**, *302*, 220–223.
52. Dedicoat, M.; Newton, R.; Alkharsah, K.R.; Sheldon, J.; Szabados, I.; Ndlovu, B.; Page, T.; Casabonne, D.; Gilks, C.F.; Cassol, S.A.; *et al.* Mother-to-child transmission of human herpesvirus-8 in south africa. *J. Infect. Dis.* **2004**, *190*, 1068–1075.

53. Brayfield, B.P.; Kankasa, C.; West, J.T.; Muyanga, J.; Bhat, G.; Klaskala, W.; Mitchell, C.D.; Wood, C. Distribution of kaposi sarcoma-associated herpesvirus/human herpesvirus 8 in maternal saliva and breast milk in zambia: Implications for transmission. *J. Infect. Dis.* **2004**, *189*, 2260–2270.

54. Boldogh, I.; Szaniszlo, P.; Bresnahan, W.A.; Flaitz, C.M.; Nichols, M.C.; Albrecht, T. Kaposi's sarcoma herpesvirus-like DNA sequences in the saliva of individuals infected with human immunodeficiency virus. *Clin. Infect. Dis.* **1996**, *23*, 406–407.

55. Koelle, D.M.; Huang, M.L.; Chandran, B.; Vieira, J.; Piepkorn, M.; Corey, L. Frequent detection of kaposi's sarcoma-associated herpesvirus (human herpesvirus 8) DNA in saliva of human immunodeficiency virus-infected men: Clinical and immunologic correlates. *J. Infect. Dis.* **1997**, *176*, 94–102.

56. Webster-Cyriaque, J.; Edwards, R.H.; Quinlivan, E.B.; Patton, L.; Wohl, D.; Raab-Traub, N. Epstein-barr virus and human herpesvirus 8 prevalence in human immunodeficiency virus-associated oral mucosal lesions. *J. Infect. Dis.* **1997**, *175*, 1324–1332.

57. Casper, C.; Krantz, E.; Selke, S.; Kuntz, S.R.; Wang, J.; Huang, M.L.; Pauk, J.S.; Corey, L.; Wald, A. Frequent and asymptomatic oropharyngeal shedding of human herpesvirus 8 among immunocompetent men. *J. Infect. Dis.* **2007**, *195*, 30–36.

58. Phipps, W.; Saracino, M.; Selke, S.; Huang, M.L.; Jaoko, W.; Mandaliya, K.; Wald, A.; Casper, C.; McClelland, R.S. Oral hhv-8 replication among women in mombasa, kenya. *J. Med. Virol.* **2014**, *86*, 1759–1765.

59. DeSantis, S.M.; Pau, C.P.; Archibald, L.K.; Nwanyanwu, O.C.; Kazembe, P.N.; Dobbie, H.; Jarvis, W.R.; Jason, J. Demographic and immune correlates of human herpesvirus 8 seropositivity in malawi, africa. *Int. J. Infect. Dis.* **2002**, *6*, 266–271.

60. Plancoulaine, S.; Abel, L.; Tregouet, D.; Duprez, R.; van Beveren, M.; Tortevoye, P.; Froment, A.; Gessain, A. Respective roles of serological status and blood specific antihuman herpesvirus 8 antibody levels in human herpesvirus 8 intrafamilial transmission in a highly endemic area. *Cancer Res.* **2004**, *64*, 8782–8787.

61. Minhas, V.; Crabtree, K.L.; Chao, A.; M'Soka, T.J.; Kankasa, C.; Bulterys, M.; Mitchell, C.D.; Wood, C. Early childhood infection by human herpesvirus 8 in zambia and the role of human immunodeficiency virus type 1 coinfection in a highly endemic area. *Am. J. Epidemiol.* **2008**, *168*, 311–320.

62. Alkharsah, K.R.; Dedicoat, M.; Blasczyk, R.; Newton, R.; Schulz, T.F. Influence of hla alleles on shedding of kaposi sarcoma-associated herpesvirus in saliva in an african population. *J. Infect. Dis.* **2007**, *195*, 809–816.

63. Olp, L.N.; Shea, D.M.; White, M.K.; Gondwe, C.; Kankasa, C.; Wood, C. Early childhood infection of kaposi's sarcoma-associated herpesvirus in zambian households: A molecular analysis. *Int. J. Cancer* **2012**, *132*, 1182–1190.

64. Wakeham, K.; Webb, E.L.; Sebina, I.; Nalwoga, A.; Muhangi, L.; Miley, W.; Johnston, W.T.; Ndibazza, J.; Whitby, D.; Newton, R.; *et al.* Risk Factors for Seropositivity to Kaposi Sarcoma-Associated Herpesvirus among Children in Uganda. *J. Acquir. Immune Defic. Syndr.* **2013**, *63*, 228–233.

65. Butler, L.M.; Were, W.A.; Balinandi, S.; Downing, R.; Dollard, S.; Neilands, T.B.; Gupta, S.; Rutherford, G.W.; Mermin, J. Human Herpesvirus 8 Infection in Children and Adults in a Population-Based Study in Rural Uganda. *J. Infect. Dis.* **2011**, *203*, 625–634.
66. Pfeiffer, R.M.; Wheeler, W.A.; Mbisa, G.; Whitby, D.; Goedert, J.J.; de The, G.; Mbulaiteye, S.M. Geographic Heterogeneity of Prevalence of the Human Herpesvirus 8 in Sub-Saharan Africa: Clues about Etiology. *Ann. Epidemiol.* **2010**, *20*, 958–963.
67. Coluzzi, M.; Manno, D.; Guzzinati, S.; Tognazzo, S.; Zambon, P.; Arca, B.; Costantini, C.; Ascoli, V. The bloodsucking arthropod bite as possible cofactor in the transmission of human herpesvirus-8 infection and in the expression of kaposi's sarcoma disease. *Parassitologia* **2002**, *44*, 123–129.
68. Lin, C.J.; Katongole-Mbidde, E.; Byekwaso, T.; Orem, J.; Rabkin, C.S.; Mbulaiteye, S.M. Intestinal parasites in kaposi sarcoma patients in uganda: Indication of shared risk factors or etiologic association. *Am. J. Trop. Med. Hyg.* **2008**, *78*, 409–412.
69. Whitby, D.; Marshall, V.A.; Bagni, R.K.; Miley, W.J.; McCloud, T.G.; Hines-Boykin, R.; Goedert, J.J.; Conde, B.A.; Nagashima, K.; Mikovits, J.; *et al.* Reactivation of kaposi's sarcoma-associated herpesvirus by natural products from kaposi's sarcoma endemic regions. *Int. J. Cancer* **2007**, *120*, 321–328.
70. Ziegler, J.L. Endemic kaposi's sarcoma in africa and local volcanic soils. *Lancet* **1993**, *342*, 1348–1351.
71. Renne, R.; Dittmer, D.; Kedes, D.; Schmidt, K.; Desrosiers, R.C.; Luciw, P.A.; Ganem, D. Experimental transmission of kaposi's sarcoma-associated herpesvirus (kshv/hhv-8) to siv-positive and siv-negative rhesus macaques. *J. Med. Primatol.* **2004**, *33*, 1–9.
72. Chang, H.; Wachtman, L.M.; Pearson, C.B.; Lee, J.S.; Lee, H.R.; Lee, S.H.; Vieira, J.; Mansfield, K.G.; Jung, J.U. Non-human primate model of kaposi's sarcoma-associated herpesvirus infection. *PLoS Pathog.* **2009**, *5*, e1000606.
73. Brehm, M.A.; Shultz, L.D.; Greiner, D.L. Humanized mouse models to study human diseases. *Curr. Opin. Endocrinol. Diabetes Obes.* **2010**, *17*, 120–125.
74. Wege, A.K.; Melkus, M.W.; Denton, P.W.; Estes, J.D.; Garcia, J.V. Functional and phenotypic characterization of the humanized blt mouse model. *Curr. Top. Microbiol. Immunol.* **2008**, *324*, 149–165.
75. Wang, L.X.; Kang, G.; Kumar, P.; Lu, W.; Li, Y.; Zhou, Y.; Li, Q.; Wood, C. Humanized-blt mouse model of kaposi's sarcoma-associated herpesvirus infection. *Proc. Natl. Acad. Sci. USA* **2014**, *111*, 3146–3151.
76. Labo, N.; Miley, W.; Marshall, V.; Gillette, W.; Esposito, D.; Bess, M.; Turano, A.; Uldrick, T.; Polizzotto, M.N.; Wyvill, K.M.; *et al.* Heterogeneity and breadth of host antibody response to kshv infection demonstrated by systematic analysis of the kshv proteome. *PLoS Pathog.* **2014**, *10*, e1004046.
77. Lennette, E.T.; Blackbourn, D.J.; Levy, J.A. Antibodies to human herpesvirus type 8 in the general population and in kaposi's sarcoma patients. *Lancet* **1996**, *348*, 858–861.
78. Kimball, L.E.; Casper, C.; Koelle, D.M.; Morrow, R.; Corey, L.; Vieira, J. Reduced levels of neutralizing antibodies to kaposi sarcoma-associated herpesvirus in persons with a history of kaposi sarcoma. *J. Infect. Dis.* **2004**, *189*, 2016–2022.

79. Inoue, N.; Spira, T.; Lam, L.; Corchero, J.L.; Luo, W. Comparison of serologic responses between kaposi's sarcoma-positive and -negative men who were seropositive for both human herpesvirus 8 and human immunodeficiency virus. *J. Med. Virol.* **2004**, *74*, 202–206.
80. Kumar, P.; Kuwa, N.Y.; Minhas, V.; Marimo, C.; Shea, D.M.; Kankasa, C.; Wood, C. Higher levels of neutralizing antibodies against KSHV in KS patients compared to asymptomatic individuals from Zambia. *PLoS One* **2013**, *8*, e71254.
81. Lepone, L.; Rappocciolo, G.; Knowlton, E.; Jais, M.; Piazza, P.; Jenkins, F.J.; Rinaldo, C.R. Monofunctional and polyfunctional CD8+ T cell responses to human Herpesvirus 8 lytic and latency proteins. *Clin. Vaccine Immunol.* **2010**, *17*, 1507–1516.
82. Robey, R.C.; Mletzko, S.; Gotch, F.M. The T-Cell immune response against kaposi's sarcoma-associated herpesvirus. *Adv. Virol.* **2010**, *2010*, 340356.
83. Carrieri, M.P.; Pradier, C.; Piselli, P.; Piche, M.; Rosenthal, E.; Heudier, P.; Durant, J.; Serraino, D. Reduced incidence of kaposi's sarcoma and of systemic non-hodgkin's lymphoma in hiv-infected individuals treated with highly active antiretroviral therapy. *Int. J. Cancer* **2003**, *103*, 142–144.
84. Franceschi, S.; Maso, L.D.; Rickenbach, M.; Polesel, J.; Hirschel, B.; Cavassini, M.; Bordoni, A.; Elzi, L.; Ess, S.; Jundt, G.; *et al.* Kaposi sarcoma incidence in the swiss hiv cohort study before and after highly active antiretroviral therapy. *Br. J. Cancer* **2008**, *99*, 800–804.
85. Ngalamika, O.; Minhas, V.; Wood, C. Kaposi's sarcoma at the university teaching hospital, Lusaka, Zambia in the antiretroviral therapy era. *Int. J. Cancer* **2014**, doi:10.1002/ijc.29184.
86. Bohlius, J.; Valeri, F.; Maskew, M.; Prozesky, H.; Garone, D.; Sengayi, M.; Fox, M.P.; Davies, M.A.; Egger, M. Kaposi's sarcoma in hiv-infected patients in south africa: Multicohort study in the antiretroviral therapy era. *Int. J. Cancer* **2014**, *135*, 2644–2652.
87. Stebbing, J.; Portsmouth, S.; Bower, M. Insights into the molecular biology and sero-epidemiology of kaposi's sarcoma. *Curr. Opin. Infect. Dis.* **2003**, *16*, 25–31.

11

PAN's Labyrinth: Molecular Biology of Kaposi's Sarcoma-Associated Herpesvirus (KSHV) PAN RNA, a Multifunctional Long Noncoding RNA

Cyprian C. Rossetto * and Gregory S. Pari

Department of Microbiology & Immunology, University of Nevada, Reno School of Medicine, Reno, NV 89557, USA; E-Mail: gpari@medicine.nevada.edu

* Author to whom correspondence should be addressed; E-Mail: crossetto@medicine.nevada.edu;

External Editor: Zhi-Ming Zheng

Abstract: Kaposi's sarcoma-associated herpesvirus (KSHV) is an oncogenic γ-herpesivrus, the causative agent of Kaposi's sarcoma and body cavity lymphomas. During infection KSHV produces a highly abundant long non-coding polyadenylated RNA that is retained in the nucleus known as PAN RNA. Long noncoding RNAs (lncRNA) are key regulators of gene expression and are known to interact with specific chromatin modification complexes, working in *cis* and *trans* to regulate gene expression. Data strongly supports a model where PAN RNA is a multifunctional regulatory transcript that controls KSHV gene expression by mediating the modification of chromatin by targeting the KSHV repressed genome.

Keywords: Kaposi's sarcoma-associated herpesvirus (KSHV); gammaherpesvirus; long noncoding RNA; PAN; histone modification; chromatin; epigenetics; gene expression

---

1. Introduction

Kaposi's sarcoma-associated herpesvirus (KSHV) was first identified as the etiological agent of Kaposi's sarcoma in 1994 [1], and it wasn't long after that before researchers first identified a highly abundant viral RNA from cells that were infected with KSHV [2,3]. Originally termed T1.1 or nut-1 RNA (U-RNA like nuclear transcript), this highly abundant RNA was identified as a 1.1-kb polyadenylated viral transcript. The coding potential of T1.1 was dismissed based on two factors. First, there are a few possible open reading frames but they are at suboptimal positions within the transcript and would code for very small proteins. The second reason T1.1 was extremely unlikely to have any coding potential was its location within the cell; it was found to be primarily located in the nucleus of infected cells. Cytoplasmic and nuclear fractionation studies of 293T cells transfected with a T1.1 expression plasmid demonstrated that it was only detectable in the nuclear fraction and *in situ* hybridization studies also localized PAN RNA to the nucleus of KSHV infected cells [2]. These same studies also quantitated the relative copy number of T1.1 transcripts to ~25,000 copies per cell based on the signal intensity of the same cells with cellular U12 RNA. During lytic infection PAN RNA is estimated to account for over 80% of the viral transcriptome. This apparent high expression level allows for the possibility that PAN RNA is a major factor with respect to contribution to KSHV replication and growth.

The PAN gene locus is found in the KSHV genome between K6 and ORF16, and slightly overlaps it's 5' end with the 3' end of the K7 ORF (see Figure 1). PAN RNA corresponds to nucleotides 28,661–29,741 (accession number U75698.1) in the KSHV genome. Molecular analysis revealed a strong K-Rta responsive promoter upstream of the PAN RNA start site, and a poly(A) signal sequence at the 3' to promote RNA stability.

More recent sequencing analysis using next generation sequencing technology has offered further insights into PAN RNA. A 2014 study by Arias *et al.* demonstrated that during lytic induction PAN RNA accounts for 65.5% of the KSHV reads as soon as 8 h post induction (hpi), at 24 hpi PAN accounts for 78.5%–92.2% of KSHV reads, and at 48 hpi PAN accounts for 84.8%–91.9% of KSHV reads [4]. There have also been reports of PAN RNA detected in latently infected, non-induced cells. Detection of PAN RNA during latency has often been attributed to spontaneous lytic reactivation [4]. Work from our laboratory demonstrated that a significant amount of PAN RNA is expressed from a virus mutant defective for expression of the major transactivator K-Rta [5]. Additionally, we showed that we could detect PAN RNA in uninduced BCBL-1 cells and PFA treated BCBL-1 cells.

There has also been a report of a transcript that is antisense to PAN and K7, termed K7.3, discovered using tiling microarray analysis [6]. Using next generation sequencing techniques along with a strand oriented RNA-seq library preparation kit we were unable to detect any appreciable amounts of the antisense PAN/K7 transcript. Due to differing isolation techniques and treatment of the RNA, it remains to be seen whether this antisense transcript is a true RNA species with viral kinetics that can be classified into a temporal category.

2. Long Noncoding RNAs (lncRNAs)

The discovery of lncRNAs has lead to increasing evidence that they play a prominent role in human disease [7]. Many examples now exist where specific lncRNAs can reprogram chromatin and promote disease or changes in gene expression [8–12]. The ability of these novel transcripts to regulate gene expression is an exciting new area for research. Although the field of lncRNAs is still developing, several mechanisms have been postulated as to how lncRNAs can impact gene expression [7,13,14]. There are four main proposed mechanisms of action for lncRNAs, these include acting as a decoy, a scaffold, a guide, or an enhancer. A decoy lncRNA acts as a "sponge" to isolate proteins, such as transcription factors or chromatin modifiers, away from their normal substrates. A scaffold lncRNA is used to assemble protein complexes; usually these types of lncRNAs will have unique domains to bind different proteins creating a complex that is functionally distinct from its individual parts. A guide lncRNA acts to localize specific proteins, often this can be a specific chromatin target to cause changes in gene expression. An enhancer lncRNA facilities the signals from transcription factors to regulate gene expression. Another interesting model of lncRNA's, which goes along with its mechanisms of action, is that they can regulate nuclear organization to orchestrate and traffic protein complexes, genes, and chromosomes to appropriate locations allowing for proper activation or repression [15].

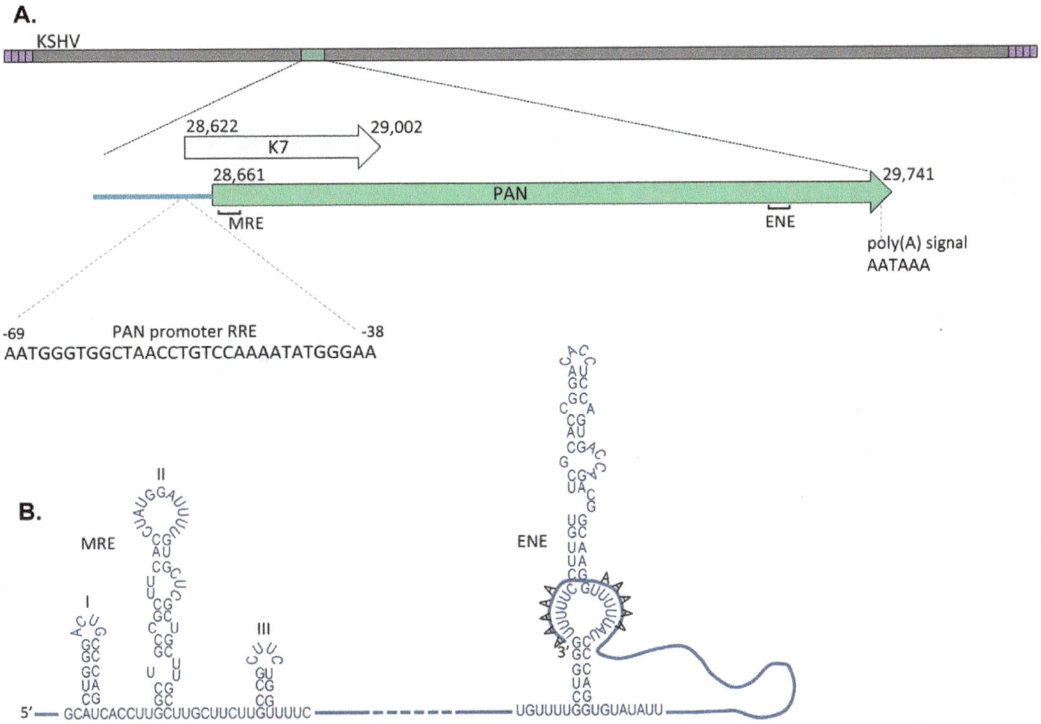

Figure 1. Schematic diagram depicting key features of the PAN locus and transcript. (**A**) The nucleotide numbering for the K7 ORF and PAN transcript corresponds to genbank accession number U75698.1 for HHV8/KSHV genome type M isolated from BC-1 cells. The numbering for the RRE within the PAN promoter and the ENE is from the transcription start site; (**B**) The secondary structure of the ENE and MRE sequence within PAN RNA to prevent degradation and increase stability.

lncRNAs associate with chromatin modifying complexes [12] and interact with components of the polycomb repressive complex 2 (PRC2). The PRC2 complex is composed of EZH2, SUZ12 and EED-1. EZH2 is a protein that adds three methyl groups to lysine 27 of histone 3 [16]. SUZ12 is a protein that contains a zinc finger domain that is the point of contact with RNA [17]. EED-1 interacts with HDAC1 and histone deacetylase and various other proteins to mediate gene repression [18]. These PRC2 proteins can mediate changes in histone modifications (methylation) and subsequent repression of gene expression from various genetic loci. Hence the interaction of lncRNAs with PRC2 can globally influence gene expression.

Although the ability of lncRNAs to repress gene expression is well documented, more recent reports suggest that they are multifunctional having the ability to activate, as well as suppress gene expression.

One interesting example of this is the lncRNA, lincRNA-Cox2, which can both activate and repress gene expression associated with immune response [19].

While thousands of lncRNAs with diverse functions have been identified in mammalian cells, the exact function or mechanisms of action for most of these transcripts are unknown. Several technological advances have been described that can be used to study lncRNAs. The use of techniques, such as Chromatin Isolation by RNA purification (ChIRP), and the more recent domain-specific chromatin isolation by RNA purification (dChIRP), are invaluable with respect to mapping specific lncRNA DNA interaction domains [20,21].

3. PAN Expression during Lytic Reactivation and Latency

The expression of PAN RNA during KSHV lytic infection is regulated by major viral transactivator K-Rta through the interaction of a specific DNA interaction domain referred to as Rta-Responsive Element (RRE) [22,23]. The RRE is located between nts −69 and −38 of the PAN RNA promoter and directly interacts with K-Rta to activate transcription [24] (Figure 1A). *In vitro* and *in vivo* studies of the interaction between the PAN promoter and K-Rta protein have demonstrated that K-Rta has an extremely high affinity for the PAN promoter, even under stringent high-salt conditions [22,24]. PAN RNA is stabilized by the interaction of ORF57 [25]. PAN RNA expression is also upregulated and stabilized by an interaction of ORF57 with PABPC1 [26]. Interestingly, in transient assays the viral encoded K-bZIP, which was shown to interact with K-Rta, suppresses transcription from many KSHV promoters has no affect on the PAN RNA promoter [27]. In addition to K-Rta mediated expression of PAN RNA, it was shown that the cellular factor CAAT/enhancer-binding protein alpha (C/EBP) cooperates with K-Rta to activate PAN RNA expression [28].

Although K-Rta activates PAN RNA expression upon reactivation, it is clear that a K-Rta independent mechanism also exists. Recombinant viruses defective for K-Rta expression still express PAN RNA, suggesting that PAN RNA is present during all phases of the KSHV life cycle [5]. It should also be noted that although PAN RNA is present at early times post infection, a significant amount of the RNA in packaged virions is PAN RNA [5]. This observation reinforces the concept that PAN RNA is a ubiquitously present regulatory factor in KSHV infected cells. Recent studies show that PAN RNA interacts with the KSHV latency associated nuclear antigen, LANA [29]. Studies show that the *N*-terminus (as 1–70) of LANA are involved in the interaction with PAN RNA. Also, the 3' end of PAN RNA contained within nucleotides 750 to 1062 was shown to be the critical interaction domains that bind to LANA. This interaction is thought to serve to sequester LANA away from viral DNA episomes to facilitate lytic reactivation.

4. *Cis*-Acting Elements, Triple Helix Structure and Localization of PAN RNA

The relative abundance and localization of PAN RNA was speculated to be attributed to specific physical characteristic unique to PAN RNA, further studies would go on to demonstrate that PAN RNA contains at least two *cis*-acting elements that contribute to it's stability and localization. One of the early insights into the function of PAN RNA comes from the elucidation of an RNA element within PAN RNA that mediates nuclear retention. PAN RNA contains an expression and nuclear retention element (ENE) that allows for formation of a triple helix, such that the Poly(A) tail of PAN RNA is sequestered, retaining the transcript in the nucleus [30–32]. There have been many interesting studies that have focused on discerning the structure of PAN RNA that allows for it's unique stability and nuclear retention.

The ENE 79-nt sequence within PAN RNA was originally described and characterized by Conrad and Steitz [31]. They reasoned that because PAN RNA was expressed at such high levels, and was not spliced or exported to the cytoplasm, they should look for the presence of a *cis*-acting element that could confer those unique characteristics to PAN RNA. They also performed studies to insert the PAN ENE within an intronless β-globin mRNA, and demonstrated that the fusion RNA transcript accumulated to high levels and was retained in the nucleus. Using the β-globin-ENE as a type of reporter RNA, they also demonstrated that tethering of export factors or splicing could overcome the retention signal that the ENE confers to the RNA. One interesting caveat was noted about the ENE, and most likely points to other *cis*-acting elements within PAN RNA, multiple copies of the ENE were required to increase the amount of the β-globin transcript retained in the nucleus and deletion of the ENE within PAN did not cause an increase in PAN RNA's cytoplasmic accumulation.

The identification of the ENE within the 3' end of PAN RNA eventually lead to physical analysis of PAN RNA's secondary structure. When the ENE was first discovered there was speculation that the unpaired stretches of U's in the ENE might hybridize with the poly(A) tail and protect the polyadenylated 3' end from exonucleolytic attack, and this would account for one of the reasons why PAN RNA was so stable [33]. Further investigation revealed that it wasn't just the amount or number of U's in the ENE that accounted for the accumulation of PAN RNA, there was a structural stem-loop component of the ENE that added to the stability of PAN RNA. The crystal structure of the ENE core with the poly(A) sequence revealed that the U-rich loop of the ENE sequestered the poly(A) tail by the formation of a major-groove triple helix thereby blocking the initiation of RNA decay [30].

Identifying the ENE and triple helix formation within PAN RNA has lead to recognizing similar ENE types of *cis*-acting elements within other lncRNA's. Using bioinformatics and computation approaches has aided in the identification of ENE *cis*-elements in other viruses such as rhesus rhadinovirus (RRV) and equine herpesvirus 2 (EHV2) [32]. Further studies looking at cellular lncRNA's have found an ENE and bipartite triple helix within metastasis-associated lung adenocarcinoma transcript 1 (MALAT1) transcript [34,35]. Secondary structure for RNA's, and lncRNA's in particular, play an important role in their regulation, stability, and functional targeting. Find additional *cis*-elements like the ENE and triple helix formation will help us better understand the molecular interactions between lncRNA's and their substrates.

Despite the presence of the ENE, more recent evidence supports stabilization of PAN RNA through a 9-nt element termed the MRE (Mta-responsive element) [25]. ORF57, the viral encoded Mta protein, was identified as a possible factor in modulating PAN RNA metabolism. The relative locations of MRE and ENE are shown in Figure 1B. Interestingly, in the absence of viral infection and other viral encoded factors, PAN RNA is stably expressed in cell lines and interacts with cellular DNA to activate or repress gene expression [5,36]. There have been many reports that support the interaction of PAN RNA with ORF57 [25,26,37–39]. Studies have described the ability of ORF57 to prevent the decay of PAN RNA even in a PAN-ΔENE mutant, and identified a larger 300 bp section called the ORF57-responsive element (ORE) that encompasses the smaller MRE [39]. Deletion studies of the ENE and MRE have clearly shown that the ENE plays only a small role in ORF57-mediated PAN stability, while the major function of the MRE is to increase the half-life of PAN RNA by its interaction with ORF57 [25]. This same study also reported that approximately 20% of PAN RNA was exported to the cytoplasm, and by using a plasmid co-transfection system of PAN RNA and ORF57, they were also able to detect 30% of PAN RNA in the cytoplasm when ORF57 was co-expressed along with PAN. While studies that used the larger 300 bp ORE were able to demonstrate that the ORE was able to confer an ORF57 depend accumulation on heterologous intronless mRNA [39]. Insertion of the MRE into other heterologous mRNA, such as vGPCR and luciferase that are normally insensitive to the effects of ORF57, showed only minimal increase in mRNA accumulation [25]. Insertion of the minimal MRE along with the ENE also showed no added increase in mRNA accumulation [25], indicating that there may be other *cis*-elements or motifs within PAN RNA that contribute to its extremely high abundance.

5. PAN RNA as a Regulatory Factor

As regulatory RNA, PAN associates with several viral and cellular factors. Early reports described the interaction of PAN RNA with KSHV encoded ORF57 [37,38]. ORF57 increases the stability of PAN RNA and increases its nuclear concentration [25,39]. The interaction of ORF57 is specific and mediated through a core MRE found within the 5'-end of PAN RNA [25,40]. The mechanism involved in ORF57 mediated stability of PAN RNA involves the recruitment of factors composed of the TREX complex [41]. PAN RNA interacts with cellular polyadenylate-binding protein C1 (PABPC1), which binds to PAN RNA through the MRE located in the 5'-end of the PAN RNA transcript and has a negative affect on PAN RNA stability [25,26].

Studies involving the elucidation of a mechanism for the increased stability and nuclear localization of PAN RNA were significant. However, none of these early endeavors unlocked the mystery with respect to the function of PAN RNA in infected cells. Proteomics approaches from the Steitz lab and our lab [42,43], using two slightly different procedures to isolated proteins interacting with PAN RNA, provided some insight into the binding partners of PAN RNA. The first studies involving a proposed function for PAN RNA revealed that depletion of PAN RNA using antisense in infected cells impaired KSHV late gene expression and decreased the amount of infectious virus [42]. Subsequent studies showed that PAN RNA also interacted with histones H1 and H2A, mitochondrial and cellular single-stranded binding proteins and interferon regulatory factor 4 (IRF4) and KSHV ORF59 [43]. Transient assays showed that PAN RNA interfered with the ability of IRF4/PU.1 to activate the interleukin-4 (IL-4) promoter. Evaluation of the expression of other cellular immune response genes also showed a decrease in expression of other factors involved in immune response [43]. These studies illuminated the first clues with respect to a possible function for PAN RNA, a modulator of immune and inflammatory response.

6. PAN RNA and Viral Replication

The ability of PAN RNA to disregulate the expression of immune response genes and those involved in formation of the inflammasome suggested that PAN RNA had a global affect on cellular and viral gene expression programs. To further investigate the role of PAN RNA in KSHV growth, we generated a recombinant KSHV BAC with a large deletion in the PAN RNA gene [44]. This viral mutant failed to express most of the viral lytic gene expression program under reactivation conditions. The over-expression of K-Rta (ORF50), the major viral transactivator protein, failed to rescue the observed gene expression defect, suggesting that PAN RNA was required for full activation of KSHV gene expression and virus production [44]. The mechanisms involved in the ability of PAN RNA to activate KSHV and cellular gene expression is directly related to the interaction of PAN RNA with chromatin modifying complexes. PAN RNA binds to UTX and JMJD3, demethylases that specifically removes the methyl groups from the repressive H3K27me3 mark, and MLL2, a histone-lysine *N*-methyltransferase, which methylates H3 at the K4 position, a mark associated with gene activation [44–46]. ChIRP-Seq showed that PAN RNA occupies much of the KSHV genome during lytic infection including the PAN RNA gene promoter itself. PAN RNA also interacts with protein components that comprise the polycomb repression complex 2 (PRC2). It was previously demonstrated that during latency the KSHV genome is composed of both activating and repressive histone marks [47]. However, most of the latent genome is associated with repressive H3K27me3 and H3K9me3 repressive marks with the ORF50 promoter and some early gene promoters associated with both repressive and H3K4me3 and acetylated H3 activation marks. It could be postulated that PAN RNA mediates latency or reactivation through its interaction with different chromatin modifying complexes, since PAN RNA is present in packaged virions and is expressed in the absence of K-Rta. Although the PAN RNA deletion mutant does establish latency in HEK293 cells, it is possible that in the appropriate target cells PAN RNA may be required to promote KSHV latency.

7. Targeting of PAN RNA to KSHV Genomic DNA

One of the least known mechanisms for lncRNAs is how these molecules are targeted to specific DNA loci. For KSHV, some insight into this targeting mechanism could lie in the observation that PAN RNA interacts with ORF59 [43]. KSHV ORF59 encodes the polymerase processivity factor [48,49]. Interestingly, the human cytomegalovirus (HCMV) lncRNA4.9 also interacts with the HCMV encoded processivity factor UL44 [50]. Although originally described as a binding partner for herpesvirus-encoded polymerases, processivity factors are emerging as having diverse functions in viral replication and gene expression [51–58]. For KSHV, ORF59 is associated with the same regions of the viral genome as PAN RNA occupancy, strongly suggesting that ORF59 is involved in the targeting mechanism or anchoring of PAN RNA to specific regions of the KSHV chromosome. Hence PAN RNA, like most lncRNAs, has a complex three-dimensional structure that allows for specific protein binding and RNA-DNA hybridization through openly configured single-stranded sequences, which cannot be predicted by the primary RNA sequence. These two characteristics of lncRNAs, protein binding and recognition of specific DNA sequences, imparts functionality.

8. PAN RNA and Cellular Gene Expression

A later investigation of the ability of PAN RNA to destabilize cellular gene expression showed that PAN RNA interacts with polycomb group proteins, which are associated with repression of gene expression [5]. Interestingly, PAN RNA was shown to globally disregulate cellular gene expression programs with a concentration on the expression of factors involved in inflammation, immune response and cell cycle. Cells that constitutively express PAN RNA grow to a higher density and have shorter doubling times consistent with an increase in cell survival [5]. This dramatic increase in cell survival mediated by PAN RNA allows for increased viral fitness. PAN RNA expressing cell lines can be efficiently infected with KSHV and are easily transfected with KSHV BACmids, suggesting that PAN RNA promotes a cellular environment highly advantageous to viral growth [59].

As mentioned previously, in addition to interacting with chromatin modifying complexes that activate gene expression, PAN RNA interacts with factors associated with the PRC2 [5]. Hence, the interaction with PCR2 factors as well as UTX. JMJD3 and MLL2 (KMTD2), a histone methyltransferase, allows for PAN RNA to both activate and repress gene expression (see Figure 2) [44]. The differential association of PAN RNA with specific chromatin modifying complexes may be key to understanding the mechanism of action for PAN RNA during different phases of the KSHV life cycle. Although the precise mechanism involved in the modulation of PAN RNA mediated activation or repression is unknown, ChIRP-Seq results clearly show that PAN RNA interacts with hundreds of cellular promoters for genes that express factors involved in a wide spectrum of host functions.

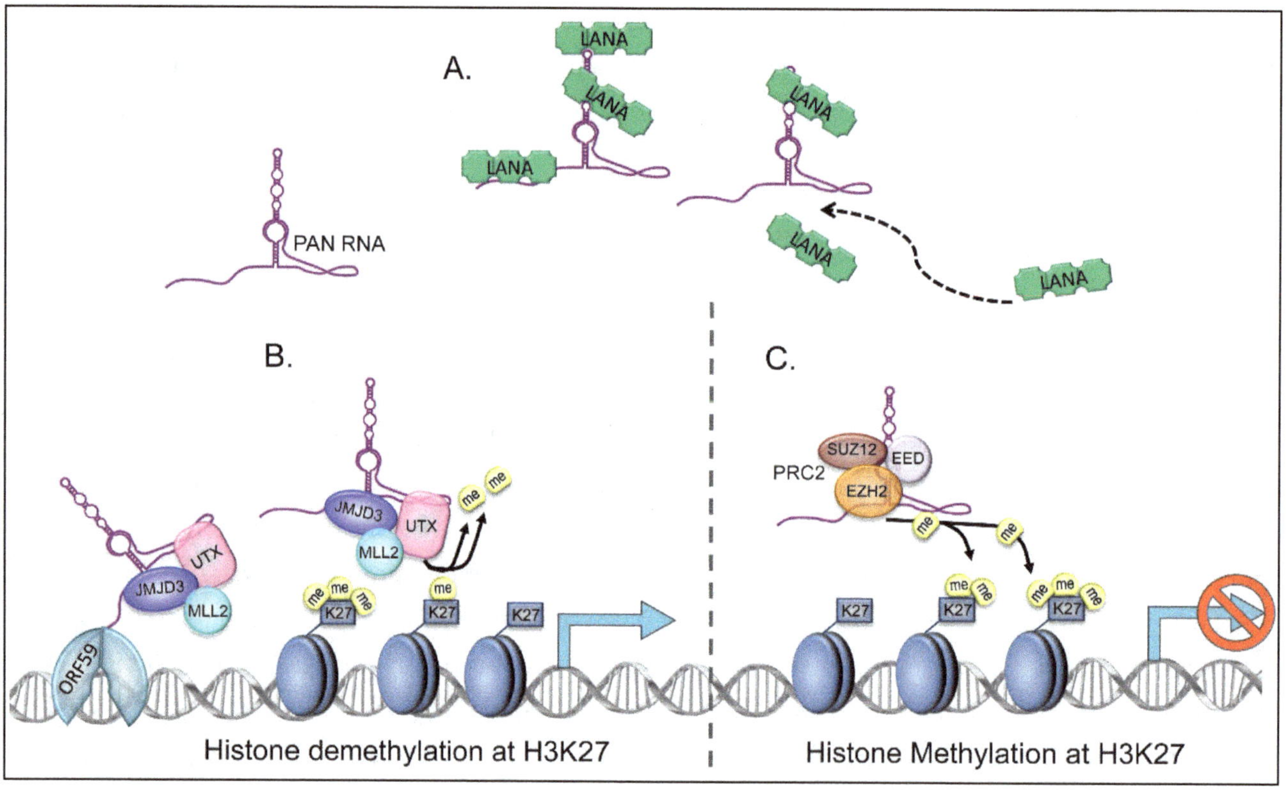

Figure 2. PAN RNA both activates and represses gene expression depending on its interaction with chromatin modifying complexes. (**A**) The interaction of PAN RNA with LANA during reactivation is partly responsible for the LANA-episome dissociation allowing for the expression of lytic transcripts. In this model PAN RNA acts as a "sponge" to keep LANA from repressing viral gene expression; (**B**) The interaction of PAN RNA with UTX/JMJD3 targets H3K37 for demethylation and acts to increase gene expression. The interaction of PAN RNA with viral encoded processivity factor ORF59 may aid in the targeting UTX/JMJD3 to viral genes during lytic gene expression; (**C**) PAN RNA's interaction with PRC2 targets H3K27 for methylation and acts to repress gene expression. The mechanism for specific targeting of PAN RNA is currently under investigation.

9. Coding Potential of PAN RNA

There has always been interest in the speculation about the coding potential of PAN RNA. In the early studies on PAN RNA, it was originally localized to the nucleus, lacked a 5' trimethylguanosine cap, and did not associated with ribosomes [3]. For these reasons, plus the fact that within PAN RNA there are stretches of sequence similar to cellular non-coding RNA U1, most researchers assumed that PAN RNA was non-coding. Proteomics LC-MS/MS analysis from latent BCBL-1, latent and lytic BC-1 cells did not detect any peptides that correspond to the potential PAN coding regions, however it was acknowledged in the same study that the proteomics is limited and may not fully characterize under-represented proteins [6]. The most recent development in the coding potential of PAN RNA comes from a study that used mRNA-Seq, along with ribosome footprinting (Ribo-Seq) and DNA-Seq, to compile a comprehensive analysis of viral gene expression [4]. The comprehensive analysis not only included a time course including during latency and following lytic reactivation, but also chemical treatment with harringtonine to arrest the progression of initiating ribosomes and cycloheximide (CHX) to arrest elongating ribosomes. Interestingly the PAN RNA followed classical ribosome protection patterns, they found initiating ribosomes accumulating at the start codons in the samples treated with harringtonine, elongating ribosomes in the body of the transcript in CHX treated samples, and releasing ribosomes at the stop codon in the samples that were not treated with an translation inhibitor. In this study the three predominant putative PAN ORFs are termed PAN1.1 (37aa), PAN1.2 (44aa), and PAN1.3 (25aa). The calculations of the ribosome release score (RSS) are comparable to those of known coding RNAs, and the low translation efficiency of the three major PAN ORFs is offset by the significant amount of PAN RNA that accumulates during the lytic cycle. This study calculated the ribosome-protected RNA corresponding to the small PAN peptides represented up to 1.7% of the total CHX Ribo-Seq reads, strongly suggesting that the putative peptide could be quite abundant.

Although the coding potential of PAN RNA is debatable, there is agreement that because PAN RNA is extremely abundant and that there is more than enough RNA to be translated and act as a long non-coding RNA. Those arguing against the coding potential of PAN would cite the fact that early studies of PAN RNA did not find it associated with a 5' trimethylguanosine cap [3], the 5' trimethylguanosine cap is most often associated with cellular small nuclear RNA's such as U1 RNA, which was used as the control in the original study. The lack of a 5' cap means that it would most likely not be exported out of the nucleus using the nuclear pore complex and if cytoplasmic localization were achieved, it would have to rely on cap-independent translation such as an internal ribosome entry site (IRES). Later studies have often noted PAN RNA has a 5' cap [30,60], but a search of relevant literature shows a lack of experimental proof that PAN RNA did not associate with a trimethylguanosine cap. Interestingly, one study focusing on developing the function of the ENE, found that the effects of the ENE were similar whether they used GpppG or m^{7}GpppG capped substrates but the data was not shown [60]. The exact state of the 5' cap on PAN RNA has not been completely confirmed or disproven, and the ambiguity may lie between not having a 5' trimethylguanosine cap [3] and possibly having a 5' monomethylguanosine cap [60]. Since PAN RNA is transcribed by RNA polymerase II, capping enzymes associated with the

RNA pol II after the CTD of RNA pol II has been phosphorylated and during the elongation phase of RNA pol II, should cap the 5' end during the transcription resulting in a 5' monomethylguanosine cap.

It has been shown that PAN RNA is polyadenylated and binds to poly(A)-binding protein C1 (PABPC1) after PABPC1 is relocalized to the nucleus during lytic infection [42]. This is particularly interesting because PABPC1 is normally found in the cytoplasm bound to poly(A) tails of mRNA and regulates mRNA stability and translation. The fact that PABPC1 binds to PAN RNA in the nucleus during a lytic infection could be indicative of the virus manipulating the host environment to favor the export and expression of viral mRNAs. A number of studies using *in situ* hybridization [3,39,42] and subcellular fractionation [2] localized PAN RNA to the nucleus. In BCBL1 cells treated with valproate to induce the virus to enter the lytic cycle, approximately 20% of PAN RNA was exported to the cytoplasm. The fact that most PAN RNA resides in the nucleus and not in the cytoplasm argues against proteins being translated from PAN RNA.

10. Conclusions

Since the discovery of PAN RNA as the most abundant viral transcript during a KSHV infection, there has been much effort put forth to uncover how PAN RNA achieves and maintains a very high steady state level of accumulation. Molecular analysis revealed PAN RNA is transcribed from a K-Rta responsive promoter, which binds K-Rta at a higher affinity than other K-Rta responsive promoters, resulting in higher levels of transcription. The other key to PAN RNA's abundance is conferred by it's unique secondary structure of the ENE which sequesters the poly(A) tail to prevent degradation and the interaction with ORF57 in infected cells. As work was going forward to discern PAN's molecular characteristics, at the same time there were theories about what functional benefit PAN RNA may impart to the virus. Our lab and other's have demonstrated that PAN RNA is essential for viral late gene expression, as shown using a virus lacking the PAN locus and alternatively using antisense oligonucleotides to degrade PAN RNA. PAN RNA can regulate cellular and viral gene expression by targeting chromatin-modifying complexes for either activation or repression. This fact cannot be overstated since PAN RNA is the most abundant transcript in lytically infected cells and is present in the absence of ORF50 (K-Rta) expression. Therefore PAN RNA has the potential to function as a master regulator of viral gene expression.

It is now well established that lncRNAs participate in a wide variety of biological processes. New and better techniques and reagents are developed to decipher the mechanism of action for lncRNAs. Nevertheless it is clear that lncRNAs have the capacity to act as decoys, scaffolds, guides, or enhancers. Viral lncRNAs are only very recently being recognized as playing a significant role in regulation of gene expression similar to their cellular counterparts. For viral lncRNAs, their high abundance compared to the expression levels of protein coding mRNAs makes them hard to ignore as modulators of gene expression and chromatin modification.

References and Notes

1. Chang, Y.; Cesarman, E.; Pessin, M.S.; Lee, F.; Culpepper, J.; Knowles, D.M.; Moore, P.S. Identification of herpesvirus-like DNA sequences in AIDS-associated Kaposi's sarcoma. *Science* **1994**, *266*, 1865–1869.
2. Zhong, W.; Wang, H.; Herndier, B.; Ganem, D. Restricted expression of Kaposi sarcoma-associated herpesvirus (human herpesvirus 8) genes in Kaposi sarcoma. *Proc. Natl. Acad. Sci. USA* **1996**, *93*, 6641–6646.
3. Sun, R.; Lin, S.F.; Gradoville, L.; Miller, G. Polyadenylylated nuclear RNA encoded by Kaposi sarcoma-associated herpesvirus. *Proc. Natl. Acad. Sci. USA* **1996**, *93*, 11883–11888.
4. Arias, C.; Weisburd, B.; Stern-Ginossar, N.; Mercier, A.; Madrid, A.S.; Bellare, P.; Holdorf, M.; Weissman, J.S.; Ganem, D. KSHV 2.0: A Comprehensive Annotation of the Kaposi's Sarcoma-Associated Herpesvirus Genome Using Next-Generation Sequencing Reveals Novel Genomic and Functional Features. *PLoS Pathog.* **2014**, *10*, e1003847.
5. Rossetto, C.C.; Tarrant-Elorza, M.; Verma, S.; Purushothaman, P.; Pari, G.S. Regulation of viral and cellular gene expression by Kaposi's sarcoma-associated herpesvirus polyadenylated nuclear RNA. *J. Virol.* **2013**, *87*, 5540–5553.
6. Dresang, L.R.; Teuton, J.R.; Feng, H.; Jacobs, J.M.; Camp, D.G., 2nd; Purvine, S.O.; Gritsenko, M.A.; Li, Z.; Smith, R.D.; Sugden, B.; *et al.* Coupled transcriptome and proteome analysis of human lymphotropic tumor viruses: Insights on the detection and discovery of viral genes. *BMC Genomics* **2011**, *12*, 625.
7. Wapinski, O.; Chang, H.Y. Long noncoding RNAs and human disease. *Trends Cell Biol.* **2011**, *21*, 354–361.
8. Gupta, R.A.; Shah, N.; Wang, K.C.; Kim, J.; Horlings, H.M.; Wong, D.J.; Tsai, M.C.; Hung, T.; Argani, P.; Rinn, J.L.; *et al.* Long non-coding RNA HOTAIR reprograms chromatin state to promote cancer metastasis. *Nature* **2010**, *464*, 1071–1076.
9. Willingham, A.T.; Orth, A.P.; Batalov, S.; Peters, E.C.; Wen, B.G.; Aza-Blanc, P.; Hogenesch, J.B.; Schultz, P.G. A strategy for probing the function of noncoding RNAs finds a repressor of NFAT. *Science* **2005**, *309*, 1570–1573.
10. Khalil, A.M.; Rinn, J.L. RNA-protein interactions in human health and disease. *Semin. Cell Dev. Biol.* **2011**, *22*, 359–365.
11. Huarte, M.; Guttman, M.; Feldser, D.; Garber, M.; Koziol, M.J.; Kenzelmann-Broz, D.; Khalil, A.M.; Zuk, O.; Amit, I.; Rabani, M.; *et al.* A large intergenic noncoding RNA induced by p53 mediates global gene repression in the p53 response. *Cell* **2010**, *142*, 409–419.
12. Khalil, A.M.; Guttman, M.; Huarte, M.; Garber, M.; Raj, A.; Rivea Morales, D.; Thomas, K.; Presser, A.; Bernstein, B.E.; van Oudenaarden, A.; *et al.* Many human large intergenic noncoding RNAs associate with chromatin-modifying complexes and affect gene expression. *Proc. Natl. Acad. Sci. USA* **2009**, *106*, 11667–11672.
13. Wang, K.C.; Yang, Y.W.; Liu, B.; Sanyal, A.; Corces-Zimmerman, R.; Chen, Y.; Lajoie, B.R.; Protacio, A.; Flynn, R.A.; Gupta, R.A.; *et al.* A long noncoding RNA maintains active chromatin to coordinate homeotic gene expression. *Nature* **2011**, *472*, 120–124.

14. Tsai, M.C.; Manor, O.; Wan, Y.; Mosammaparast, N.; Wang, J.K.; Lan, F.; Shi, Y.; Segal, E.; Chang, H.Y. Long noncoding RNA as modular scaffold of histone modification complexes. *Science* **2010**, *329*, 689–693.
15. Batista, P.J.; Chang, H.Y. Long noncoding RNAs: Cellular address codes in development and disease. *Cell* **2013**, *152*, 1298–1307.
16. Cao, R.; Wang, L.; Wang, H.; Xia, L.; Erdjument-Bromage, H.; Tempst, P.; Jones, R.S.; Zhang, Y. Role of histone H3 lysine 27 methylation in Polycomb-group silencing. *Science* **2002**, *298*, 1039–1043.
17. Rinn, J.L.; Kertesz, M.; Wang, J.K.; Squazzo, S.L.; Xu, X.; Brugmann, S.A.; Goodnough, L.H.; Helms, J.A.; Farnham, P.J.; Segal, E.; *et al.* Functional demarcation of active and silent chromatin domains in human HOX loci by noncoding RNAs. *Cell* **2007**, *129*, 1311–1323.
18. Van der Vlag, J.; Otte, A.P. Transcriptional repression mediated by the human polycomb-group protein EED involves histone deacetylation. *Nat. Genet.* **1999**, *23*, 474–478.
19. Carpenter, S.; Aiello, D.; Atianand, M.K.; Ricci, E.P.; Gandhi, P.; Hall, L.L.; Byron, M.; Monks, B.; Henry-Bezy, M.; Lawrence, J.B.; *et al.* A long noncoding RNA mediates both activation and repression of immune response genes. *Science* **2013**, *341*, 789–792.
20. Chu, C.; Qu, K.; Zhong, F.L.; Artandi, S.E.; Chang, H.Y. Genomic maps of long noncoding RNA occupancy reveal principles of RNA-chromatin interactions. *Mol. Cell* **2011**, *44*, 667–678.
21. Quinn, J.J.; Ilik, I.A.; Qu, K.; Georgiev, P.; Chu, C.; Akhtar, A.; Chang, H.Y. Revealing long noncoding RNA architecture and functions using domain-specific chromatin isolation by RNA purification. *Nat. Biotechnol.* **2014**, *32*, 933–940.
22. Song, M.J.; Brown, H.J.; Wu, T.T.; Sun, R. Transcription activation of polyadenylated nuclear RNA by rta in human herpesvirus 8/Kaposi's sarcoma-associated herpesvirus. *J. Virol.* **2001**, *75*, 3129–3140.
23. Chang, P.J.; Shedd, D.; Gradoville, L.; Cho, M.S.; Chen, L.W.; Chang, J.; Miller, G. Open reading frame 50 protein of Kaposi's sarcoma-associated herpesvirus directly activates the viral PAN and K12 genes by binding to related response elements. *J. Virol.* **2002**, *76*, 3168–3178.
24. Song, M.J.; Li, X.; Brown, H.J.; Sun, R. Characterization of interactions between RTA and the promoter of polyadenylated nuclear RNA in Kaposi's sarcoma-associated herpesvirus/human herpesvirus 8. *J. Virol.* **2002**, *76*, 5000–5013.
25. Massimelli, M.J.; Kang, J.G.; Majerciak, V.; Le, S.Y.; Liewehr, D.J.; Steinberg, S.M.; Zheng, Z.M. Stability of a Long Noncoding Viral RNA Depends on a 9-nt Core Element at the RNA 5' End to Interact with Viral ORF57 and Cellular PABPC1. *Int. J. Biol. Sci.* **2011**, *7*, 1145–1160.
26. Massimelli, M.J.; Majerciak, V.; Kruhlak, M.; Zheng, Z.M. Interplay between polyadenylate-binding protein 1 and Kaposi's sarcoma-associated herpesvirus ORF57 in accumulation of polyadenylated nuclear RNA, a viral long noncoding RNA. *J. Virol.* **2013**, *87*, 243–256.
27. Izumiya, Y.; Lin, S.F.; Ellison, T.; Chen, L.Y.; Izumiya, C.; Luciw, P.; Kung, H.J. Kaposi's Sarcoma-Associated Herpesvirus K-bZIP Is a Coregulator of K-Rta: Physical Association and Promoter-Dependent Transcriptional Repression. *J. Virol.* **2003**, *77*, 1441–1451.
28. Wang, S.E.; Wu, F.Y.; Yu, Y.; Hayward, G.S. CCAAT/Enhancer-Binding Protein-alpha is induced during the Early Stages of Kaposi's Sarcoma-Associated Herpesvirus (KSHV) Lytic Cycle Reactivation and Together with the KSHV Replication and Transcription Activator (RTA) Cooperatively Stimulates the Viral RTA, MTA, and PAN Promoters. *J. Virol.* **2003**, *77*, 9590–9612.

29. Campbell, M.; Kim, K.Y.; Chang, P.C.; Huerta, S.; Shevchenko, B.; Wang, D.H.; Izumiya, C.; Kung, H.J.; Izumiya, Y. A Lytic Viral Long Non-coding RNA Modulates the Function of a Latent Protein. *J. Virol.* **2014**, *88*, 1843–1848.
30. Mitton-Fry, R.M.; deGregorio, S.J.; Wang, J.; Steitz, T.A.; Steitz, J.A. Poly(A) tail recognition by a viral RNA element through assembly of a triple helix. *Science* **2010**, *330*, 1244–1247.
31. Conrad, N.K.; Steitz, J.A. A Kaposi's sarcoma virus RNA element that increases the nuclear abundance of intronless transcripts. *Embo J.* **2005**, *24*, 1831–1841.
32. Tycowski, K.T.; Shu, M.D.; Borah, S.; Shi, M.; Steitz, J.A. Conservation of a triple-helix-forming RNA stability element in noncoding and genomic RNAs of diverse viruses. *Cell Rep.* **2012**, *2*, 26–32.
33. Conrad, N.K.; Shu, M.D.; Uyhazi, K.E.; Steitz, J.A. Mutational analysis of a viral RNA element that counteracts rapid RNA decay by interaction with the polyadenylate tail. *Proc. Natl. Acad. Sci. USA* **2007**, *104*, 10412–10417.
34. Brown, J.A.; Bulkley, D.; Wang, J.; Valenstein, M.L.; Yario, T.A.; Steitz, T.A.; Steitz, J.A. Structural insights into the stabilization of MALAT1 noncoding RNA by a bipartite triple helix. *Nat. Struct. Mol. Biol.* **2014**, *21*, 633–640.
35. Conrad, N.K. The emerging role of triple helices in RNA biology. *Wiley Interdiscip. Rev. RNA* **2014**, *5*, 15–29.
36. Rossetto, C.C.; Pari, G.S. KSHV noncoding PAN RNA interacts with virus and cellular-encoded proteins and suppresses expression of genes involved in immune modulation. *J. Virol.* **2011**, doi:10.1128/JVI.05886-11.
37. Majerciak, V.; Yamanegi, K.; Nie, S.H.; Zheng, Z.M. Structural and functional analyses of Kaposi sarcoma-associated herpesvirus ORF57 nuclear localization signals in living cells. *J. Biol. Chem.* **2006**, *281*, 28365–28378.
38. Kang, J.G.; Pripuzova, N.; Majerciak, V.; Kruhlak, M.; Le, S.Y.; Zheng, Z.M. Kaposi's sarcoma-associated herpesvirus ORF57 promotes escape of viral and human interleukin-6 from microRNA-mediated suppression. *J. Virol.* **2011**, *85*, 2620–2630.
39. Sahin, B.B.; Patel, D.; Conrad, N.K. Kaposi's sarcoma-associated herpesvirus ORF57 protein binds and protects a nuclear noncoding RNA from cellular RNA decay pathways. *PLoS Pathog.* **2010**, *6*, e1000799.
40. Sei, E.; Conrad, N.K. Delineation of a core RNA element required for Kaposi's sarcoma-associated herpesvirus ORF57 binding and activity. *Virology* **2011**, *419*, 107–116.
41. Stubbs, S.H.; Hunter, O.V.; Hoover, A.; Conrad, N.K. Viral factors reveal a role for REF/Aly in nuclear RNA stability. *Mol. Cell. Biol.* **2012**, *32*, 1260–1270.
42. Borah, S.; Darricarrere, N.; Darnell, A.; Myoung, J.; Steitz, J.A. A Viral Nuclear Noncoding RNA Binds Re-localized Poly(A) Binding Protein and is Required for Late KSHV Gene Expression. *PLoS Pathog.* **2011**, *7*, e1002300.
43. Rossetto, C.C.; Pari, G.S. Kaposi's sarcoma-associated herpesvirus noncoding polyadenylated nuclear RNA interacts with virus- and host cell-encoded proteins and suppresses expression of genes involved in immune modulation. *J. Virol.* **2011**, *85*, 13290–13297.
44. Rossetto, C.C.; Pari, G. KSHV PAN RNA Associates with Demethylases UTX and JMJD3 to Activate Lytic Replication through a Physical Interaction with the Virus Genome. *PLoS Pathog.* **2012**, *8*, e1002680.

45. Agger, K.; Cloos, P.A.; Christensen, J.; Pasini, D.; Rose, S.; Rappsilber, J.; Issaeva, I.; Canaani, E.; Salcini, A.E.; Helin, K. UTX and JMJD3 are histone H3K27 demethylases involved in HOX gene regulation and development. *Nature* **2007**, *449*, 731–734.
46. Issaeva, I.; Zonis, Y.; Rozovskaia, T.; Orlovsky, K.; Croce, C.M.; Nakamura, T.; Mazo, A.; Eisenbach, L.; Canaani, E. Knockdown of ALR (MLL2) reveals ALR target genes and leads to alterations in cell adhesion and growth. *Mol. Cell. Biol.* **2007**, *27*, 1889–1903.
47. Toth, Z.; Maglinte, D.T.; Lee, S.H.; Lee, H.R.; Wong, L.Y.; Brulois, K.F.; Lee, S.; Buckley, J.D.; Laird, P.W.; Marquez, V.E.; Jung, J.U. Epigenetic analysis of KSHV latent and lytic genomes. *PLoS Pathog.* **2010**, *6*, e1001013.
48. Chan, S.R.; Bloomer, C.; Chandran, B. Identification and characterization of human herpesvirus-8 lytic cycle-associated ORF 59 protein and the encoding cDNA by monoclonal antibody. *Virology* **1998**, *240*, 118–126.
49. Chan, S.R.; Chandran, B. Characterization of human herpesvirus 8 ORF59 protein (PF-8) and mapping of the processivity and viral DNA polymerase-interacting domains. *J. Virol.* **2000**, *74*, 10920–10929.
50. Rossetto, C.C.; Tarrant-Elorza, M.; Pari, G.S. Cis and Trans Acting Factors Involved in Human Cytomegalovirus Experimental and Natural Latent Infection of CD14 (+) Monocytes and CD34 (+) Cells. *PLoS Pathog* **2013**, *9*, e1003366.
51. Rossetto, C.C.; Susilarini, N.K.; Pari, G.S. Interaction of Kaposi's Sarcoma-Associated Herpesvirus ORF59 with oriLyt Is Dependent on Binding with K-Rta. *J. Virol.* **2011**, *85*, 3833–3841.
52. McDowell, M.E.; Purushothaman, P.; Rossetto, C.C.; Pari, G.S.; Verma, S.C. Phosphorylation of Kaposi's Sarcoma-Associated Herpesvirus Processivity Factor ORF59 by a Viral Kinase Modulates Its Ability To Associate with RTA and oriLyt. *J. Virol.* **2013**, *87*, 8038–8052.
53. Strang, B.L.; Bender, B.J.; Sharma, M.; Pesola, J.M.; Sanders, R.L.; Spector, D.H.; Coen, D.M. A Mutation Deleting Sequences Encoding the Amino Terminus of human Cytomegalovirus UL84 impairs Interaction with UL44 and Capsid Localization. *J. Virol.* **2012**, *86*, 11066–11077.
54. Strang, B.L.; Geballe, A.P.; Coen, D.M. Association of human cytomegalovirus proteins IRS1 and TRS1 with the viral DNA polymerase accessory subunit UL44. *J. Gen. Virol.* **2010**, *91*, 2167–2175.
55. Strang, B.L.; Sinigalia, E.; Silva, L.A.; Coen, D.M.; Loregian, A. Analysis of the association of the human cytomegalovirus DNA polymerase subunit UL44 with the viral DNA replication factor UL84. *J. Virol.* **2009**, *83*, 7581–7589.
56. Gao, Y.; Colletti, K.; Pari, G.S. Identification of human cytomegalovirus UL84 virus- and cell-encoded binding partners by using proteomics analysis. *J. Virol.* **2008**, *82*, 96–104.
57. Nakayama, S.; Murata, T.; Murayama, K.; Yasui, Y.; Sato, Y.; Kudoh, A.; Iwahori, S.; Isomura, H.; Kanda, T.; Tsurumi, T. Epstein-Barr virus polymerase processivity factor enhances BALF2 promoter transcription as a coactivator for the BZLF1 immediate-early protein. *J. Biol. Chem.* **2009**, *284*, 21557–21568.
58. Zhang, Q.; Holley-Guthrie, E.; Ge, J.Q.; Dorsky, D.; Kenney, S. The Epstein-Barr virus (EBV) DNA polymerase accessory protein, BMRF1, activates the essential downstream component of the EBV oriLyt. *Virology* **1997**, *230*, 22–34.
59. Rossetto, C.; Pari, G. University of Nevada, Reno School of Medicine, Reno, NV, USA. Unpublished work, 2014.

60. Conrad, N.K.; Mili, S.; Marshall, E.L.; Shu, M.D.; Steitz, J.A. Identification of a rapid mammalian deadenylation-dependent decay pathway and its inhibition by a viral RNA element. *Mol. Cell* **2006**, *24*, 943–953.

12

Long Non-Coding RNA and Epigenetic Gene Regulation of KSHV

Mel Campbell [1,2,*], Hsing-Jien Kung [2,3,4,5] and Yoshihiro Izumiya [1,2,3]

1 Department of Dermatology, University of California, Davis, CA 95616, USA; E-Mail: yizumiya@ucdavis.edu

2 UC Davis Comprehensive Cancer Center, University of California, Davis, CA 95616, USA; E-Mail: hkung@ucdavis.edu

3 Department of Biochemistry and Molecular Medicine, University of California, Davis, CA 95616, USA

4 National Health Research Institutes, Maioli 35053, Taiwan

5 Taipei Medical University, Taipei 110, Taiwan

* Author to whom correspondence should be addressed; E-Mail: mcampbell@ucdavis.edu;

External Editor: Zhi-Ming Zheng

Abstract: Kaposi's sarcoma-associated herpesvirus (KSHV/human herpesvirus 8) is a γ-herpesvirus linked to Kaposi's sarcoma (KS) and two lymphoproliferative disorders, primary effusion lymphoma (PEL or body-cavity B-lymphoma [BCBL]) and a subset of Multicentric Castleman's Disease. During lytic growth, pervasive viral transcription generating a variety of transcripts with uncertain protein-coding potential has been described on a genome-wide scale in β- and γ-herpesviruses. One class of such RNAs is called long non-coding RNAs (lncRNAs). KSHV encodes a viral lncRNA known as polyadenylated nuclear RNA (PAN RNA), a copious early gene product. PAN RNA has been implicated in KSHV gene expression, replication, and immune modulation. PAN RNA expression is required for optimal expression of the entire KSHV lytic gene expression program. Latent KSHV episomes are coated with viral latency-associated nuclear antigen (LANA). LANA rapidly dissociates from episomes during reactivation. Here we review recent studies suggesting that PAN RNA may function as a viral lncRNA, including a role in the facilitation of LANA-episomal dissociation during lytic replication.

Keywords: viruses; long non-coding RNA; epigenetics; viral latency; viral reactivation

1. Introduction

Recent new developments in epigenetic control of transcriptional programs by chromatin modifiers are highlighted by findings that long non-coding RNA (lncRNA), previously thought to be "transcription noise", interact with chromatin-associated proteins to modulate their functions [1,2]. lncRNA interactions have been reported with most classes of proteins that associate with chromatin including transcription factors [3], chromatin remodelers [4], and histone methylases and demethylases [5,6]. Non-coding (nc) RNAs can be classified as either housekeeping or regulatory ncRNA, and based on transcript size, regulatory ncRNA can be further grouped into two subclasses; small non-coding RNA (20–200 nt) and long non-coding RNA (lncRNA, >200 nt). Although the role of small regulatory RNAs (microRNAs and siRNAs [small interfering RNA]) in gene silencing is well defined, the biological function of lncRNAs is still unclear. Like proteins, diverse biological functions of lncRNAs have been identified/proposed, including the structural integrity of the nucleus, regulation of gene expression, chromatin remodeling, transcription, and posttranscriptional processing [7,8]. The actions of several lncRNAs and their mechanisms are summarized in the Table 1.

Table 1. Mechanisms of lncRNA action.

Name of lncRNA	Nature of lncRNA/ Mode of Action	Consequence of the Regulation	Reference
ANRIL, HOTAIR, XIST, H19, KCNQ1OT1	Scaffold molecule/ Histone modification	Epigenetic gene silencing	[6,9–12]
NEAT1	Scaffold molecule/ Protein assembly	Paraspeckles formation	[13]
MALAT1	Modifier of alternative splicing/Relocation of splicing factor	Modulation of alternative splicing	[14]
BACE1AS	Regulation of mRNA stability/Modification of mRNA stability	Increased translation of BACE1	[15]
Pseudo-NOS	Translational control/Displacement of ribosome	Repressed translation of nNOS	[16]
GAS5	Decoy for DNA binding/Inhibition of DNA binding	Repressed GR mediated gene activation	[17]

Currently, the most dominant function explored in lncRNA studies relates to epigenetic regulation of target genes. This role is typically associated with gene repression, which has been studied in many cellular lncRNAs including ANRIL, HOTAIR, H19, and XIST [6,9–12]. These lncRNAs exhibit their repression function through interactions with histone modifying enzymes. The most common protein partners of lncRNAs are the PRC1 and PRC2 polycomb repressive complexes. These complexes transfer repressive posttranslational modification marks (H2AK119ub or H3K27me3) to histone tails, thereby facilitating chromatin compaction to repress gene transcription [1]. It is estimated that nearly 20% of all lncRNAs bind PRC2 [18], although the biological meaning of this observation remains unclear. ANRIL, HOTAIR, H19, KCNQ1OT1, and XIST have been shown to interact with the PRC2 complex, and in all except H19, direct binding has been observed between PRC2 proteins and the lncRNAs ([6,9–12], reviewed in [19]). The programmers of chromatin remodeling are enzymes involved in histone modifications, namely histone methylases and demethylases, thus regulation of such enzymes by lncRNAs through recruitment or assembly of specific complexes may have profound local effects on chromatin modification.

2. KSHV Epigenetics

While the KSHV genome is histone-free in virions, the viral genome adopts a highly organized chromatin structure in infected cells that have established latency [20,21] and following *de novo* infection [22]. During *de novo* infection there is a rapid association of KSHV genomic DNA with histones, followed by a biphasic period of chromatinization [22]. These early events include an initial transient enrichment of H3K4me3 and H3K27ac activating histone marks on the viral chromatin that is followed by a decline of activating marks and the transition to a heterochromatic state enriched for H3K27me3/H2AK119ub marks. This transition is dependent on PRC1 and PRC2 complexes and ultimately results in the inhibition of lytic gene expression, however, these events differ among cell types [22]. During latency, the promoter of the master lytic switch regulator, K-Rta, is characterized by a bivalent chromatin structure, consisting of both activating and repressive histone marks, including H3K27me3 [20,21]. Upon reactivation, immediate early/early gene expression, including K-Rta, is accompanied by decreased local levels of repressive H3K27me3 marks while activating histone marks such as acetylated H3 and H3K4me3 are simultaneously increased. This shift is important for optimal K-Rta production and activation of the entire KSHV lytic program [20,21]. In all of these studies [20–22], the deposition of histone marks on viral chromatin is site-specific, with certain histone marks enriched only on specific viral genomic regions. Although the exact mechanisms of targeted recruitment of histone-modifying machinery are unclear, it is likely that cellular and viral DNA-binding factors may be important for recruitment [22]. In addition, the lncRNA function of PAN RNA has been suggested to be important in this regard, perhaps serving as a guide RNA to deliver factors to specific viral genomic locations [23] (see below). We have been interested in how viral gene products can change the epigenetic landscape of the viral and host genome. We, and others, have found that cellular histone lysine methyltransferases (KMT), protein arginine methyltransferases (PRMT), and lysine demethylases (KDM) are targets of viral proteins and important for the viral life cycle [21,24–28]. The potential significance of histone methylases for KSHV biology was also highlighted by the recent finding that histone modifying enzymes, but not modified histones *per se*, act as epigenetic marks for inheritance [29]. In the study, Petruk *et al.* showed that methylated histones were not detected during S-phase of the cell cycle, however Trithorax and Ezh2, which are H3K4 and H3K27 methylases were continuously associated with their response elements on the newly replicated DNA which thus re-establishes the histone marks on newly assembled unmethylated histones. This finding suggests that dislodging PRC1/PRC2 polycomb complexes from the KSHV genome may be important for effective viral replication, because the KSHV genome has been shown to be heavily loaded by polycomb complexes [20–22]. This further supports the notion that removal of enzymatic complexes that deposit or maintain a repressive KSHV chromatin state is required for optimal expression of the lytic program. As deletion studies have suggested a generalized repressive role for LANA in lytic gene expression [30], it is tempting to speculate that sequestering polycomb complexes and/or LANA from KSHV genomes is important to allow late gene expression. Persistence of these complexes might allow them to bind newly synthesized DNA [29] and continue to silence viral gene expression. In the following sections, we propose that one of the lncRNA-like functions of PAN RNA may be in the titration of the repressive complexes away from the KSHV genome during reactivation.

3. KSHV PAN RNA

KSHV encodes a 1.1 kb viral lncRNA known as polyadenylated nuclear RNA or PAN RNA, an abundant early gene product. High level expression of PAN RNA is directly regulated by a master switch gene K-Rta (replication and transcription activator [31–33]. While PAN RNA was first described 18 years ago [31,34,35] as a non-coding RNA, its discovery predated the widespread recognition of long non-coding RNAs and their role in gene regulation. The structure, unique stability, and nuclear retention properties of PAN RNA has been exquisitely characterized by the groups of Conrad, Steitz, and Zheng [36–41]. Despite this knowledge, the role of PAN RNA in KSHV lifecycle is still unclear. New studies by Steitz, Pari and colleagues have begun to address this question, and PAN RNA has been demonstrated to play a role in KSHV gene expression, replication, and immune modulation [23,42–44]. With the recent discovery of thousands of lncRNAs, PAN RNA has been re-examined under the premise that one of its functions may be that of a viral lncRNA. Recent reports have demonstrated that PAN RNA binds the transcription factor IRF4 [43], the lysine demethylase JMJD3 [44], and the lysine methylase Ezh2, [23] supporting the notion that similarly to cellular lncRNAs, PAN RNA may function in epigenetic gene regulation. Although PAN RNA is classified as a lytic transcript, prior reports have demonstrated its presence in virions [45] and virion packaging has been further demonstrated by others [23]. Moreover, recent studies have also reported that PAN RNA is expressed in uninduced KSHV positive cell lines [23,46] suggesting that PAN RNA is a transcript present (at variable levels) during both latency and lytic infection. Thus, the current view of PAN RNA function depicts broad effects of PAN RNA expression acting as a multifunctional regulator, which influences both viral and host transcriptional programs.

3.1. LANA Is Released from KSHV Genome in a Binding Site-Specific Manner

Our discovery of PAN RNA involvement in the regulation of LANA began with the observation that LANA is released from the KSHV genome immediately after triggering reactivation by K-Rta expression [25]. For the study, we took four different time points to examine the dynamics of LANA recruitment sites on the KSHV genome. Initially, we expected that LANA binds to different sites during the course of reactivation as LANA was reported to interact with K-Rta [47]. Unexpectedly, we found that LANA is dissociated from the KSHV genome almost immediately after induction of lytic replication by K-Rta expression. In our experiments, we used K-Rta inducible cells instead of chemicals to synchronize and induce viral lytic replication in almost all cells [48]; this strategy significantly reduced the background caused by non-reactivating cells. Accordingly, the negative effects (reduction of recruitment) could be observed. Importantly, dissociation of LANA is only seen along the unique region of the KSHV genome, and not at the terminal repeat region (TR), where LANA was still enriched nearly 40-fold over input DNA in our tiling array analyses at all time points [25].This may be due to the combination of tightness of LANA binding to the TR region through the DNA binding domain as well as copy numbers of TR elements in the infected cells. Dissociation of LANA after reactivation was further confirmed with independent ChIP experiments with qt-PCR [49]. As LANA is generally viewed as a key factor for maintaining latency, sequestration of LANA from the KSHV genome may play a role in regulating lytic gene expression. Ganem and colleagues have also reported that the association of LANA with viral chromatin is disrupted in cells during the lytic cycle [50].

3.2. PAN RNA Interacts with LANA

What molecular mechanisms would underlie LANA dissociation from KSHV episomes? As our lab and another group independently found; LANA associates with multiple RNA-binding proteins *in vivo* [26,51]. We hypothesized that LANA might possess the property of RNA binding and we speculated that robust expression of PAN RNA might have a role in dissociation of LANA during lytic replication. In fact, interactions between lncRNAs and transcriptional factors have been reported elsewhere [3]. Could PAN RNA function in an lncRNA-like manner to reduce LANA's occupancy on chromatin by acting as a protein sponge? To test the hypothesis, a series of experiments were conducted to examine if LANA associates with PAN RNA *in vitro* and *in vivo*. The results clearly demonstrated that LANA has a property to bind PAN RNA [49]. LANA may recognize RNA secondary structure in addition to sequence as several experiments using different salt concentrations repeatedly showed that antisense PAN RNA also bound to LANA, albeit weaker than PAN RNA in the sense orientation [49]. Subsequent mapping experiments with a series of PAN RNA deletions showed that the nucleotides towards the 3' end of PAN RNA are responsible for the interaction [49]. By using RNA immunoprecipitation (RIP) analyses with anti-LANA antibody, LANA was shown to be associated with PAN RNA in naturally infected cells during the course of reactivation [49]. The RNA-binding domain of LANA was mapped with both RNA- and GST-pull-down analyses. Intriguingly, results showed that the RNA-binding domain completely overlapped with the previously identified LANA histone-binding domain (residues 1–20) [49]; this may be the mechanistic explanation of why LANA dissociates from the unique region, but not TR region of the KSHV chromosome during reactivation, as TR binding requires a DNA-binding domain, located at the C-terminal region of LANA [52,53]. It is exciting to speculate that PAN RNA may have evolved as a viral "Aptamer" for LANA to counteract its function and allow for effective lytic replication.

3.3. PAN RNA Is Responsible for LANA Dissociation from KSHV Genomes

To study if PAN RNA/LANA association is responsible for dissociation of LANA from KSHV genomes, we adapted the strategy reported by the Steitz lab [42] which utilizes modified antisense oligonucleotides and endogenous RNase H to knock-down PAN RNA expression during reactivation. Similar to their research design, we also designed antisense oligonucleotides for the K7 transcript as a control, because K7 partially overlaps with PAN RNA and co-terminates with PAN RNA transcripts. The antisense oligonucleotides were electroporated into latent KSHV positive cell lines prior to reactivation. The effects on LANA binding to the KSHV genome by the partial ablation of PAN RNA expression were then examined by ChIP assay with anti-LANA antibody. The results showed that knocking-down of PAN RNA expression during reactivation resulted in retained LANA occupancy on the KSHV genome [49]. This results support the idea that PAN RNA plays a role in the sequestration of the LANA complex from the KSHV genome; however, how binding of LANA on the KSHV chromosome affects the landscape of KSHV histone modifications remains to be seen. Thus, one function of the lytic lncRNA PAN RNA includes an influence on the activity of a latent viral DNA-binding protein (LANA). Partial inhibition of PAN RNA production has been shown to decrease late gene expression significantly [42]. PRC2 complex components interact with both PAN RNA and purified LANA *in vitro* [61] and *in vivo* [23]. Together, these results suggest that PAN RNA is important for optimal lytic gene expression, in part through mechanisms involving sequestration of repressive protein complexes away from KSHV genomes undergoing reactivation.

3.4. Other Functions of PAN RNA

PAN RNA has been reported to interact with both viral and cellular proteins suggestive of a multifunctional role for PAN RNA in the KSHV lifecycle. This interaction list includes KSHV ORFs 26, 57, and 59 [40,41,43,54,55], as well as host-encoded factors such as histones, single-stranded DNA binding proteins, transcription factors, ribonucleoproteins, and histone methylases and demethylases [23,40,42–44]. Follow up studies have been performed for several of these interactions. PAN RNA interaction with both ORF57 and the cellular ribonucleoprotein PAPBC1 has been reported to be involved in the abundant nuclear accumulation of PAN RNA [40,42,54]. PAN RNA interaction with the transcription factor IRF4 [43], the histone demethylases UTX and JMJD3, the histone methylase MLL2 [44], and components of the polycomb complex 2 (PRC2) [23], have suggested that PAN RNA has potential widespread effects on both KSHV and cellular gene expression and epigenetic states. Moreover, using chromatin isolation by RNA purification (ChIRP) analysis [56], Rossetto *et al.* [23] have shown PAN RNA interacts with many regions of both the KSHV genome and cellular chromosomes. These interactions influence almost the entire KSHV transcriptional program as well as host genes controlling the cell cycle, inflammation, and immune responses. Finally, similar to what has been previously observed with cellular lncRNAs [57], a recent ribosomal profiling study of KSHV infected cells has found a fraction of PAN RNA to be associated with ribosomes, suggesting that in addition to its role as a non-coding RNA, PAN may also serve as a *bona fide* mRNA with the potential to produce several small viral peptides [58].

4. Concluding Remarks

Pathogen-encoded lncRNAs have been reported for several virus groups, including herpesviruses (see [59] for a recent review). For KSHV, several studies have described properties of PAN RNA, which are consistent with its role as a multifunctional viral lncRNA. In view of the dependence of viral replication on the host cell machinery, it is perhaps not surprising that viral and cellular lncRNAs are increasingly linked to viral processes. It should be noted that our studies have focused on PAN RNA as a molecular sink or decoy for LANA or other repressive protein complexes in order to facilitate lytic replication [49]. In addition to that aspect of the model (Figure 1), the results of Rossetto, Pari and colleagues [23,43,44] suggests that PAN RNA is capable of functioning in a lncRNA-like manner using other archetypes of lncRNA molecular mechanisms as put forth by Wang and Chang, including lncRNA guides, scaffolds and signals [1]. For example, the widespread distribution of PAN RNA at both KSHV and cellular genomic loci suggests that PAN RNA may act as a guide or scaffold to target chromatin-modifying complexes to specific locations with PAN RNA serving as a conduit for either gene activation or repression [23,44]. This paradigm is illustrated by the detection of PAN RNA interactions at the ORF50 promoter and with JMJD3, UTX, and MLL2; these interactions are consistent with the changes in local histone modifications observed at this locus during viral reactivation [20,21]. Interestingly, a recombinant bacmid lacking PAN RNA expression does not produce virus, nor could viral production by this mutant be rescued by K-Rta overexpression [23]. Although PAN RNA is considered a K-Rta target gene [32,33] these surprising results indicate that a certain level of PAN RNA is needed for the activation of the entire KSHV lytic expression program. While PAN RNA is often considered a lytic transcript, it is important to keep in mind that PAN RNA is packaged into virions [45] and the transcript is also detected in latent cells [23], Thus PAN RNA has the potential to influence virus-host interactions at all stages of infection. One curious aspect that arises when attempting to invoke lncRNA mechanisms for PAN RNA concerns the extreme abundance of the transcript. Early studies had suggested a number of up to ~500,000 copies of PAN RNA per induced BC-1 cell representing ~80% of the total poly(A) selected RNA pool [31,32]. This abundance has been confirmed by next generation mRNA-sequencing results where PAN RNA represented up to 90% of the KSHV reads at 72 h post-induction [58]. This expression level contrasts that of cellular noncoding RNAs, which have been reported to often be expressed at significantly lower levels than coding RNA [60]. While one could propose that this high level of PAN RNA expression (a) creates a better sponge or decoy; (b) is necessary to ensure its incorporation into virions; or (c) assures perturbation of host cell transcriptional programs, it will be exciting to see how specificity or targeting is accomplished under these conditions. A working model that outlines our current view of PAN RNA lncRNA function is presented in Figure 1.

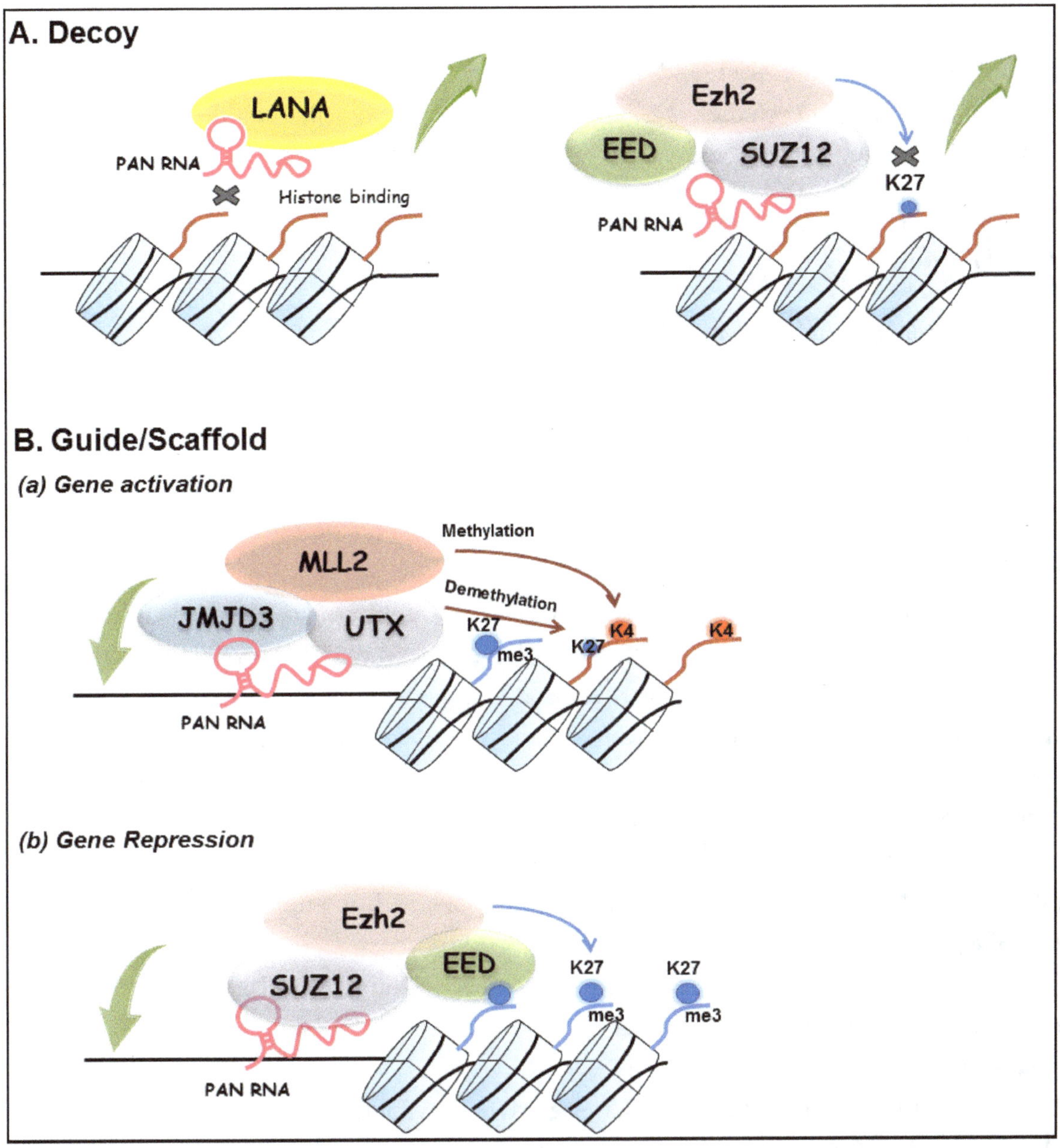

Figure 1. Working model of KSHV PAN lncRNA action. (**A**) Decoy. Expression of PAN RNA sequesters LANA (left) and chromatin modifying complexes such as PRC2 (right) from the KSHV genome that reduces or prevents H3K27me3 mark deposition [1]. (**B**) Guide/Scaffold. PAN RNA aids in the targeting of regulatory factor complexes to specific loci of viral or cellular genes, which regulate gene expression to suit viral needs requiring (**a**) activation (*i.e.*, ORF50 locus during reactivation) (**b**) repression (*i.e.*, cellular response to viral infection or reactivation [23,43,44]. Mechanisms of PAN RNA targeting to specific loci is currently unknown. Not shown is the potential for PAN RNA to act in *cis* as a lncRNA signal [1]. Large curved green arrows indicate complex eviction (upward arrow) or deposition (downward arrow) from a genomic locus. Histones are depicted as cylinders with histone tails emanating, (red tail, activation; blue tail, repression) K27 (demethylated histone H3K27) and K27me3 (histone H3K27 tri-methylated; repressive mark); K4 (histone H3K4, substrate for methylation, activating mark).

Acknowledgments

The authors wish to thank past and present members of Izumiya and Kung labs who have contributed to our studies. Our studies are supported by grants from the National Institutes of Health (CA14779) and American Cancer Society (RSG-13-383-01-MPC) to YI.

Author Contributions

Mel Campbell, Hsing-Jien Kung and Yoshihiro Izumiya wrote the manuscript.

References and Notes

1. Wang, K.C.; Chang, H.Y. Molecular mechanisms of long noncoding RNAs. *Mol. Cell* **2011**, *43*, 904–914.
2. Wang, K.C.; Yang, Y.W.; Liu, B.; Sanyal, A.; Corces-Zimmerman, R.; Chen, Y.; Lajoie, B.R.; Protacio, A.; Flynn, R.A.; Gupta, R.A.; *et al.* A long noncoding RNA maintains active chromatin to coordinate homeotic gene expression. *Nature* **2011**, *472*, 120–124.
3. Jeon, Y.; Lee, J.T. YY1 tethers Xist RNA to the inactive X nucleation center. *Cell* **2011**, *146*, 119–133.
4. Prensner, J.R.; Iyer, M.K.; Sahu, A.; Asangani, I.A.; Cao, Q.; Patel, L.; Vergara, I.A.; Davicioni, E.; Erho, N.; Ghadessi, M.; *et al.* The long noncoding RNA SChLAP1 promotes aggressive prostate cancer and antagonizes the SWI/SNF complex. *Nat. Genet.* **2013**, *45*, 1392–1398.
5. Tsai, M.C.; Manor, O.; Wan, Y.; Mosammaparast, N.; Wang, J.K.; Lan, F.; Shi, Y.; Segal, E.; Chang, H.Y. Long noncoding RNA as modular scaffold of histone modification complexes. *Science* **2010**, *329*, 689–693.
6. Zhao, J.; Sun, B.K.; Erwin, J.A.; Song, J.J.; Lee, J.T. Polycomb proteins targeted by a short repeat RNA to the mouse X chromosome. *Science* **2008**, *322*, 750–756.
7. Mercer, T.R.; Dinger, M.E.; Mattick, J.S. Long non-coding RNAs: Insights into functions. *Nat. Rev. Genet.* **2009**, *10*, 155–159.
8. Wilusz, J.E.; Sunwoo, H.; Spector, D.L. Long noncoding RNAs: Functional surprises from the RNA world. *Genes Dev.* **2009**, *23*, 1494–1504.
9. Kotake, Y.; Nakagawa, T.; Kitagawa, K.; Suzuki, S.; Liu, N.; Kitagawa, M.; Xiong, Y. Long non-coding RNA ANRIL is required for the PRC2 recruitment to and silencing of p15(INK4B) tumor suppressor gene. *Oncogene* **2011**, *30*, 1956–1962.
10. Rinn, J.L.; Kertesz, M.; Wang, J.K.; Squazzo, S.L.; Xu, X.; Brugmann, S.A.; Goodnough, L.H.; Helms, J.A.; Farnham, P.J.; Segal, E.; Chang, H.Y. Functional demarcation of active and silent chromatin domains in human HOX loci by noncoding RNAs. *Cell* **2007**, *129*, 1311–1323.
11. Li, T.; Hu, J.F.; Qiu, X.; Ling, J.; Chen, H.; Wang, S.; Hou, A.; Vu, T.H.; Hoffman, A.R. CTCF regulates allelic expression of Igf2 by orchestrating a promoter-polycomb repressive complex 2 intrachromosomal loop. *Mol. Cell Biol.* **2008**, *28*, 6473–6482.

12. Pandey, R.R.; Mondal, T.; Mohammad, F.; Enroth, S.; Redrup, L.; Komorowski, J.; Nagano, T.; Mancini-Dinardo, D.; Kanduri, C. Kcnq1ot1 antisense noncoding RNA mediates lineage-specific transcriptional silencing through chromatin-level regulation. *Mol. Cell* **2008**, *32*, 232–246.
13. Clemson, C.M.; Hutchinson, J.N.; Sara, S.A.; Ensminger, A.W.; Fox, A.H.; Chess, A.; Lawrence, J.B. An architectural role for a nuclear noncoding RNA: NEAT1 RNA is essential for the structure of paraspeckles. *Mol. Cell* **2009**, *33*, 717–726.
14. Tripathi, V.; Ellis, J.D.; Shen, Z.; Song, D.Y.; Pan, Q.; Watt, A.T.; Freier, S.M.; Bennett, C.F.; Sharma, A.; Bubulya, P.A.; *et al.* The nuclear-retained noncoding RNA MALAT1 regulates alternative splicing by modulating SR splicing factor phosphorylation. *Mol. Cell* **2010**, *39*, 925–938.
15. Faghihi, M.A.; Modarresi, F.; Khalil, A.M.; Wood, D.E.; Sahagan, B.G.; Morgan, T.E.; Finch, C.E.; St. Laurent, G., 3rd; Kenny, P.J.; Wahlestedt, C. Expression of a noncoding RNA is elevated in Alzheimer's disease and drives rapid feed-forward regulation of beta-secretase. *Nat. Med.* **2008**, *14*, 723–730.
16. Korneev, S.A.; Park, J.H.; O'Shea, M. Neuronal expression of neural nitric oxide synthase (nNOS) protein is suppressed by an antisense RNA transcribed from an NOS pseudogene. *J. Neurosci.* **1999**, *19*, 7711–7720.
17. Kino, T.; Hurt, D.E.; Ichijo, T.; Nader, N.; Chrousos, G.P. Noncoding RNA gas5 is a growth arrest- and starvation-associated repressor of the glucocorticoid receptor. *Sci. Signal* **2010**, *3*, ra8.
18. Zhao, J.; Ohsumi, T.K.; Kung, J.T.; Ogawa, Y.; Grau, D.J.; Sarma, K.; Song, J.J.; Kingston, R.E.; Borowsky, M.; Lee, J.T. Genome-wide identification of polycomb-associated RNAs by RIP-seq. *Mol. Cell* **2010**, *40*, 939–953.
19. Brockdorff, N. Noncoding RNA and polycomb recruitment. *RNA* **2013**, *19*, 429–442.
20. Toth, Z.; Maglinte, D.T.; Lee, S.H.; Lee, H.R.; Wong, L.Y.; Brulois, K.F.; Lee, S.; Buckley, J.D.; Laird, P.W.; Marquez, V.E.; *et al.* Epigenetic analysis of KSHV latent and lytic genomes. *PLoS Pathog.* **2010**, *6*, e1001013.
21. Gunther, T.; Grundhoff, A. The epigenetic landscape of latent Kaposi sarcoma-associated herpesvirus genomes. *PLoS Pathog.* **2010**, *6*, e1000935.
22. Toth, Z.; Brulois, K.; Lee, H.R.; Izumiya, Y.; Tepper, C.; Kung, H.J.; Jung, J.U. Biphasic euchromatin-to-heterochromatin transition on the KSHV genome following de novo infection. *PLoS Pathog.* **2013**, *9*, e1003813.
23. Rossetto, C.C.; Tarrant-Elorza, M.; Verma, S.; Purushothaman, P.; Pari, G.S. Regulation of viral and cellular gene expression by Kaposi's sarcoma-associated herpesvirus polyadenylated nuclear RNA. *J. Virol.* **2013**, *87*, 5540–5553.
24. Chang, P.C.; Fitzgerald, L.D.; Hsia, D.A.; Izumiya, Y.; Wu, C.Y.; Hsieh, W.P.; Lin, S.F.; Campbell, M.; Lam, K.S.; Luciw, P.A.; *et al.* Histone demethylase JMJD2A regulates Kaposi's sarcoma-associated herpesvirus replication and is targeted by a viral transcriptional factor. *J. Virol.* **2011**, *85*, 3283–3293.
25. Campbell, M.; Chang, P.C.; Huerta, S.; Izumiya, C.; Davis, R.; Tepper, C.G.; Kim, K.Y.; Shevchenko, B.; Wang, D.H.; Jung, J.U.; *et al.* Protein arginine methyltransferase 1-directed methylation of Kaposi sarcoma-associated herpesvirus latency-associated nuclear antigen. *J. Biol. Chem.* **2012**, *287*, 5806–5818.

26. Kim, K.Y.; Huerta, S.B.; Izumiya, C.; Wang, D.H.; Martinez, A.; Shevchenko, B.; Kung, H.J.; Campbell, M.; Izumiya, Y. Kaposi's sarcoma-associated herpesvirus (KSHV) latency-associated nuclear antigen regulates the KSHV epigenome by association with the histone demethylase KDM3A. *J. Virol.* **2013**, *87*, 6782–6793.
27. Jong, J.E.; Cha, S.; Jang, J.H.; Seo, T. Alteration of histone H3 lysine 4 trimethylation on putative lytic gene promoters by human Set1 complex during reactivation of Kaposi's sarcoma-associated herpesvirus. *Intervirology* **2013**, *56*, 91–103.
28. Hu, J.; Yang, Y.; Turner, P.C.; Jain, V.; McIntyre, L.M.; Renne, R. LANA binds to multiple active viral and cellular promoters and associates with the H3K4methyltransferase hSET1 complex. *PLoS Pathog.* **2014**, *10*, e1004240.
29. Petruk, S.; Sedkov, Y.; Johnston, D.M.; Hodgson, J.W.; Black, K.L.; Kovermann, S.K.; Beck, S.; Canaani, E.; Brock, H.W.; Mazo, A. TrxG and PcG proteins but not methylated histones remain associated with DNA through replication. *Cell* **2012**, *150*, 922–933.
30. Li, Q.; Zhou, F.; Ye, F.; Gao, S.J. Genetic disruption of KSHV major latent nuclear antigen LANA enhances viral lytic transcriptional program. *Virology* **2008**, *379*, 234–244.
31. Sun, R.; Lin, S.F.; Gradoville, L.; Miller, G. Polyadenylylated nuclear RNA encoded by Kaposi sarcoma-associated herpesvirus. *Proc. Natl. Acad. Sci. USA* **1996**, *93*, 11883–11888.
32. Song, M.J.; Brown, H.J.; Wu, T.T.; Sun, R. Transcription activation of polyadenylated nuclear rna by rta in human herpesvirus 8/Kaposi's sarcoma-associated herpesvirus. *J. Virol.* **2001**, *75*, 3129–3140.
33. Song, M.J.; Li, X.; Brown, H.J.; Sun, R. Characterization of interactions between RTA and the promoter of polyadenylated nuclear RNA in Kaposi's sarcoma-associated herpesvirus/human herpesvirus 8. *J. Virol.* **2002**, *76*, 5000–5013.
34. Zhong, W.; Wang, H.; Herndier, B.; Ganem, D. Restricted expression of Kaposi sarcoma-associated herpesvirus (human herpesvirus 8) genes in Kaposi sarcoma. *Proc. Natl. Acad. Sci. USA* **1996**, *93*, 6641–6646.
35. Zhong, W.; Ganem, D. Characterization of ribonucleoprotein complexes containing an abundant polyadenylated nuclear RNA encoded by Kaposi's sarcoma-associated herpesvirus (human herpesvirus 8). *J. Virol.* **1997**, *71*, 1207–1212.
36. Conrad, N.K.; Steitz, J.A. A Kaposi's sarcoma virus RNA element that increases the nuclear abundance of intronless transcripts. *EMBO J.* **2005**, *24*, 1831–1841.
37. Conrad, N.K.; Mili, S.; Marshall, E.L.; Shu, M.D.; Steitz, J.A. Identification of a rapid mammalian deadenylation-dependent decay pathway and its inhibition by a viral RNA element. *Mol. Cell* **2006**, *24*, 943–953.
38. Conrad, N.K.; Shu, M.D.; Uyhazi, K.E.; Steitz, J.A. Mutational analysis of a viral RNA element that counteracts rapid RNA decay by interaction with the polyadenylate tail. *Proc. Natl. Acad. Sci. USA* **2007**, *104*, 10412–10417.
39. Mitton-Fry, R.M.; DeGregorio, S.J.; Wang, J.; Steitz, T.A.; Steitz, J.A. Poly(A) tail recognition by a viral RNA element through assembly of a triple helix. *Science* **2010**, *330*, 1244–1247.
40. Massimelli, M.J.; Kang, J.G.; Majerciak, V.; Le, S.Y.; Liewehr, D.J.; Steinberg, S.M.; Zheng, Z.M. Stability of a long noncoding viral RNA depends on a 9-nt core element at the RNA 5' end to interact with viral ORF57 and cellular PABPC1. *Int. J. Biol. Sci.* **2011**, *7*, 1145–1160.

41. Massimelli, M.J.; Majerciak, V.; Kruhlak, M.; Zheng, Z.M. Interplay between polyadenylate-binding protein 1 and Kaposi's sarcoma-associated herpesvirus ORF57 in accumulation of polyadenylated nuclear RNA, a viral long noncoding RNA. *J. Virol.* **2013**, *87*, 243–256.
42. Borah, S.; Darricarrere, N.; Darnell, A.; Myoung, J.; Steitz, J.A. A viral nuclear noncoding RNA binds re-localized poly(A) binding protein and is required for late KSHV gene expression. *PLoS Pathog.* **2011**, *7*, e1002300.
43. Rossetto, C.C.; Pari, G.S. Kaposi's sarcoma-associated herpesvirus noncoding polyadenylated nuclear RNA interacts with virus- and host cell-encoded proteins and suppresses expression of genes involved in immune modulation. *J. Virol.* **2011**, *85*, 13290–13297.
44. Rossetto, C.C.; Pari, G. KSHV PAN RNA associates with demethylases UTX and JMJD3 to activate lytic replication through a physical interaction with the virus genome. *PLoS Pathog.* **2012**, *8*, e1002680.
45. Bechtel, J.; Grundhoff, A.; Ganem, D. RNAs in the virion of Kaposi's sarcoma-associated herpesvirus. *J. Virol.* **2005**, *79*, 10138–10146.
46. Dresang, L.R.; Teuton, J.R.; Feng, H.; Jacobs, J.M.; Camp, D.G., 2nd; Purvine, S.O.; Gritsenko, M.A.; Li, Z.; Smith, R.D.; Sugden, B.; *et al.* Coupled transcriptome and proteome analysis of human lymphotropic tumor viruses: Insights on the detection and discovery of viral genes. *BMC Genomics* **2011**, *12*, 625.
47. Lan, K.; Kuppers, D.A.; Verma, S.C.; Robertson, E.S. Kaposi's sarcoma-associated herpesvirus-encoded latency-associated nuclear antigen inhibits lytic replication by targeting Rta: A potential mechanism for virus-mediated control of latency. *J. Virol.* **2004**, *78*, 6585–6594.
48. Izumiya, Y.; Ellison, T.J.; Yeh, E.T.; Jung, J.U.; Luciw, P.A.; Kung, H.J. Kaposi's sarcoma-associated herpesvirus K-bZIP represses gene transcription via SUMO modification. *J. Virol.* **2005**, *79*, 9912–9925.
49. Campbell, M.; Kim, K.Y.; Chang, P.C.; Huerta, S.; Shevchenko, B.; Wang, D.H.; Izumiya, C.; Kung, H.J.; Izumiya, Y. A lytic viral long noncoding RNA modulates the function of a latent protein. *J. Virol.* **2014**, *88*, 1843–1848.
50. Mercier, A.; Arias, C.; Madrid, A.S.; Holdorf, M.M.; Ganem, D. Site-Specific Association with Host and Viral Chromatin by Kaposi's Sarcoma-Associated Herpesvirus LANA and Its Reversal during Lytic Reactivation. *J. Virol.* **2014**, *88*, 6762–6777.
51. Chen, W.; Dittmer, D.P. Ribosomal protein S6 interacts with the latency-associated nuclear antigen of Kaposi's sarcoma-associated herpesvirus. *J. Virol.* **2011**, *85*, 9495–9505.
52. Schwam, D.R.; Luciano, R.L.; Mahajan, S.S.; Wong, L.; Wilson, A.C. Carboxy terminus of human herpesvirus 8 latency-associated nuclear antigen mediates dimerization, transcriptional repression, and targeting to nuclear bodies. *J. Virol.* **2000**, *74*, 8532–8540.
53. Shinohara, H.; Fukushi, M.; Higuchi, M.; Oie, M.; Hoshi, O.; Ushiki, T.; Hayashi, J.; Fujii, M. Chromosome binding site of latency-associated nuclear antigen of Kaposi's sarcoma-associated herpesvirus is essential for persistent episome maintenance and is functionally replaced by histone H1. *J. Virol.* **2002**, *76*, 12917–12924.
54. Sahin, B.B.; Patel, D.; Conrad, N.K. Kaposi's sarcoma-associated herpesvirus ORF57 protein binds and protects a nuclear noncoding RNA from cellular RNA decay pathways. *PLoS Pathog.* **2010**, *6*, e1000799.

55. Majerciak, V.; Uranishi, H.; Kruhlak, M.; Pilkington, G.R.; Massimelli, M.J.; Bear, J.; Pavlakis, G.N.; Felber, B.K.; Zheng, Z.M. Kaposi's sarcoma-associated herpesvirus ORF57 interacts with cellular RNA export cofactors RBM15 and OTT3 to promote expression of viral ORF59. *J. Virol.* **2011**, *85*, 1528–1540.
56. Chu, C.; Qu, K.; Zhong, F.L.; Artandi, S.E.; Chang, H.Y. Genomic maps of long noncoding RNA occupancy reveal principles of RNA-chromatin interactions. *Mol. Cell* **2011**, *44*, 667–678.
57. Ingolia, N.T.; Lareau, L.F.; Weissman, J.S. Ribosome profiling of mouse embryonic stem cells reveals the complexity and dynamics of mammalian proteomes. *Cell* **2011**, *147*, 789–802.
58. Arias, C.; Weisburd, B.; Stern-Ginossar, N.; Mercier, A.; Madrid, A.S.; Bellare, P.; Holdorf, M.; Weissman, J.S.; Ganem, D. KSHV 2.0: A comprehensive annotation of the Kaposi's sarcoma-associated herpesvirus genome using next-generation sequencing reveals novel genomic and functional features. *PLoS Pathog.* **2014**, *10*, e1003847.
59. Zhang, Q.; Jeang, K.T. Long non-coding RNAs (lncRNAs) and viral infections. *Biomed. Pharmacother.* **2013**, *3*, 34–42.
60. Djebali, S.; Davis, C.A.; Merkel, A.; Dobin, A.; Lassmann, T.; Mortazavi, A.; Tanzer, A.; Lagarde, J.; Lin, W.; Schlesinger, F.; *et al.* Landscape of transcription in human cells. *Nature* **2012**, *489*, 101–108.
61. Campbell, M.; Izumiya, Y. UC Davis, Sacramento, CA, USA. Unpublished work, 2013.

Permissions

All chapters in this book were first published by MDPI; hereby published with permission under the Creative Commons Attribution License or equivalent. Every chapter published in this book has been scrutinized by our experts. Their significance has been extensively debated. The topics covered herein carry significant findings which will fuel the growth of the discipline. They may even be implemented as practical applications or may be referred to as a beginning point for another development.

The contributors of this book come from diverse backgrounds, making this book a truly international effort. This book will bring forth new frontiers with its revolutionizing research information and detailed analysis of the nascent developments around the world.

We would like to thank all the contributing authors for lending their expertise to make the book truly unique. They have played a crucial role in the development of this book. Without their invaluable contributions this book wouldn't have been possible. They have made vital efforts to compile up to date information on the varied aspects of this subject to make this book a valuable addition to the collection of many professionals and students.

This book was conceptualized with the vision of imparting up-to-date information and advanced data in this field. To ensure the same, a matchless editorial board was set up. Every individual on the board went through rigorous rounds of assessment to prove their worth. After which they invested a large part of their time researching and compiling the most relevant data for our readers.

The editorial board has been involved in producing this book since its inception. They have spent rigorous hours researching and exploring the diverse topics which have resulted in the successful publishing of this book. They have passed on their knowledge of decades through this book. To expedite this challenging task, the publisher supported the team at every step. A small team of assistant editors was also appointed to further simplify the editing procedure and attain best results for the readers.

Apart from the editorial board, the designing team has also invested a significant amount of their time in understanding the subject and creating the most relevant covers. They scrutinized every image to scout for the most suitable representation of the subject and create an appropriate cover for the book.

The publishing team has been an ardent support to the editorial, designing and production team. Their endless efforts to recruit the best for this project, has resulted in the accomplishment of this book. They are a veteran in the field of academics and their pool of knowledge is as vast as their experience in printing. Their expertise and guidance has proved useful at every step. Their uncompromising quality standards have made this book an exceptional effort. Their encouragement from time to time has been an inspiration for everyone.

The publisher and the editorial board hope that this book will prove to be a valuable piece of knowledge for researchers, students, practitioners and scholars across the globe.

List of Contributors

Timsy Uppal and Subhash C. Verma
Department of Microbiology and Immunology, University of Nevada, Reno, School of Medicine, 1664 N Virginia Street, MS 320, Reno, NV 89557, USA

Sagarika Banerjee, Zhiguo Sun and Erle S. Robertson
Department of Microbiology and the Tumor Virology Program of the Abramson Cancer Center, Perelman School of Medicine at the University of Pennsylvania, 201E Johnson Pavilion, 3610 Hamilton Walk, Philadelphia, PA 19104, USA

Shuhei Sakakibara
Department of Molecular Immunology, Research Institute for Microbial Diseases, Osaka University, Suita, Osaka 565-0871, Japan
World Premier International Immunology Frontier Research Center, Osaka University, Suita, Osaka 565-0871, Japan

Giovanna Tosato
Laboratory of Cellular Oncology, National Cancer Institute, National Institutes of Health, Bethesda, MD 20982, USA

Maria Julia Massimelli, Vladimir Majerciak, Jeong-Gu Kang and Zhi-Ming Zheng
Tumor Virus RNA Biology Section, Gene Regulation and Chromosome Biology Laboratory, Center for Cancer Research, National Cancer Institute, National Institutes of Health, 1050 Boyles Street, Frederick, MD 21702, USA

David J. Liewehr and Seth M. Steinberg
Biostatistics & Data Management Section, Center for Cancer Research, National Cancer Institute, National Institutes of Health, Bethesda, MD 20892, USA

Tiejun Zhang, Jun Wang and Na He
Department of Epidemiology, School of Public Health, Fudan University, and The Key Laboratory of Public Health Safety of Ministry of Education, Shanghai 200032, China

Ying Liu
Department of Epidemiology, School of Public Health, Fudan University, and The Key Laboratory of Public Health Safety of Ministry of Education, Shanghai 200032, China
Putuo District Center for Disease Control and Prevention, Shanghai 200032, China

Yuyan Zhang
Putuo District Center for Disease Control and Prevention, Shanghai 200032, China

Karlie Plaisance-Bonstaff, Hong Seok Choi, Tyler Beals, Brian J. Krueger, Isaac W. Boss, Lauren A. Gay, Irina Haecker and Jianhong Hu
Department of Molecular Genetics and Microbiology, University of Florida, Gainesville, FL 32610, USA

Rolf Renne
Department of Molecular Genetics and Microbiology, University of Florida, Gainesville, FL 32610, USA
UF Health Cancer Center, University of Florida, Gainesville, FL 32610, USA
UF Institute of Genetics, University of Florida, Gainesville, FL 32610, USA

Natacha Coen, Sophie Duraffour, Robert Snoeck and Graciela Andrei
Rega Institute for Medical Research, KU Leuven, B-3000 Leuven, Belgium

Hui Wu, Jun Xiao, Mengdie Chi, Yixiao Qu and Hao Feng
Key Laboratory of Protein Chemistry and Developmental Biology of Ministry of Education of China, College of Life Science, Hunan Normal University, Changsha 410081, China

Liqun Liu
Division of Pediatric Neurology, Children's Medical Center, The Second Xiangya Hospital, Central South University, Changsha 410011, China

Pravinkumar Purushothaman
Department of Microbiology and Immunology, University of Nevada, Reno, School of Medicine, 1664 N Virginia Street, MS 320, Reno, NV 89557, USA

Xinxing Ouyang, Bishi Fu and Minhua Luo
State Key Laboratory of Virology, Wuhan Institute of Virology, Chinese Academy of Sciences, Wuhan 430071, China

Yan Zeng
Laboratory of Xinjiang Endemic and Ethnic Diseases, Shihezi University, Shihezi 832002, China

Xiaowu Wang, Wei Chen and Yuan Fang
Department of Microbiology, Anhui Medical University, Hefei 230032, China

Linding Wang
Department of Microbiology, Anhui Medical University, Hefei 230032, China
Central Laboratory of Molecular and Cellular Biology, School of Basic Medicine, Anhui Medical University, Hefei, Anhui 230032, China

Veenu Minhas and Charles Wood
Nebraska Center for Virology, School of Biological Sciences, University of Nebraska, Morrison Center, 4240 Fair Street, Lincoln, NE 68583, USA

Cyprian C. Rossetto and Gregory S. Pari
Department of Microbiology & Immunology, University of Nevada, Reno School of Medicine, Reno, NV 89557, USA

Mel Campbell
Department of Dermatology, University of California, Davis, CA 95616, USA
UC Davis Comprehensive Cancer Center, University of California, Davis, CA 95616, USA

Hsing-Jien Kung
UC Davis Comprehensive Cancer Center, University of California, Davis, CA 95616, USA
Department of Biochemistry and Molecular Medicine, University of California, Davis, CA 95616, USA National Health Research Institutes, Maioli 35053, Taiwan
Taipei Medical University, Taipei 110, Taiwan

Yoshihiro Izumiya
Department of Dermatology, University of California, Davis, CA 95616, USA
UC Davis Comprehensive Cancer Center, University of California, Davis, CA 95616, USA
Department of Biochemistry and Molecular Medicine, University of California, Davis, CA 95616, USA

Index

www.ingramcontent.com/pod-product-compliance
Lightning Source LLC
LaVergne TN
LVHW081028210926
848485LV00004B/971